Discrete-Valued Time Series

Discrete-Valued Time Series

Editor

Christian H. Weiss

Basel • Beijing • Wuhan • Barcelona • Belgrade • Novi Sad • Cluj • Manchester

Editor
Christian H. Weiss
Helmut-Schmidt-University
Hamburg
Germany

Editorial Office
MDPI
St. Alban-Anlage 66
4052 Basel, Switzerland

This is a reprint of articles from the Special Issue published online in the open access journal *Entropy* (ISSN 1099-4300) (available at: https://www.mdpi.com/journal/entropy/special_issues/Discrete_Valued_Time_Series).

For citation purposes, cite each article independently as indicated on the article page online and as indicated below:

Lastname, A.A.; Lastname, B.B. Article Title. *Journal Name* **Year**, *Volume Number*, Page Range.

ISBN 978-3-7258-0477-1 (Hbk)
ISBN 978-3-7258-0478-8 (PDF)
doi.org/10.3390/books978-3-7258-0478-8

© 2024 by the authors. Articles in this book are Open Access and distributed under the Creative Commons Attribution (CC BY) license. The book as a whole is distributed by MDPI under the terms and conditions of the Creative Commons Attribution-NonCommercial-NoDerivs (CC BY-NC-ND) license.

Contents

About the Editor vii

Christian H. Weiß
Discrete-Valued Time Series
Reprinted from: *Entropy* **2023**, *25*, 1576, doi:10.3390/e25121576 1

Kaizhi Yu and Tielai Tao
An Observation-Driven Random Parameter INAR(1) Model Based on the Poisson Thinning Operator
Reprinted from: *Entropy* **2023**, *25*, 859, doi:10.3390/e25060859 5

Huaping Chen, Jiayue Zhang and Xiufang Liu
A Conway–Maxwell–Poisson-Binomial AR(1) Model for Bounded Time Series Data
Reprinted from: *Entropy* **2023**, *25*, 126, doi:10.3390/e25010126 35

Isabel Silva, Maria Eduarda Silva, Isabel Pereira and Brendan McCabe
Time Series of Counts under Censoring: A Bayesian Approach
Reprinted from: *Entropy* **2023**, *25*, 549, doi:10.3390/e25040549 51

Mengya Liu, Fukang Zhu, Jianfeng Li and Chuning Sun
A Systematic Review of INGARCH Models for Integer-Valued Time Series
Reprinted from: *Entropy* **2023**, *25*, 922, doi:10.3390/e25060922 67

Yue Xu, Qi Li and Fukang Zhu
A Modified Multiplicative Thinning-Based INARCH Model: Properties, Saddlepoint Maximum Likelihood Estimation, and Application
Reprinted from: *Entropy* **2023**, *25*, 207, doi:10.3390/e25020207 94

Sidratul Moontaha, Bert Arnrich and Andreas Galka
State Space Modeling of Event Count Time Series
Reprinted from: *Entropy* **2023**, *25*, 1372, doi:10.3390/e25101372 112

Christian H. Weiß, Boris Aleksandrov, Maxime Faymonville and Carsten Jentsch
Partial Autocorrelation Diagnostics for Count Time Series
Reprinted from: *Entropy* **2023**, *25*, 105, doi:10.3390/e25010105 133

Lihong Guan and Xiaohong Wang
Ruin Analysis on a New Risk Model with Stochastic Premiums and Dependence Based on Time Series for Count Random Variables
Reprinted from: *Entropy* **2023**, *25*, 698, doi:10.3390/e25040698 154

Manuel Cabral Morais
Two Features of the GINAR(1) Process and Their Impact on the Run-Length Performance of Geometric Control Charts
Reprinted from: *Entropy* **2023**, *25*, 444, doi:10.3390/e25030444 179

Maria Papapetrou, Elsa Siggiridou and Dimitris Kugiumtzis
Adaptation of Partial Mutual Information from Mixed Embedding to Discrete-Valued Time Series
Reprinted from: *Entropy* **2022**, *24*, 1505, doi:10.3390/e24111505 191

About the Editor

Christian H. Weiss

Christian H. Weiss is a Professor in the Department of Mathematics and Statistics at the Helmut Schmidt University in Hamburg, Germany. He obtained his Doctoral Degree in Mathematical Statistics from the University of Würzburg, Germany. His research areas include time series analysis, statistical quality control, and computational statistics. He is an author of several textbooks and has published his work in international scientific journals such as *Bernoulli*, *Entropy*, *Journal of the American Statistical Association*, *Journal of Multivariate Analysis*, *Journal of Nonparametric Statistics*, *Journal of Quality Technology*, *Journal of the Royal Statistical Society*, *Journal of Time Series Analysis*, *Spatial Statistics*, and *Technometrics*.

Editorial

Discrete-Valued Time Series

Christian H. Weiß

Department of Mathematics and Statistics, Helmut Schmidt University, 22043 Hamburg, Germany; weissc@hsu-hh.de; Tel.: +49-40-6541-2779

Citation: Weiß, C.H. Discrete-Valued Time Series. *Entropy* **2023**, *25*, 1576. https://doi.org/10.3390/e25121576

Received: 14 November 2023
Accepted: 22 November 2023
Published: 23 November 2023

Copyright: © 2023 by the author. Licensee MDPI, Basel, Switzerland. This article is an open access article distributed under the terms and conditions of the Creative Commons Attribution (CC BY) license (https://creativecommons.org/licenses/by/4.0/).

Time series are sequentially observed data in which important information about the phenomenon under consideration is contained not only in the individual observations themselves, but also in the way these observations follow one another. Therefore, the stochastic counterpart of time series data, the stochastic process, and the methods and models for time series analysis, carefully consider this sequential information. The first approaches to the analysis and modeling of time series were developed about 100 years ago, and they have formed an entire research discipline at least since the publication of the famous textbook by Box and Jenkins [1] about 50 years later. This double anniversary recently gave rise to the Special Issue "Time Series Modeling" in *Entropy* [2], where numerous contributions illustrated the strong research interest and the various research directions in this area. Here, it is interesting to note that these contributions could be roughly split into two halves. The first half deals with real-valued time series, i.e., time series having a continuous range consisting of real numbers or real vectors. Indeed, time series of this kind have been the main subject of time series analysis since the beginning, and for most of the last 100 years. Remarkably, however, the second half of the papers deals with a rather young subfield of time series analysis that now seems to be attracting a great deal of research interest: discrete-valued time series. The first papers on discrete-valued time series appeared in the 1980s, but it was not until the 2000s that a rapid increase in research activity could be observed. It is only in the last few years that a certain maturity and consolidation of this research area can be observed, which is manifested in, among other things, the textbooks by Davis et al. [3], Weiß [4], and in a number of recent survey articles, such as Davis et al. [5], Fokianos [6], Armillotta et al. [7], Karlis and Mamode Khan [8], and Li et al. [9], as well as in contribution 4 by Liu et al.

Discrete-valued time series can be of several types (see Weiß [4] for a comprehensive discussion). Undoubtedly, the most popular are count time series, where the range is quantitative and consists of non-negative integers [5–8]. However, truly integer-valued time series (where the range also includes negative integers) are increasingly being considered [9], while categorical time series, where the range is qualitative (symbols), are still somewhat neglected. Finally, the discretization of a real-valued time series has to be mentioned in this context, e.g., when methods based on ordinal patterns are used for its analysis (see Bandt [10] for a recent overview). Motivated by these diverse research directions, this Special Issue on "Discrete-valued Time Series" was initiated, which was actually very well received. It was possible to collect articles on a wide range of topics in this area, covering stochastic models for discrete-valued time series, as well as methods for their analysis, univariate and multivariate discrete-valued time series, and various applications of discrete time-series methods. The remainder of this Editorial provides a brief summary of the contributions to this Special Issue, grouping the articles thematically.

A large class of models for count time series make use of so-called thinning operators [4]. Such thinnings are used as integer substitutes of the multiplication, and allow us to adapt the classical autoregressive moving-average (ARMA) models to the count-data case. Therefore, the resulting integer ARMA-type models are abbreviated as INARMA models [4]. The observation-driven random parameter INAR(1) model proposed by Yu and Tao (contribution 1) belongs to this class, and it is characterized by using the Poisson thinning operator together with a state-dependent thinning parameter (i.e., the thinning parameter at time t depends on the count observation at time $t-1$). Furthermore, the authors introduce further flexibility into the model by allowing the thinning parameter to be chosen randomly (given the previous observation). The AR(1)-type model developed by Chen et al. (contribution 2) belongs to the INARMA family as well, but differs from the aforementioned model of Yu and Tao in several respects. First, Chen et al. focus on bounded counts (i.e., the observed counts have the range $\{0, \ldots, n\}$ with specified $n \in \mathbb{N} = \{1, 2, \ldots\}$), whereas Yu and Tao's INAR(1) model assumes the unbounded range $\mathbb{N}_0 = \{0, 1, \ldots\}$. Second, Chen et al. use Conway–Maxwell–Poisson-binomial thinning operators, thus allowing the resulting model to handle counts exhibiting either underdispersion, equidispersion, or overdispersion with respect to a binomial distribution. Further, the article by Silva et al. (contribution 3) refers to the INARMA modeling of count time series, but under a different perspective. In many application contexts, it is not possible to observe the true count value at a time t, but only a censored version of it. Neglecting censoring might cause biased and inconsistent model estimates, so Silva et al. developed two Bayesian algorithms to explicitly account for the censoring of data while fitting an INAR(1) model.

Another large class of models for count time series adapt an ARMA-like structure via a conditional regression approach. Although there is an ongoing debate on how these models should be referred to [4], most authors use the term integer-valued generalized autoregressive conditional heteroskedasticity (INGARCH). Compared to the INARMA class, INGARCH models are better able to handle higher-order models, and by using a feedback term within the model recursion, they also allow for an intensified memory of the count process. For these reasons, among others, INGARCH models have become very popular in recent years, and a comprehensive survey about the state of the art is presented by Liu et al. (contribution 4). It is worth noting that their survey article is not restricted to univariate counts only (where both the cases of unbounded and bounded counts are considered), but it also covers INGARCH models for multivariate counts, as well as models for "$\mathbb{Z}$-valued time series". The latter refers to time series with range $\mathbb{Z} = \{\ldots, -1, 0, 1, \ldots\}$, i.e., where also negative integer outcomes are possible (also see Li et al. [9] for further information). At this point, the article by Xu et al. (contribution 5) has to be mentioned, where a particular type of INARCH model is developed. The proposal adapts the multiplicative error model to the count-data case by combining the INARCH approach with the binomial thinning operator. Consequently, the resulting model might be understood as a hybrid model that combines features from the INARMA and INGARCH classes. Finally, the paper of Moontaha et al. (contribution 6) is also concerned with the modeling of (possibly multivariate) count time series, but the proposed approach does not fall into either the INARMA or the INGARCH class. On the contrary, although this may seem contradictory at first, Moontaha et al. use the Gaussian linear state space model together with the Kalman filter. However, to ensure that non-negative integer outcomes are generated, the model is equipped with special observation functions.

The remaining contributions to this Special Issue are not about the mere modeling of discrete-valued time series, but consider miscellaneous topics in this research area. In Weiß et al. (contribution 7), a particular type of analyzing count time series is discussed, namely hypothesis tests based on the partial autocorrelation function (PACF). Such PACF tests are commonly used for identifying the model order of AR-type count processes, but their asymptotics have been derived for Gaussian data. Therefore, different ways of getting more reliable test implementations are investigated in a comparative study. Guan and Wang (contribution 8) present an application of INARMA models to the risk analysis of an insurance portfolio. More precisely, an INAR(1) model is used to describe the temporal dependence among the premium numbers, and an INMA(1) model for the temporal dependence among the claim numbers. Another application of INARMA models is presented by Morais (contribution 9), namely statistical process monitoring by control charts. Morais derives two stochastic ordering results regarding the geometric INAR(1) process which, in turn, can be utilized to conclude on the properties of the geometric control chart's run length distribution. Last but not least, Papapetrou et al. (contribution 10) discuss the causality analysis of discrete-valued multivariate time series. In contrast to the aforementioned contributions, these time series do not need to consist of quantitative count values, but might also be of qualitative nature. Papapetrou et al. propose a discrete type of partial mutual information from mixed embedding, and investigate its performance in a simulation study.

Acknowledgments: The Guest Editor is grateful to the authors for their contributions to this Special Issue, to the anonymous peer reviewers for carefully reading the submissions as well as for their constructive feedback, and to the MDPI editorial staff for their support during this work.

Conflicts of Interest: The author declares no conflict of interest.

List of Contributions

1. Yu, K.; Tao, T. An observation-driven random parameter INAR(1) model based on the Poisson thinning operator. *Entropy* **2023**, *25*, 859. https://doi.org/10.3390/e25060859.
2. Chen, H.; Zhan, J.; Liu, X. A Conway-Maxwell-Poisson-binomial AR(1) model for bounded time series data. *Entropy* **2023**, *25*, 126. https://doi.org/10.3390/e25010126.
3. Silva, I.; Silva, M.E.; Pereira, I.; McCabe, B. Time series of counts under censoring: A Bayesian approach. *Entropy* **2023**, *25*, 549. https://doi.org/10.3390/e25040549.
4. Liu, M.; Zhu, F.; Li, J.; Sun, C. A systematic review of INGARCH models for integer-valued time series. *Entropy* **2023**, *25*, 922. https://doi.org/10.3390/e25060922.
5. Xu, Y.; Li, Q.; Zhu, F. A modified multiplicative thinning-based INARCH model: properties, saddlepoint maximum likelihood estimation, and application. *Entropy* **2023**, *25*, 207. https://doi.org/10.3390/e25020207.
6. Moontaha, S.; Arnrich, B.; Galka, A. State space modeling of event count time series. *Entropy* **2023**, *25*, 1372. https://doi.org/10.3390/e25101372.
7. Weiß, C.H.; Aleksandrov, B.; Faymonville, M.; Jentsch, C. Partial autocorrelation diagnostics for count time series. *Entropy* **2023**, *25*, 105. https://doi.org/10.3390/e25010105.
8. Guan, L.; Wang, X. Ruin analysis on a new risk model with stochastic premiums and dependence based on time series for count random variables. *Entropy* **2023**, *25*, 698. https://doi.org/10.3390/e25040698.
9. Morais, M.C. Two features of the GINAR(1) process and their impact on the run-length performance of geometric control charts. *Entropy* **2023**, *25*, 444. https://doi.org/10.3390/e25030444.
10. Papapetrou, M.; Siggiridou, E.; Kugiumtzis, D. Adaptation of partial mutual information from mixed embedding to discrete-valued time series. *Entropy* **2022**, *24*, 1505. https://doi.org/10.3390/e24111505.

References

1. Box, G.E.P.; Jenkins, G.M. *Time Series Analysis: Forecasting and Control*, 1st ed.; Holden-Day: San Francisco, CA, USA, 1970.
2. Weiß, C.H. (Ed.) *Time Series Modelling*; MDPI: Basel, Switzerland, 2021. [CrossRef]
3. Davis, R.; Holan, S.; Lund, R.; Ravishanker, N. (Eds.) *Handbook of Discrete-Valued Time Series*; Chapman & Hall/CRC Press: Boca Raton, FL, USA, 2015.
4. Weiß, C.H. *An Introduction to Discrete-Valued Time Series*, 1st ed.; John Wiley & Sons, Inc.: Chichester, UK, 2018.
5. Davis, R.A.; Fokianos, K.; Holan, S.H.; Joe, H.; Livsey, J.; Lund, R.; Pipiras, V.; Ravishanker, N. Count time series: A methodological review. *J. Am. Stat. Assoc.* **2021**, *116*, 1533–1547. [CrossRef]
6. Fokianos, K. Multivariate count time series modelling. *Econom. Stat.* **2021**, in press. [CrossRef]
7. Armillotta, M.; Luati, A.; Lupparelli, M. An overview of ARMA-like models for count and binary data. In *Trends and Challenges in Categorical Data Analysis: Statistical Modelling and Interpretation*; Kateri, M., Moustaki, I., Eds.; Springer International Publishing: Cham, Switzerland, 2023; pp. 233–274.
8. Karlis, D.; Mamode Khan, N. Models for integer data. *Annu. Rev. Stat. Its Appl.* **2023**, *10*, 297–323. [CrossRef]

9. Li, Q.; Chen, H.; Zhu, F. Z-valued time series: models, properties and comparison. *J. Stat. Plan. Inference* **2023**, 106099, *in press*. [CrossRef]
10. Bandt, C. Statistics and contrasts of order patterns in univariate time series. *Chaos Interdiscip. J. Nonlinear Sci.* **2023**, *33*, 033124. [CrossRef] [PubMed]

Disclaimer/Publisher's Note: The statements, opinions and data contained in all publications are solely those of the individual author(s) and contributor(s) and not of MDPI and/or the editor(s). MDPI and/or the editor(s) disclaim responsibility for any injury to people or property resulting from any ideas, methods, instructions or products referred to in the content.

Article

An Observation-Driven Random Parameter INAR(1) Model Based on the Poisson Thinning Operator

Kaizhi Yu and Tielai Tao *

School of Statistics, Southwestern University of Finance and Economics, Chengdu 611130, China; yukz@swufe.edu.cn
* Correspondence: 118020209005@smail.swufe.edu.cn

Abstract: This paper presents a first-order integer-valued autoregressive time series model featuring observation-driven parameters that may adhere to a particular random distribution. We derive the ergodicity of the model as well as the theoretical properties of point estimation, interval estimation, and parameter testing. The properties are verified through numerical simulations. Lastly, we demonstrate the application of this model using real-world datasets.

Keywords: integer-valued time series; thinning operator; observation-driven; ergodicity; interval estimation

1. Introduction

Integer-valued time series data are prevalent in both scientific research and various socioeconomic contexts. Examples of such data encompass the annual number of companies listed on stock exchanges, the monthly usage of hospital beds in specific departments, and the yearly frequency of major earthquakes or tsunamis. However, traditional continuous-valued time series models are unable to precisely capture the unique characteristics of integer-valued data, resulting in only approximations through continuous-valued models. This shortcoming may lead to model mis-specification, posing challenges in statistical inference. Consequently, the modeling and analysis of integer-valued time series data have increasingly gained attention within academia. Amongst the extensive range of integer-valued time series models, thinning operator models have attracted considerable interest from scholars due to their resemblance to Autoregressive Moving Average (ARMA) models in continuous-valued time series theory. Thinning operator models replace multiplication in ARMA models with the binomial thinning operator, which was initially introduced by Steutel and Van Harn [1]:

$$\phi \circ Y_i = \sum_{i=1}^{Y_i} B_i, \qquad (1)$$

where $\{Y_i\}$ refers to a count series and $\{B_i\}$ represents a Bernoulli random variable sequence that independent of $\{Y_i\}$, satisfying the condition $P(B_i = 1) = 1 - P(B_i = 0) = \phi$. Building upon this concept, Al-Osh and Alzaid [2] developed the first-order Integer-valued Autoregressive (INAR(1)) model, for $t \in \mathbb{N}^+$:

$$Y_t = \phi \circ Y_{t-1} + Z_t, \qquad (2)$$

where Z_t is considered the innovation term entering the model during period t. Its marginal distribution aligns with a Poisson distribution, exhibiting an expected value of λ, thereby giving rise to the nomenclature of the Poisson INAR(1) model. An intuitive interpretation of this model is that, within a hospital setting, the number of in-patients in period t comprises patients from period $t-1$ who have not yet been discharged, along with patients newly admitted in period t. Given that B_i adheres to a Bernoulli distribution, the binomial thinning operator can exclusively express the $\{0,1\}$ to $\{0,1\}$ excitation states. However, the binomial thinning operator does not represent the sole available option for thinning operators. Latour [3] expanded the distribution of B_i in Equation (1) to encompass any non-negative integer-valued random variable, thus establishing the Generalized Integer-valued Autoregressive (GINAR) model and providing conditions for model stationarity. Furthermore, the ϕ in Equation (1) need not be a fixed constant. Joe [4] and Zheng, Basawa, and Datta [5] constructed the Random Coefficient Thinning Operator (RCINAR(1)) model by permitting the parameter ϕ in the INAR(1) model to follow a specified random distribution. Gomes and Castro [6] generalized the thinning operator in RCINAR(1) to GINAR(1) model, culminating in the development of the Random Coefficient Generalized Integer-valued Autoregressive model. Weiß and Jentsch [7] proposed a bootstrap estimation method based on the INAR model to facilitate the introduction of semi-parametric structures within the INAR model, in turn reducing model assumptions and augmenting model generalization capabilities. Kang, Wang, and Yang [8] mixed the binomial thinning operator with the operator introduced by Pegram [9], resulting in the development of a novel INAR model capable of addressing equi-dispersed, under-dispersed, over-dispersed, zero-inflated, and multi-modal integer-valued time series data. Salinas, Flunkert, Gasthaus, and Januschowski [10] proposed a new method for time series forecasting based on autoregressive recurrent neural network models. Huang, Zhu, and Deng [11] mixed quasi-binomial distribution operators with generalized Poisson operators, thus equipping the INAR model with the ability to describe structural changes in the data generation processes. Mohammadi, Sajjadnia, Bakouch, and Sharafi [12] incorporated innovation terms conforming to the Poisson-Lindley distribution, thereby enhancing the INAR(1) model's capacity to capture $\{0,1\}$ inflated integer-valued time series data. For further discussion on thinning operator models, Scotto, Weiß, and Gouveia [13] provide a comprehensive review article.

The thinning operator models previously mentioned presuppose that ϕ is independent of other variables, thereby neglecting the dynamic features of the coefficient ϕ in INAR models. To tackle this limitation, Zheng and Basawa [14] proposed a first-order observation-driven integer-valued autoregressive process. Triebsch [15] introduced the first-order Functional Coefficient Integer-valued Time Series model based on the thinning operator, in which the coefficient ϕ_t during period t is a measurable function of the previous observation Y_{t-1}. Furthermore, Montriro, Scotto, and Pereira [16] presented the Self-Exciting Threshold Integer-valued Time Series model (SETINAR) in which the coefficient ϕ_t during period t assumes diverse values contingent on the varying observations in prior limited periods. Building on the geometric thinning operator (alternatively known as the negative binomial thinning operator) proposed by Ristić, Bakouch, and Nastić [17], Yu, Wang, and Yang [18] introduced an INAR(1) model encompassing observation-driven parameters.

With respect to integer-valued time series models featuring observation-driven parameters, existing studies primarily focus on binomial and geometric thinning operators. However, the binomial thinning operator cannot represent one-to-many excitation states, and both binomial and geometric thinning operators exhibit limited descriptive capacity for locally non-stationary phenomena and extreme values in real data. Consequently, this paper employs a Poisson thinning operator, defined as follows:

$$\phi_t \ominus Y_t = \sum_{i=1}^{Y_t} B_i^{(t)}, \qquad (3)$$

where, $\left\{B_i^{(t)}\right\}$ is independent of Y_t and constitutes an independent and identically distributed Poisson random variable sequence with an intensity parameter $\phi_t > 0$. The probability mass function is expressed by:

$$\mathbb{P}\left(B_i^{(t)} = x\right) = \frac{\phi_t^x}{x!} \exp(-\phi_t),$$

where $\left\{B_i^{(t)}\right\}$ and Y_t are mutually independent. Leveraging this thinning operator, the INAR(1) model in this study is formulated as follows:

$$Y_t = \phi_t \ominus Y_{t-1} + Z_t,$$

where the sequence $\{Z_t\}$ comprises independent and identically distributed non-negative integer-valued random variables, which are independent of $\left\{B_i^{(t)}\right\}$ and $\{Y_s\}_{s<t}$. Furthermore, diverging from the parameters set forth by Yu, Wang, and Yang [18], we posit that ϕ_t correlates with the previous observation Y_{t-1}, and given Y_{t-1}, $\phi_t | Y_{t-1}$ may still conform to a specific non-negative probability distribution. In Section 2, we will demonstrate that if the expectation of this non-negative discrete probability distribution falls below 1, it does not affect the model's ergodicity. Simultaneously, due to instances where $\phi_t | Y_{t-1}$ occasionally exceeds 1, the autoregressive model exhibits non-stationary features or generates extreme values within specific periods—all without compromising its overall stationarity. In comparison to existing research, this setting offers the advantage of simultaneously illustrating one-to-many excitation states and observation-driven and time-varying parameter structures, as well as localized non-stationary features or extreme values. For example, in public health, a patient with an infectious disease may not transmit the illness to others or could potentially infect one or multiple individuals, indicating one-to-many excitation states. As the number of infections fluctuates, local epidemic prevention policies may undergo changes, consequently modifying the disease's transmissibility and reflecting the time-varying and observation-driven characteristics of the coefficient. During particular periods of rapid infectious disease spread, the majority of infected individuals are likely to infect more than one other person, resulting in infection data that exhibit extreme values or localized non-stationary characteristics.

The organization of this paper is as follows: in Section 2, we introduce the integer-valued time series model featuring observation-driven coefficients under investigation and outline its essential statistical properties. In Section 3, we describe the estimation and testing methods pertinent to this model and present asymptotic results. Section 4 provides numerical simulation outcomes of these techniques, elaborating on the performance of the estimation and testing approaches across diverse settings and sample conditions. Section 5 demonstrates the application of the proposed model using real-world data. Finally, Section 6 offers a summary and discussion.

2. Model Construction and Basic Properties

For the time series $\{Y_t\}$, consider the following data generating process:

$$Y_t = \phi_t \ominus Y_{t-1} + Z_t \qquad (4)$$

Given Y_{t-1}, ϕ_t may be fixed as:

$$\frac{\exp[\nu(Y_{t-1};\beta)]}{1+\exp[\nu(Y_{t-1};\beta)]}.$$

Alternatively, $\{\phi_t\}$ could represent an independent random variable sequence with a conditional expectation of:

$$\mathbb{E}(\phi_t | Y_{t-1}) = \frac{\exp[\nu(Y_{t-1};\beta)]}{1+\exp[\nu(Y_{t-1};\beta)]}, \qquad (5)$$

where β is an ℓ-dimensional parameter vector, the function $\nu(\cdot;\cdot)$ belongs to a specific parametric family of functions $\mathcal{G}\{\nu(Y_{t-1};\beta); \beta \in \Theta\}$, and Θ is a compact subset of $\mathbb{R}^\ell$. β is an interior point of Θ and $\nu(y;\beta)$ is thrice continuously differentiable with respect to β. The conditional variance is given by $Var(\phi | Y_{t-1}) = \sigma^2_{\phi_t | Y_{t-1}}$. Additionally, $\{Z_t\}$ comprises an independent and identically distributed non-negative integer-valued random variable sequence with a probability mass function f_z with expectation $\mathbb{E}(Z_t) = \lambda < \infty$ and variance $Var(Z_t) = \sigma_Z^2 < \infty$. Furthermore, $\{Z_t\}$ is independent of $\{Y_t\}$.

Remark 1. *Integer-valued probability distributions that align with the settings of Z_t are common, with typical examples being Poisson and geometric distributions. This paper employs a Poisson distribution in the numerical simulation section.*

Remark 2. *There are numerous functions that align with the setting of $\nu(\cdot;\cdot)$, with the most typical being the linear function $\nu(Y_{t-1};\beta) = \beta_0 + \beta_1 Y_{t-1}$. In this paper's numerical simulation section, a linear function setting will be adopted.*

Remark 3. *From model (4), it is evident that $\{Y_t\}$ is a Markov chain defined on the set of natural numbers $\mathbb{N}$, with a one-step-ahead transition probability:*

$$\mathbb{P}(Y_t = y_t | Y_{t-1} = y_{t-1}) = \int \mathbb{P}(Y_t = y_t | Y_{t-1} = y_{t-1}, \phi_t = \phi) \mathbb{P}(\phi_t = \phi | Y_{t-1} = y_{t-1}) d\phi$$
$$= \int \sum_{k=0}^{y_t} \frac{(\phi y_{t-1})^k}{k!} \exp(-\phi y_{t-1}) f_z(y_t - k) \mathbb{P}(\phi_t = \phi | Y_{t-1} = y_{t-1}) d\phi \quad (6)$$

Based on the above model construction, we can obtain the conditional moments for Model (4). Starting from these conditional moments, we can construct estimating equations to estimate the unknown parameters in the model:

Property 1. *for $t \geq 1$*

(i) $\mathbb{E}(Y_t | Y_{t-1}) = \frac{\exp[\nu(Y_{t-1};\beta)]}{1+\exp[\nu(Y_{t-1};\beta)]} Y_{t-1} + \lambda$,

(ii) $Var(Y_t | Y_{t-1}) = \frac{\exp[\nu(Y_{t-1};\beta)]}{1+\exp[\nu(Y_{t-1};\beta)]} Y_{t-1} + \sigma_Z^2 + \sigma_{\phi_t | Y_{t-1}}^2$,

(iii) $Cov(Y_t, Y_{t-1}) = \mathbb{E}\left(\frac{\exp[\nu(Y_{t-1};\beta)]}{1+\exp[\nu(Y_{t-1};\beta)]} Y_{t-1}^2\right) - \mathbb{E}\left(\frac{\exp[\nu(Y_{t-1};\beta)]}{1+\exp[\nu(Y_{t-1};\beta)]} Y_{t-1}\right) \mathbb{E}(Y_{t-1})$.

Ergodicity is crucial for the convergence of parameter estimation, as presented in the following property:

Property 2. *If $\sup_{y \in \mathbb{N}} \nu(y;\beta) < \infty, \beta \in \Theta$, then the data generating process $\{Y_t\}$ defined by (4) is an ergodic Markov chain.*

Remark 4. *In Property 2, since the form of the function ν is not determined, we cannot directly provide the conditions for the ergodicity of $\{Y_t\}$. However, for specific cases, such as $\nu(Y_{t-1};\beta) = \beta_0 + \beta_1 Y_{t-1}$, we can intuitively see that the stationary and ergodic property of the data generating process requires $\beta_1 \leq 0$ at the very least, making the expected value of ϕ_t lower when Y_t is higher and vice versa. From the proof of Property A1 in Appendix A, it can be observed that the ergodicity of $\{Y_t\}$ requires the existence of a constant $0 < m < 1$ such that $\frac{\exp(\beta_0 + \beta_1 Y_{t-1})}{1+\exp(\beta_0 + \beta_1 Y_{t-1})} < m$; however, if $\beta_1 > 0$, then $\frac{\exp(\beta_0 + \beta_1 Y_{t-1})}{1+\exp(\beta_0 + \beta_1 Y_{t-1})}$ will increase with the rise of Y_t, making it impossible to determine a constant m that meets requirements.*

3. Parameter Estimation and Hypothesis Testing

In this section, we assume that the time series $\{Y_t\}_{t=1}^T$ satisfies the data-generating process defined by Equation (4), with $\theta_0 = (\beta_0', \lambda_0)$ as the true parameter vector of this process and $\theta = (\beta', \lambda)$ as the unknown parameter vector to be estimated. In this paper, our primary focus is on two estimation methods: Conditional Least Squares (CLS) and Conditional Maximum Likelihood (CML). Additionally, we attempt to establish observation-driven interval estimation through estimating equations in CLS and observation-driven hypothesis testing through the framework of Empirical Likelihood (EL). Here, we first make assumptions about the data-generating process $\{Y_t\}$ and the function $v(y;\beta)$, assuming the existence of a neighborhood B of β_0 and a positive integrable function $N(y)$, such that:

(A1) $\{Y_t\}$ is a strictly stationary and ergodic sequence.

(A2) $1 \leq i, j \leq \ell$, $\left|\frac{\partial v(y;\beta)}{\partial \beta_i}\right|$ and $\left|\frac{\partial^2 v(y;\beta)}{\partial \beta_i \partial \beta_j}\right|$ are continuous with respect to β and dominated by $N(y)$ on B, where $N(y)$ is a positive integrable function.

(A3) $1 \leq i, j, k \leq \ell$, $\left|\frac{\partial^3 v(y;\beta)}{\partial \beta_i \partial \beta_j \partial \beta_k}\right|$ are continuous with respect to β and dominated by $N(y)$ on B, where $N(y)$ is a positive integrable function.

(A4) $\exists \delta > 0$, such that $\mathbb{E}|Y_t|^{8+\delta} < \infty$, $\mathbb{E}|N(Y_t)|^{8+\delta} < \infty$.

(A5) $\mathbb{E}\left(\frac{\partial v(y;\beta)}{\partial \beta} \cdot \frac{\partial v(y;\beta)}{\partial \beta'}\right)$ is a full-rank matrix, i.e., of rank ℓ.

(A6) The parameters of $v(y;\beta)$ are identifiable, that is, if $\beta \neq \beta_0$, then $P_{v(Y_t;\beta)} \neq P_{v(Y_t;\beta_0)}$, where $P_{v(Y_t;\beta)}$ represents the marginal probability measure of $v(Y_t;\beta)$.

3.1. Conditional Least Squares Estimation

Let $S(\theta) = \sum_{t=2}^{T}(Y_t - \mathbb{E}(Y_t|Y_{t-1}))^2 = \sum_{t=2}^{T}\left(Y_t - \frac{\exp[v(Y_{t-1};\beta)]}{1+\exp[v(Y_{t-1};\beta)]}Y_{t-1} - \lambda\right)^2$, where $\theta = (\beta', \lambda)$. The CLS estimator is then given by:

$$\hat{\theta}_{CLS} = \mathrm{argmin}_\theta(S(\theta)).$$

Let $S_t(\theta) = (Y_t - \mathbb{E}(Y_t|Y_{t-1}))^2$. The first-order condition equation is represented as follows:

$$-\frac{1}{2}\frac{\partial S_t(\theta)}{\partial \theta} = 0 = M_t(\theta) = \left(m_{t1}(\theta), m_{t2}(\theta), \ldots, m_{t(\ell+1)}(\theta)\right)', \tag{7}$$

where

$$m_{ti}(\theta) = \left(Y_t - \frac{\exp[v(Y_{t-1};\beta)]}{1+\exp[v(Y_{t-1};\beta)]}Y_{t-1} - \lambda\right)\frac{\exp[v(Y_{t-1};\beta)]}{(1+\exp[v(Y_{t-1};\beta)])^2}\frac{\partial v(y;\beta)}{\partial \beta_i}Y_{t-1}, \quad 1 \leq i \leq \ell,$$

$$m_{t(\ell+1)}(\theta) = Y_t - \frac{\exp[v(Y_{t-1};\beta)]}{1+\exp[v(Y_{t-1};\beta)]}Y_{t-1} - \lambda.$$

Thus, the estimating equation is given by $\sum_{t=1}^{T} M_t(\theta) = 0$. Solving this equation provides the CLS estimate $\hat{\theta}_{CLS}$ for the parameter vector $\theta = (\beta, \lambda)$.

Theorem 1. *Under assumptions (A1) to (A5), the CLS estimator $\hat{\theta}_{CLS}$ is a consistent estimator for the true parameter θ_0, and it has an asymptotic distribution:*

$$\sqrt{T}(\hat{\theta}_{CLS} - \theta_0) \xrightarrow{d} N\left(0, V^{-1}(\theta_0)W(\theta_0)V^{-1}(\theta_0)\right),$$

where

$$W(\theta_0) = \mathbb{E}\left(M_t(\theta_0)M_t'(\theta_0)\right),$$

$$V(\theta_0) = \mathbb{E}\left(\frac{\partial \mathbb{E}(Y_t|Y_{t-1})}{\partial \theta} \cdot \frac{\partial \mathbb{E}(Y_t|Y_{t-1})}{\partial \theta'}\right) - \mathbb{E}\left(u_t(\theta_0)\frac{\partial^2 \mathbb{E}(Y_t|Y_{t-1})}{\partial \theta \partial \theta'}\right),$$

$$u_t(\theta_0) = Y_t - \mathbb{E}\left(Y_t\big|Y_{(t-1)}\right).$$

3.2. Interval Estimation

Based on the estimating equations from the CLS estimation, we can construct observation-driven interval estimation and hypothesis testing. Let:

$$H(\theta) = \left(\sum_{t=2}^{T} M_t(\theta)\right)' \left(\sum_{t=2}^{T} M_t(\theta) M_t(\theta)'\right)^{-1} \left(\sum_{t=2}^{T} M_t(\theta)\right).$$

We can then obtain the following theorem:

Theorem 2. *Under assumptions (A1)–(A5), as $T \to \infty$,*

$$H(\theta_0) \xrightarrow{d} \chi^2(\ell+1). \tag{8}$$

Remark 5. *From Equation (8), we can construct an interval estimation for θ_0:*

$$\{\theta | H(\theta) \leq C_\alpha\},$$

where C_α satisfies that for $0 < \alpha < 1$, $\mathbb{P}\left(\chi^2_{\ell+1} \leq C_\alpha\right) = \alpha$. From the perspective of hypothesis testing, this serves as an acceptance region for testing the null hypothesis $\mathbb{H}_0 : \theta_0 = \theta$. If $H(\theta) > C_\alpha$, then the null hypothesis is rejected.

3.3. Empirical Likelihood Test

In the following, we introduce hypothesis testing based on empirical likelihood estimation. First, we provide a brief introduction to the empirical likelihood (EL) method. Initially proposed by Owen [19] for providing interval estimations for expectation, the EL method was later extended to estimating equation estimation by Qin and Lawless [20]. For T observations $y_1, y_2, \ldots, y_T$ of a random variable Y with distribution F, the empirical likelihood ratio is defined as:

$$R(F) = \frac{L(F)}{L(F_T)} = \prod_{t=1}^{T} T p_t,$$

where $L(F) = \prod_{t=1}^{T} p_t$ is the nonparametric likelihood function, $p_t = dF(y_t) = \mathbb{P}(Y = y_t)$, and $F_T(y) = \frac{1}{T} \sum_{t=1}^{T} 1_{\{y_t \leq y\}}$ is the empirical distribution function of the random variable Y, $dF_T = \frac{1}{T}, \forall t \in T$. Under constraints $\sum_{t=1}^{T} p_t = 1$ and $p_t \geq 0, \forall t$, F_T maximizes $L(F)$, so $R(F) \leq 1$.

Suppose we are interested in the parameter vector θ, which satisfies the estimating equation $\mathbb{E}(M_t(\theta)) = 0$. We need to add a new constraint for p_t: $\sum_{t=1}^{T} p_t M_t(\theta) = 0$. Based on this, we can establish the profile empirical likelihood ratio function:

$$\mathcal{R}(\theta) = \sup\left\{\prod_{t=1}^{T} T p_t : p_t \geq 0, \sum_{t=1}^{T} p_t = 1, \sum_{t=1}^{T} p_t M_t(\theta) = 0\right\}.$$

The profile empirical likelihood ratio function can be solved using the Lagrange multiplier method. Let:

$$\mathcal{L}(\theta) = \sum_{t=1}^{T} \log(p_t) + \hbar \left(\sum_{t=1}^{T} p_t - 1\right) + \gamma' T \sum_{t=1}^{T} p_t M_t(\theta),$$

where k and γ are Lagrange multipliers. It can be proved that when $\mathcal{L}(\theta)$ is maximized, $k = T$, and:

$$p_t = \frac{1}{T} \cdot \frac{1}{\gamma' M_t(\theta)}.$$

Here, as a function of θ, $\gamma = \gamma(\theta)$ is the solution to the following equation:

$$\sum_{t=1}^{T} \frac{M_t(\theta)}{1 + \gamma' M_t(\theta)} = 0, \qquad (9)$$

substituting this into p_t and $R(F)$, we find:

$$R(F) = \prod_{t=1}^{T} \frac{1}{1 + \gamma(\theta)' M_t(\theta)}.$$

Thus, the log empirical likelihood ratio function can be defined as:

$$\mathcal{L}_E(\theta) = -\log(\mathcal{R}(\theta)) = \sum_{t=1}^{T} \log\left[1 + \gamma(\theta)' M_t(\theta)\right].$$

The empirical likelihood estimate is then given by:

$$\hat{\theta}_{EL} = \operatorname{argmin}_\theta(\mathcal{L}_E(\theta)).$$

The corresponding γ is denoted by $\hat{\gamma}(\hat{\theta}_{EL})$.

Remark 6. *Given that $0 \leq p_t \leq 1$ for all $t \in T$, it can be deduced that $\mathcal{L}_E(\theta) = -\log\left(\prod_{t=1}^{T} p_t\right) \geq 0$.*

Remark 7. *Since the number of estimating equations matches the number of parameters to be estimated (also known as just-identified in some econometrics literature), and $\hat{\theta}_{CLS}$ is the solution to the estimating equation $\sum_{t=1}^{T} M_t(\theta) = 0$, it follows from Chen and Keilegom [21] that:*

$$\hat{\theta}_{EL} = \hat{\theta}_{CLS}.$$

Therefore, we will omit empirical likelihood estimation in the point estimation segment in the numerical simulation section.

Theorem 3. *Under assumptions (A1)–(A5), let $\theta = \left(\theta_1', \theta_2'\right)'$, where θ_1 and θ_2 are $q \times 1$ and $(\ell + 1 - q) \times 1$-dimensional parameter vectors to be estimated, respectively. For the hypothesis $\mathbb{H}_0: \theta^{(1)} = \theta_0^{(1)}$, a test statistic can be constructed as follows:*

$$\mathcal{L}_E\left(\theta_0^{(1)}, \widetilde{\theta}_{EL}^{(2)}\right) - \mathcal{L}_E\left(\hat{\theta}_{EL}^{(1)}, \hat{\theta}_{EL}^{(2)}\right) \xrightarrow{d} \chi^2(q),$$

where $\left(\hat{\theta}_{EL}^{(1)}, \hat{\theta}_{EL}^{(2)}\right) = \hat{\theta}_{EL}$, and $\widetilde{\theta}_{EL}^{(2)}$ is the estimate obtained by minimizing $\mathcal{L}_E\left(\theta_0^{(1)}, \theta^{(2)}\right)$ concerning $\theta^{(2)}$.

Remark 8. *As Remark 7 indicates, in a just-identified situation, $\hat{\theta}_{EL} = \hat{\theta}_{CLS}$ and $\mathcal{L}_E(\hat{\theta}_{CLS}) = 0$. Thus, the conclusion of Theorem 3 can be further simplified as:*

$$\mathcal{L}_E\left(\theta_0^{(1)}, \widetilde{\theta}_{EL}^{(2)}\right) \xrightarrow{d} \chi^2(q).$$

3.4. Conditional Maximum Likelihood Estimation

It is straightforward to derive the log-likelihood function $logL(\theta)$ from the one-step-ahead transition probability (6) of model (4). In time-series models, the probability distribution of the first observation Y_1 is unknown, and its influence on the likelihood function is minimal when the sample size T is sufficiently large. Thus, we focus only on the conditional likelihood function. Given that the log-conditional likelihood function is a nonlinear function of the parameter vector $\theta = (\beta, \lambda)$, we employ numerical methods to solve:

$$\hat{\theta}_{CML} = argmin_\theta(logL(\theta)).$$

To obtain the asymptotic distribution of $\hat{\theta}_{CML}$, we need to verify the regularity conditions presented in Billingsley [22]. The satisfaction of these conditions can be directly observed from the model-building process in Section 2 and the assumptions provided in Section 3. Therefore, the proof is omitted. We arrive at the following theorem:

Theorem 4. *Under assumptions (A1)–(A6), the conditional maximum likelihood estimator $\hat{\theta}_{CML}$ consistently estimates the true parameter θ_0 and exhibits an asymptotic distribution:*

$$\sqrt{T}(\hat{\theta}_{CML} - \theta_0) \xrightarrow{d} N(0, E^{-1}),$$

where $E = \mathbb{E}\left(\frac{\partial log(\mathbb{P}(X_1|X_0))}{\partial \theta} \cdot \frac{\partial log(\mathbb{P}(X_1|X_0))}{\partial \theta'}\right)$ represents the Fisher information matrix.

Remark 9. *Achieving CML estimation requires making specific assumptions about the probability distribution of Z_t. In this paper, we assume Z_t follows a Poisson distribution with parameter λ. This strong assumption can result in significant errors or even inconsistency in statistical inference based on the CML method if the assumed model does not represent the true data-generating process. This constitutes the primary drawback of CML estimation. The impact of model mis-specification on CML estimation will be examined in the following numerical simulation section.*

4. Numerical Simulation

In this section, we set the function ν as a linear function, considering the following data-generating process:

$$Y_t = \phi_t \ominus Y_{t-1} + Z_t, \tag{10}$$

$$\mathbb{E}(\phi_t | Y_t) = \frac{\exp(\beta_0 + \beta_1 Y_{t-1})}{1 + \exp(\beta_0 + \beta_1 Y_{t-1})}. \tag{11}$$

Here, $\{Z_t\}$ represents an independently and identically distributed Poisson random variable sequence with a mean of λ. In subsequent numerical simulation studies, we mainly concentrate on three aspects: parameter estimation, interval estimation, and empirical likelihood ratio testing. All numerical simulations are conducted based on 1000 repeated sampling.

4.1. Parameter Estimation

We generate data using the above model and apply the CLS and CML methods to estimate parameters. Moreover, we define three statistical measures for evaluating estimation performance (using λ as an example):

$$\text{Sample bias}: \text{Bias} = \bar{\lambda} - \lambda,$$

$$\text{Root mean square error}: \text{RMSE} = \sqrt{\frac{1}{1000}\sum_{i=1}^{1000}(\hat{\lambda}_i - \lambda)^2},$$

Mean absolute percentage error: $\text{MAPE} = \frac{1}{1000}\sum_{i=1}^{1000}\left|\frac{\hat{\lambda}_i - \lambda}{\lambda}\right|$.

In CML estimation, the score function is defined as:

$$\sum_{t=1}^{T}\frac{\left(\frac{\partial}{\partial\theta}\right)\left\{\int \sum_{k=0}^{y_t}\frac{(\phi y_{t-1})^k}{k!}\exp(-\phi y_{t-1})f_z(y_t-k)\mathbb{P}(\phi_t=\phi|Y_{t-1}=y_{t-1})d\phi\right\}}{\int \sum_{k=0}^{y_t}\frac{(\phi y_{t-1})^k}{k!}\exp(-\phi y_{t-1})f_z(y_t-k)\mathbb{P}(\phi_t=\phi|Y_{t-1}=y_{t-1})d\phi} = 0.$$

In the CML estimation, we primarily consider four distribution cases for $\phi_t|Y_{t-1}$ when Z_t follows a Poisson distribution. Let the variable $A_t = \frac{\exp(\beta_0+\beta_1 Y_{t-1})}{1+\exp(\beta_0+\beta_1 Y_{t-1})}$, and the function $dpois(x,l) = \frac{l^x}{x!}\exp(-l), l \geq 0, x \in \mathbb{N}$. Then:

(i) $\phi_t|Y_{t-1}$ is fixed at A_t, without any randomness. In this case, the log-likelihood function is:

$$logL(\theta) = -\sum_{t=2}^{T}\log\left(\sum_{k=0}^{y_t}(dpois(k,y_{t-1}A_t)\cdot dpois(y_t-k,\lambda))\right).$$

(ii) $\phi_t|Y_{t-1}$ follows a uniform distribution with mean A_t, minimum value 0, and maximum value $2A_t$. In this case, the log-likelihood function is:

$$logL(\theta) = -\sum_{t=2}^{T}\log\left(\sum_{k=0}^{y_t}\frac{dpois(y_t-k,\lambda)}{2k!y_{t-1}A_t}\cdot(\Gamma(k+1,0)-\Gamma(k+1,2A_t))\right).$$

where $\Gamma(\alpha,x) = \int_{x}^{\infty}t^{\alpha-1}\exp(-t)dt$.

(iii) $\phi_t|Y_{t-1}$ follows an exponential distribution with mean A_t. In this case, the log-likelihood function is:

$$logL(\theta) = -\sum_{t=2}^{T}\log\left(\sum_{k=0}^{y_t}\frac{A_t}{(A_t+y_{t-1})^{k+1}}\cdot y_{t-1}^k\cdot dpois(y_t-k,\lambda)\right).$$

(iv) $\phi_t|Y_{t-1}$ follows a chi-square distribution with the mean A_t. Specifically, the density function of $\phi_t|Y_{t-1}$ is:

$$\mathbb{P}(\phi_t=\phi|Y_{t-1}=y_{t-1}) = \frac{1}{2^{A_t}\Gamma(A_t/2)}\phi^{\frac{A_t}{2}-1}\exp\left(-\frac{\phi}{2}\right).$$

Although A_t is not an integer, we still call it a chi-square distribution. In this case, the log-likelihood function is:

$$logL(\theta) = -\sum_{t=2}^{T}\log\left(\sum_{k=0}^{y_t}\frac{y_{t-1}}{k!}\cdot dpois(y_t-k,\lambda)\cdot\frac{1}{2^{\frac{A_t}{2}}\Gamma\left(\frac{A_t}{2},0\right)}\cdot\frac{\Gamma\left(\frac{A_t+2k}{2},0\right)}{\left(\frac{1}{2}+y_{t-1}\right)^{\frac{A_t+2k}{2}}}\right).$$

The specific simulation results are shown in the table below:

From Table 1, we can observe that for both CLS and CML estimators, as the sample size T gradually increases, BIAS, RMSE, and MADE all decline, indicating the consistency of these estimators. Notably, both CLS and CML yield satisfactory parameter estimates. In large samples, CLS and CML estimates are approximately equal, while in small samples, under the premise of a correctly specified model, CML tends to provide superior estimation precision. Furthermore, we present an additional set of parameter estimation simulation results in the Appendix A, as shown in Table A1.

Table 1. Parameter Estimation Simulation Results.

Sample Size	$\beta_0^{(CLS)}$	$\beta_0^{(CML)}$	$\beta_1^{(CLS)}$	$\beta_1^{(CML)}$	$\lambda^{(CLS)}$	$\lambda^{(CML)}$
Parameter: $\beta_0 = 1$, $\beta_1 = -0.6$, $\lambda = 1.2$ $\phi_t\|y_{t-1}$ is fixed.						
T = 300						
BIAS	0.0571	0.0471	−0.0321	−0.0287	0.0051	0.0059
RMSE	0.7399	0.6983	0.2096	0.2008	0.1368	0.1337
MAPE	0.5636	0.5486	0.2691	0.2619	0.0909	0.0886
T = 500						
BIAS	0.0506	0.0407	−0.0251	−0.0221	0.0033	0.0042
RMSE	0.5678	0.5562	0.1556	0.1523	0.1113	0.1091
MAPE	0.4443	0.4346	0.1978	0.1946	0.0738	0.0721
T = 800						
BIAS	0.0349	0.0246	−0.0152	−0.0127	−0.0011	0.0004
RMSE	0.4165	0.4076	0.1188	0.1163	0.0828	0.0817
MAPE	0.3327	0.3254	0.1587	0.1554	0.0546	0.0535
T = 1200						
BIAS	0.0139	0.0071	−0.0074	−0.0055	0.0009	0.0017
RMSE	0.3471	0.3393	0.0951	0.0931	0.0697	0.0688
MAPE	0.2726	0.2686	0.1252	0.1234	0.0465	0.0459
T = 2000						
BIAS	0.0112	0.0085	−0.0058	−0.0053	0.0017	0.0023
RMSE	0.2719	0.2711	0.0732	0.0728	0.0533	0.0525
MAPE	0.2195	0.2176	0.0981	0.0978	0.0352	0.0347
Parameter: $\beta_0 = 1$, $\beta_1 = -0.6$, $\lambda = 1.2$ $\phi_t\|y_{t-1}$ follows a uniform distribution.						
T = 300						
BIAS	0.0865	0.0428	−0.0456	−0.0354	0.0046	0.0121
RMSE	0.8065	0.7395	0.2301	0.2163	0.1454	0.1361
MAPE	0.6267	0.5773	0.2865	0.2696	0.0964	0.0903
T = 500						
BIAS	0.0312	0.0076	−0.0228	−0.0169	0.0043	0.0082
RMSE	0.5636	0.5288	0.1567	0.1488	0.1052	0.0997
MAPE	0.4493	0.4239	0.2046	0.1968	0.0703	0.0657
T = 800						
BIAS	0.0292	0.0062	−0.0165	−0.0113	0.0038	0.0079
RMSE	0.4503	0.4233	0.1244	0.1191	0.0852	0.0793
MAPE	0.3587	0.3373	0.1651	0.1575	0.0563	0.0525

Table 1. *Cont.*

Sample Size	$\beta_0^{(CLS)}$	$\beta_0^{(CML)}$	$\beta_1^{(CLS)}$	$\beta_1^{(CML)}$	$\lambda^{(CLS)}$	$\lambda^{(CML)}$
T = 1200						
BIAS	0.0249	0.0133	−0.0127	−0.0108	0.0003	0.0031
RMSE	0.3513	0.3295	0.0971	0.0923	0.0689	0.0639
MAPE	0.2815	0.2627	0.1289	0.1249	0.0464	0.0428
T = 2000						
BIAS	0.0062	−0.0019	−0.0041	−0.0023	0.0016	0.0032
RMSE	0.2735	0.2529	0.0749	0.0719	0.0529	0.0483
MAPE	0.2165	0.1997	0.0983	0.0942	0.0353	0.0323

Parameter : $\beta_0 = 1$, $\beta_1 = -0.6$, $\lambda = 1.2$
$\phi_t|y_{t-1}$ follows an exponential distribution.

T = 300						
BIAS	0.1165	0.0594	−0.0594	−0.0491	0.0048	0.0135
RMSE	0.8356	0.7986	0.2648	0.2541	0.1407	0.1138
MAPE	0.6249	0.5392	0.3071	0.2785	0.0931	0.0752
T = 500						
BIAS	0.0174	−0.0175	−0.0195	−0.0116	0.0019	0.0088
RMSE	0.5929	0.5009	0.1649	0.1507	0.1059	0.0871
MAPE	0.4677	0.3955	0.2133	0.1932	0.0701	0.0582
T = 800						
BIAS	0.0389	0.0125	−0.0177	−0.0119	−0.0008	0.0042
RMSE	0.4646	0.3871	0.1267	0.1149	0.0839	0.0657
MAPE	0.3673	0.3052	0.1644	0.1486	0.0563	0.0438
T = 1200						
BIAS	0.0236	0.0014	−0.0103	−0.0057	0.0016	0.0057
RMSE	0.3709	0.3109	0.0997	0.0903	0.0687	0.0542
MAPE	0.2879	0.2472	0.1299	0.1201	0.0451	0.0362
T = 2000						
BIAS	0.0196	0.0074	−0.0091	−0.0072	−0.0021	0.0009
RMSE	0.2837	0.2493	0.0795	0.0746	0.0527	0.0427
MAPE	0.2261	0.1991	0.1047	0.0983	0.0356	0.0286

Parameter : $\beta_0 = 1$, $\beta_1 = -0.6$, $\lambda = 1.2$
$\phi_t|y_{t-1}$ follows a chi−square distribution.

T = 300						
BIAS	0.9382	0.2286	−0.3652	−0.1152	−0.0292	0.0041
RMSE	3.7397	1.2307	1.7201	0.5974	0.1471	0.0955
MAPE	1.3992	0.7657	0.8326	0.4475	0.0945	0.0636
T = 500						
BIAS	0.3437	0.1486	−0.1325	−0.0738	−0.0213	0.0007
RMSE	1.0769	0.7791	0.3808	0.2767	0.1129	0.0737
MAPE	0.7455	0.5794	0.4262	0.3345	0.0738	0.0493

Table 1. Cont.

Sample Size	$\beta_0^{(CLS)}$	$\beta_0^{(CML)}$	$\beta_1^{(CLS)}$	$\beta_1^{(CML)}$	$\lambda^{(CLS)}$	$\lambda^{(CML)}$
T = 800						
BIAS	0.1769	0.0771	−0.0628	−0.0339	−0.0139	−0.0006
RMSE	0.7215	0.5257	0.2459	0.1844	0.0889	0.0556
MAPE	0.5301	0.4118	0.2954	0.2363	0.0586	0.0374
T = 1200						
BIAS	0.0883	0.0452	−0.0322	−0.0216	−0.0054	0.0012
RMSE	0.5649	0.4368	0.1849	0.1498	0.0703	0.0455
MAPE	0.4353	0.3445	0.2367	0.1949	0.0469	0.0299
T = 2000						
BIAS	0.0766	0.0269	−0.0292	−0.0128	−0.0057	0.0005
RMSE	0.4163	0.3267	0.1345	0.1103	0.0542	0.0371
MAPE	0.3256	0.2585	0.1706	0.1441	0.0361	0.0246

Figure 1 showcases the typical trajectory of data generated by models (10) and (11) with parameters $\beta_0 = 1$, $\beta_1 = -0.6$, and $\lambda = 1.2$. In this figure, "fixed" represents $\phi_t|y_{t-1}$ as a fixed parameter given y_{t-1}, "uniform" denotes $\phi_t|y_{t-1}$ following a uniform distribution, "exponential" signifies $\phi_t|y_{t-1}$ following an exponential distribution, and "chi-square" indicates $\phi_t|y_{t-1}$ following a chi-square distribution. Figure 1 reveals that some extreme values are present in the sample paths when $\phi_t|y_{t-1}$ follows either an exponential or chi-square distribution, with the latter capable of generating even higher extreme values. This suggests that these two distribution settings for $\phi_t|y_{t-1}$ contain a certain descriptive ability concerning the extreme values in the data.

As pointed out in Section 3.4, the CML method depends upon correct model specification. To evaluate the effects of model misspecification on parameter estimation, we consider $\{Z_t\}$ as an independently and identically distributed geometric random-variable sequence with a mean of λ within the data generation process (10) and (11). Subsequently, we employ both the CLS and CML methods for estimation, presenting the results in the table below.

From Table 2, we can observe that the three statistical measures BIAS, RMSE, and MAPE for the CML estimator have noticeably increased compared to the CLS estimator. This indicates that model misspecification significantly impacts CML estimation, necessitating appropriate model selection efforts before employing the CML estimation method. As long as the conditional expectation $\mathbb{E}(Y_t|Y_{t-1})$ is correctly specified, CLS estimation will be more robust than CML estimation. Moreover, we provide the parameter estimation simulation results obtained under the misspecification of the $\phi_t|y_{t-1}$ distribution in the Appendix A, as shown in Table A2.

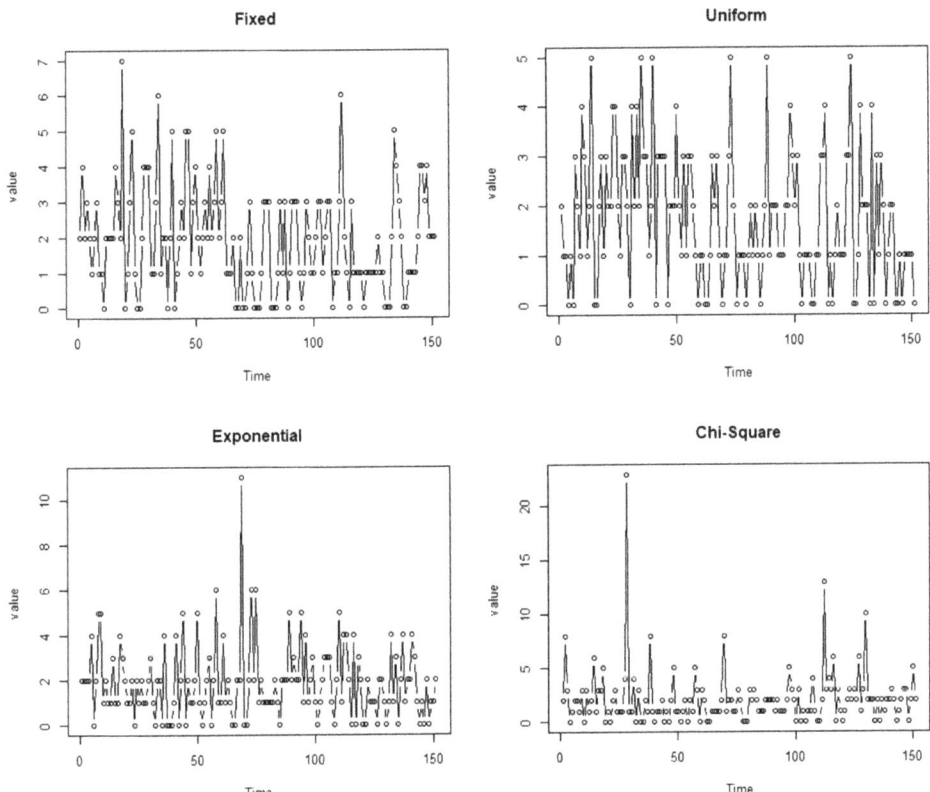

Figure 1. Typical trajectory of the model with $\beta_0 = 1$, $\beta_1 = -0.6$, and $\lambda = 1.2$.

Table 2. Parameter Estimation Simulation Results under Model Misspecification.

Sample Size	$\beta_0^{(CLS)}$	$\beta_0^{(CML)}$	$\beta_1^{(CLS)}$	$\beta_1^{(CML)}$	$\lambda^{(CLS)}$	$\lambda^{(CML)}$
Parameter : $\beta_0 = 1$, $\beta_1 = -0.6$, $\lambda = 1.2$						
$\phi_t \vert y_{t-1}$ follows a uniform distribution, Z_t follows a geometric distribution.						
T = 300						
BIAS	0.1823	0.8261	−0.1275	−0.1563	−0.0057	−0.1187
RMSE	1.2279	1.5387	0.7587	0.4665	0.1577	0.1798
MAPE	0.8237	1.0551	0.4661	0.4531	0.1027	0.1257
T = 500						
BIAS	0.0931	0.7121	−0.0632	−0.1079	−0.0009	−0.1198
RMSE	0.7686	1.0457	0.4375	0.2613	0.1198	0.1574
MAPE	0.5752	0.8394	0.3016	0.3169	0.0786	0.1118
T = 800						
BIAS	0.0858	0.6913	−0.0346	−0.0914	0.0001	−0.1199
RMSE	0.5812	0.9088	0.1651	0.1954	0.1006	0.1468
MAPE	0.4509	0.7538	0.2049	0.2451	0.0657	0.1069

Table 2. Cont.

Sample Size	$\beta_0^{(CLS)}$	$\beta_0^{(CML)}$	$\beta_1^{(CLS)}$	$\beta_1^{(CML)}$	$\lambda^{(CLS)}$	$\lambda^{(CML)}$
T = 1200						
BIAS	0.0193	0.6389	−0.0132	−0.0732	0.0043	−0.01191
RMSE	0.4427	0.7848	0.1234	0.1503	0.0829	0.1385
MAPE	0.3495	0.6687	0.1607	0.1913	0.0545	0.1027
T = 2000						
BIAS	0.0224	0.6386	−0.0116	−0.0711	0.0021	−0.1213
RMSE	0.3576	0.7362	0.0951	0.1243	0.0625	0.1321
MAPE	0.2796	0.6517	0.1234	0.1612	0.0416	0.1016

4.2. Interval Estimation

We perform a numerical simulation study on the coverage frequency of the interval estimation, as proposed in Theorem 2 and Remark 5, for the true values in the model. We consider parameter settings of $\beta_0 = 1$, $\beta_1 = -0.6$, and $\lambda = 1.2$. The nominal levels considered are 0.90 and 0.95, with the specific simulation results presented in the following table:

From Table 3, we can observe that as the sample size T increases, the coverage frequency of interval estimation gradually approaches the nominal level. Even with smaller sample sizes, the coverage frequency of the interval estimation for the true values remains satisfactory. This result suggests that the data-driven interval estimation has achieved commendable performance.

Table 3. Coverage Frequency of Interval Estimation.

Parameter: $\beta_0 = 1, \beta_1 = -0.6, \lambda = 1.2$ $\phi_t\|y_{t-1}$ is fixed.					
T	300	500	800	1200	2000
0.95	0.941	0.957	0.957	0.953	0.956
0.9	0.897	0.908	0.912	0.908	0.905
Parameter: $\beta_0 = 1, \beta_1 = -0.6, \lambda = 1.2$ $\phi_t\|y_{t-1}$ follows a uniform distribution.					
T	300	500	800	1200	2000
0.95	0.949	0.959	0.961	0.949	0.954
0.9	0.89	0.913	0.899	0.904	0.903
Parameter: $\beta_0 = 1, \beta_1 = -0.6, \lambda = 1.2$ $\phi_t\|y_{t-1}$ follows an exponential distribution.					
T	300	500	800	1200	2000
0.95	0.942	0.938	0.951	0.955	0.953
0.9	0.891	0.894	0.906	0.910	0.909
Parameter: $\beta_0 = 1, \beta_1 = -0.6, \lambda = 1.2$ $\phi_t\|y_{t-1}$ follows a chi−square distribution.					
T	300	500	800	1200	2000
0.95	0.905	0.917	0.918	0.92	0.939
0.9	0.854	0.853	0.856	0.864	0.881

4.3. Empirical Likelihood Test

Lastly, we perform a numerical simulation study on the empirical likelihood test (EL test). For the observation-driven parameter model defined by data generation processes (10) and (11), we aim to test whether β_1 equals 0. If $\beta_1 = 0$, our model's parameters are not driven by observations. We employ models (10) and (11) to generate sequences, assuming $\phi_t | y_{t-1}$ is a fixed parameter, and perform estimation under the null hypothesis. Then, we compare the test statistic proposed in Theorem 3 with the upper 0.90 and 0.95 quantiles of the corresponding chi-square distribution; if the EL test statistic exceeds the critical value, we reject the null hypothesis.

Initially, we investigate scenarios in which the true value of β_1 for the data generation process equals 0, considering the following hypotheses:

$$\mathbb{H}_0 : \beta_1 = b \neq 0 \qquad \mathbb{H}_1 : \beta_1 \neq b.$$

where b is a nonnegative constant, the simulation results for the test power are presented below (the simulation results for $\mathbb{H}_0 : \beta_1 = 0$ represent the frequency of Type I errors).

Next, we examine the scenarios where the true value of β_1 in the data generation process is not equal to 0, considering the following hypotheses:

$$\mathbb{H}_0 : \beta_1 = 0 \qquad \mathbb{H}_1 : \beta_1 \neq 0.$$

The simulation results for the test power are as follows.

From Tables 4 and 5, we observe that the Type I error frequency of the EL test gradually diminishes to the corresponding confidence level as the sample size T increases, while the test power concurrently ascends to 1. Notably, in small sample scenarios, when the true value of β_1 is 0, the test power level for $\mathbb{H}_0 : \beta_1 = -0.1$ is relatively low. Likewise, when the true value of β_1 is -0.1, the test power for $\mathbb{H}_0 : \beta_1 = 0$ exhibits a similar pattern. Overall, however, the EL test performs satisfactorily when the gap between the true and hypothesized values of β_1 is relatively large, or in cases involving large samples. Owing to space constraints, we include in the Appendix A, the EL test simulation results for the parameter λ under $\phi_t | y_{t-1}$ following four distinct random distributions, as shown in Table A3.

Table 4. Empirical Likelihood Test for β_1 with a True Value of 0.

Parameter : $\beta_0 = 1, \beta_1 = 0, \lambda = 1.2$ $\phi_t \| y_{t-1}$ is fixed, significance level 0.05.					
T	300	500	800	1200	2000
$\mathbb{H}_0 : \beta_1 = 0$ (true)	0.096	0.073	0.065	0.057	0.046
$\mathbb{H}_0 : \beta_1 = -0.1$	0.296	0.386	0.658	0.823	0.935
$\mathbb{H}_0 : \beta_1 = -0.2$	0.707	0.802	0.941	0.984	1
$\mathbb{H}_0 : \beta_1 = -0.3$	0.778	0.837	0.988	1	1
$\mathbb{H}_0 : \beta_1 = -0.4$	0.822	0.861	0.997	1	1
Parameter : $\beta_0 = 1, \beta_1 = 0, \lambda = 1.2$ $\phi_t \| y_{t-1}$ is fixed, significance level 0.10.					
T	300	500	800	1200	2000
$\mathbb{H}_0 : \beta_1 = 0$ (true)	0.146	0.126	0.110	0.103	0.107
$\mathbb{H}_0 : \beta_1 = -0.1$	0.399	0.447	0.716	0.874	0.976
$\mathbb{H}_0 : \beta_1 = -0.2$	0.784	0.883	0.969	1	1
$\mathbb{H}_0 : \beta_1 = -0.3$	0.823	0.904	0.993	1	1
$\mathbb{H}_0 : \beta_1 = -0.4$	0.875	0.921	1	1	1

Table 5. Empirical Likelihood Test for β_1 with True Value Not Equal to 0.

Parameter: $\beta_0 = 1, \mathbb{H}_0 : \beta_1 = 0, \lambda = 1.2$ $\phi_t\|y_{t-1}$ is fixed, significance level 0.05.					
T	300	500	800	1200	2000
$\beta_1 = -0.1$ (true)	0.363	0.536	0.608	0.751	0.907
$\beta_1 = -0.2$ (true)	0.647	0.806	0.936	0.988	1
$\beta_1 = -0.3$ (true)	0.768	0.935	1	1	1
$\beta_1 = -0.4$ (true)	0.875	0.945	1	1	1
Parameter: $\beta_0 = 1, \mathbb{H}_0 : \beta_1 = 0, \lambda = 1.2$ $\phi_t\|y_{t-1}$ is fixed, significance level 0.10.					
T	300	500	800	1200	2000
$\beta_1 = -0.1$ (true)	0.439	0.705	0.767	0.859	0.966
$\beta_1 = -0.2$ (true)	0.751	0.877	0.96	1	1
$\beta_1 = -0.3$ (true)	0.835	0.99	1	1	1
$\beta_1 = -0.4$ (true)	0.941	0.997	1	1	1

It is crucial to note that the estimation equation employed in the empirical likelihood test solely reflects the linear mean structure inherent in the data-generating process. For more intricate and nonlinear coefficient random distributions, the test exhibits limited descriptive capacity. As a result, we advise against utilizing the empirical likelihood test in cases where $\phi_t|y_{t-1}$ is stochastic. In Appendix A, we present numerical simulation results pertaining to the empirical likelihood test when $\phi_t|y_{t-1}$ adheres to an exponential distribution. As evidenced by Table A4, the empirical likelihood test demonstrates a very high frequency of Type I errors when $\phi_t|y_{t-1}$ conforms to an exponential distribution. Consequently, we discourage the use of the empirical likelihood test in such circumstances.

5. Real Data Application

In this section, we analyze the daily download count data for the software CWB TeXpert, covering the period from 1 June 2006, to 28 February 2007, resulting in a sample size of T = 267. This dataset is made available on the Supplementary webpage associated with Weiß [23].

From the sample path in Figure 2, we observe that this data contains a considerable number of extreme values. Simultaneously, the ACF and PACF plots suggest that the sample might have originated from a first-order autoregressive data-generating process. We proceed to analyze this data using the models introduced in this paper. For the CML estimation, CML_{fix} in the table below represents $\phi_t|y_{t-1}$ as a fixed parameter, CML_{unif} denotes $\phi_t|y_{t-1}$ following a uniform distribution, CML_{exp} signifies $\phi_t|y_{t-1}$ following an exponential distribution, and CML_{chi} indicates $\phi_t|y_{t-1}$ following a chi-square distribution. Additionally, for comparison purposes, we applied the model proposed by Yu et al. [18] to this dataset, which is denoted as CML_{geom} in the subsequent table:

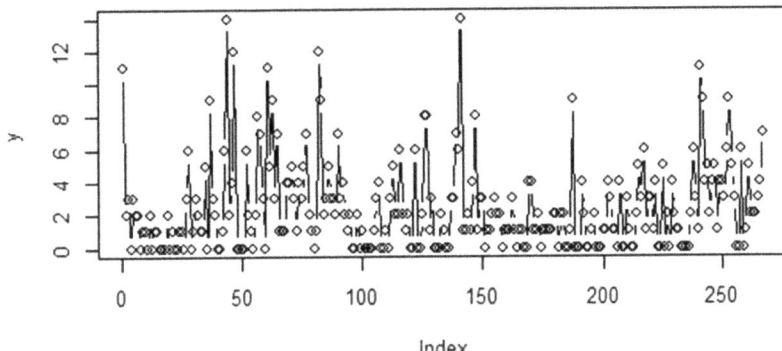

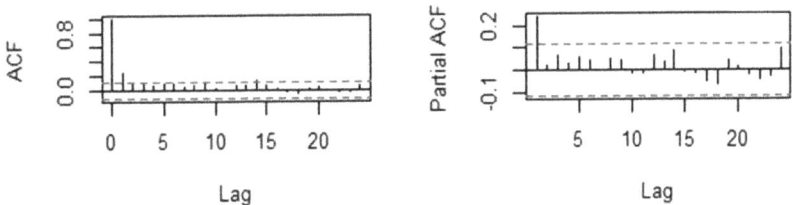

Figure 2. Sample Path of Software Download Data and Corresponding ACF and PACF Plots.

The estimation results are displayed in Table 6, where we provide AIC and BIC values for the four distributions that $\phi_t|y_{t-1}$ may follow. Based on these two information criteria, we show a preference for models in which $\phi_t|y_{t-1}$ follows either a chi-square distribution or an exponential distribution. This preference might be attributable to the presence of extreme values in the sample path, as anticipated. As observed in Figure 1 in Section 4, models with $\phi_t|y_{t-1}$ following either a chi-square or exponential distribution prove more effective in capturing data characterized by extreme values.

Table 6. Model Estimation Results.

	CLS	CML_{fix}	CML_{unif}	CML_{exp}	CML_{chi}	CML_{geom}
β_0	0.302	0.209	1.379	1.305	0.658	1.244
β_1	−0.151	−0.143	−0.227	−0.244	−0.097	−0.231
λ	1.463	1.493	1.201	1.196	1.359	1.166
AIC	-	1243.986	1189.377	1151.465	1143.669	1184.96
BIC	-	1254.748	1200.138	1162.227	1154.431	1195.322

6. Discussion and Conclusions

In this paper, we propose a first-order integer-valued autoregressive time series model based on the Poisson thinning operator. The parameters of this model are observation-driven and may follow specific random distributions, resulting in time-varying autoregressive coefficients. We established the ergodicity of this model and performed estimation and hypothesis testing using conditional least squares (CLS), conditional maximum likelihood (CML), and empirical likelihood (EL) methods. Additionally, we provided a data-driven interval estimation.

In the numerical simulation study, we compared the parameter estimation performance of CLS and CML, verified the coverage frequency of the interval estimation for the true parameter values in the data generation process, and conducted corresponding simulation studies for the EL test. The simulation study reveals that the properties of the CML estimation depend on the correct model specification, while the CLS estimation demonstrates a degree of robustness against model misspecifications.

In future research, observation-driven parameter integer-valued time series models offer numerous promising avenues for development. In this discussion, a brief overview of some of these directions is provided:

(1) Combining observation-driven parameters with self-driven parameters, namely self-exciting threshold models: the SETINAR model proposed by Montriro, Scotto, and Pereira [16] is defined as follows:

$$Y_t = \begin{cases} \sum_{i=1}^{p^{(1)}} \alpha_i^{(1)} \circ Y_{t-i} + Z_t^{(1)}, & Y_{t-d} \leq R, \\ \sum_{i=1}^{p^{(2)}} \alpha_i^{(2)} \circ Y_{t-i} + Z_t^{(2)}, & Y_{t-d} > R, \end{cases} \quad (12)$$

in this model, $p^{(1)}$ and $p^{(2)}$ represent given positive integers, with $\sum_{i=1}^{p^{(j)}} \alpha_i^{(j)} \in (0,1)$ for $j = 1, 2$. Additionally, the innovation series $\{Z_t^{(1)}\}$ and $\{Z_t^{(2)}\}$ possess probability distributions F_1 and F_2 on the set of natural numbers $\mathbb{N}_0$, respectively. The constant R represents the threshold value responsible for the structural transition in the lagged d-period observation excitation model. Montriro, Scotto, and Pereira [16] demonstrated that model 6.1 possesses a strictly stationary solution when $p^{(1)} = p^{(2)} = 1$. By effectively combining observation-driven parameter models with self-driven parameter models and flexibly selecting thinning operators, a more diverse range of integer-valued time series models can be characterized.

(2) Expanding upon current observation-driven models to incorporate higher-order models: Du and Li [24] introduced the INAR(p) model:

$$Y_t = \alpha_1 \circ Y_{t-1} + \cdots + \alpha_p \circ Y_{t-p} + Z_t, \quad (13)$$

in this model, $\sum_{i=1}^{p} \alpha_i < 1$, and $\{Z_t\}$ represents a sequence of integer-valued random variables defined on the set of natural numbers $\mathbb{N}_0$. Existing observation-driven models are primarily first-order models. By extending these models to higher-order versions, the capability to describe more intricate and complex parameter dynamics can be achieved. It is important to note that when progressing to higher-order models, the technique utilized in the proof of Property 2. is no longer applicable for establishing the model's ergodicity. As a result, new proof methods need to be sought from related Markov chain theories.

(3) Extending the observation-driven parameter setting to Integer-valued Autoregressive Conditional Heteroskedasticity (INARCH) models: Fokianos, Rahbek, and Tjøstheim [25] proposed the INARCH model (which they referred to as Poisson Autoregressive) as follows:

$$\begin{aligned} Y_t|\mathcal{F}_{t-1} &\sim Poisson(\lambda_t), \\ \lambda_t &= d + \alpha \lambda_{t-1} + \beta Y_{t-1}, \end{aligned} \quad (14)$$

where $\alpha \geq 0$, $\beta \geq 0$, and $\alpha + \beta < 1$. This model is a natural extension of the generalized linear model and helps to capture the fluctuating changes of observed variables over time. Another advantage of this model is its simplicity, which makes it easy to establish the likelihood function of the INARCH model. Extending the observation-driven parameter setting to integer-valued autoregressive conditional heteroskedasticity models allows the model to describe the driving effect of the fluctuations of observed variables on the parameters. However, the challenge in doing

(4) Forecasting Integer-Valued Time Series: In time series research, it is common to employ h-step forward conditional expectations for forecasting:

$$\hat{Y}_{t+h} = \mathbb{E}(Y_{t+h}|Y_t) \tag{15}$$

Nonetheless, this approach does not guarantee that the predicted values will be integers, and such predictions primarily describe the expected characteristics of the model, without capturing potentially time-varying coefficients or other features, as illustrated in Figure A1. Furthermore, Freeland and McCabe [26] highlighted that utilizing conditional medians or conditional modes for forecasting could be misleading. Consequently, it is essential to adopt innovative forecasting methods for integer-valued time series analysis. The rapid advancement of machine learning and deep learning in recent years has offered numerous new perspectives, such as the deep autoregressive model based on autoregressive recurrent neural network proposed by Salinas, Flunkert, Gasthaus, and Januschowski [10], which may hold significant potential for widespread application in the domain of integer-valued time series.

Supplementary Materials: The following supporting information can be downloaded at: https://www.mdpi.com/article/10.3390/e25060859/s1.

Author Contributions: Conceptualization, K.Y. and T.T.; methodology, T.T.; software, T.T.; validation, K.Y. and T.T.; formal analysis, T.T.; investigation, T.T.; resources, K.Y.; data curation, K.Y.; writing—original draft preparation, T.T.; writing—review and editing, K.Y.; visualization, T.T.; supervision, K.Y.; project administration, K.Y.; funding acquisition, K.Y. All authors have read and agreed to the published version of the manuscript.

Funding: This research was funded by the National Social Science Fund of China (No. 18BTJ039).

Data Availability Statement: The following supporting data can be downloaded at: http://www.wiley.com/go/weiss/discrete-valuedtimeseries, (accessed on 27 April 2023). The code has been uploaded as Supplementary File of this paper. Interested readers are also encouraged to request the relevant data and code from the authors directly through e-mail.

Conflicts of Interest: The authors declare no conflict of interest.

Appendix A

Appendix A.1. Proofs

Property A1.

(i) Given the data generation process (4), the following can be proved using the law of iterated expectation:

$$\mathbb{E}(Y_t|Y_{t-1};\phi_t) = \phi_t Y_{t-1} + \lambda,$$

$$Var(Y_t|Y_{t-1};\phi_t) = \phi_t Y_{t-1} + \sigma_Z^2.$$

Using the formula $Var(Y) = Var(\mathbb{E}(Y|X)) + \mathbb{E}(Var(Y|X))$, the result can be proved

(ii) By the law of iterated expectation, we know:

$$\mathbb{E}(Y_t Y_{t-1}) = \mathbb{E}(Y_{t-1}\mathbb{E}(Y_t|Y_{t-1})) = \mathbb{E}\left(\phi_{t-1}Y_{t-1}^2 + Y_{t-1}\lambda\right),$$

$$\mathbb{E}(Y_t)\mathbb{E}(Y_{t-1}) = \mathbb{E}\left(\frac{\exp[\nu(Y_{t-1};\beta)]}{1+\exp[\nu(Y_{t-1};\beta)]}Y_{t-1} + \lambda\right)\mathbb{E}(Y_{t-1}).$$

From this, it follows that:

$$\mathrm{Cov}(Y_t, Y_{t-1}) = \mathbb{E}\left(\frac{\exp[\nu(Y_{t-1};\beta)]}{1+\exp[\nu(Y_{t-1};\beta)]}Y_{t-1}^2\right) - \mathbb{E}\left(\frac{\exp[\nu(Y_{t-1};\beta)]}{1+\exp[\nu(Y_{t-1};\beta)]}Y_{t-1}\right)\mathbb{E}(Y_{t-1}).$$

Property A2. *According to Theorem 1 in Tweedie [27] (also see Meyn and Tweedie [28]), the sufficient condition for $\{Y_t\}$ to be an ergodic Markov chain is the existence of a set K and a measurable function g in the state space $\mathcal{Y}$ of $\{Y_t\}$ such that:*

$$\int_\mathcal{Y} P(x,dy)g(y) \leq g(x) - 1, \quad x \in K^c.$$

and for a constant B:

$$\int_\mathcal{Y} P(x,dy)g(y) = \lambda(x) \leq B < \infty, \quad x \in K.$$

where $P(x,A) = \mathbb{P}(Y_t \in A | Y_{t-1} = x)$.

The state space $\mathcal{Y}$ of $\{Y_t\}$ is the set of natural numbers $\mathbb{N} = \{0,1,2,3,\ldots\}$. Let $g(y) = y$, then we have:

$$\begin{aligned}\int_\mathbb{N} P(x,dy)g(y) &= \sum_{y=0}^\infty \mathbb{P}(Y_t = y | Y_{t-1} = x) = \mathbb{E}(Y_t | Y_{t-1} = x)\\ &= \frac{\exp[\nu(Y_{t-1};\beta)]}{1+\exp[\nu(Y_{t-1};\beta)]}x + \lambda.\end{aligned}$$

Since $\sup_{y \in \mathbb{N}} \nu(y;\beta) < \infty$, then:

$$\frac{\exp[\nu(Y_{t-1};\beta)]}{1+\exp[\nu(Y_{t-1};\beta)]} = \frac{1}{1+\exp[-\nu(x;\beta)]} \leq \frac{1}{1+\exp\left[-\sup_{y\in\mathbb{N}}\nu(y;\beta)\right]} < 1.$$

Therefore, we can choose a constant $0 < m < 1$, such that $\frac{\exp[\nu(Y_{t-1};\beta)]}{1+\exp[\nu(Y_{t-1};\beta)]} < m$. Let $N = \left\lfloor \frac{\lambda+1}{1-m} \right\rfloor + 1$, where $\lfloor c \rfloor$ represents the floor function of c. Defining $K = \{0,1,2,\ldots,N-1\}$, we know:

$$\int_\mathbb{N} P(x,dy)g(y) = \mathbb{E}(Y_t | Y_{t-1} = x) < mx + \lambda < x - 1 = g(x) - 1, \quad x \in K^c,$$

$$\int_\mathbb{N} P(x,dy)g(y) = \mathbb{E}(Y_t | Y_{t-1} = x) < x + \lambda < N + \lambda < \infty, \quad x \in K.$$

Hence, the data generation process $\{Y_t\}$ is ergodic.

Theorem A1. *According to Theorems 5 and 6 in Klimko and Nelson [29], let $g = \mathbb{E}\left(Y_t \middle| Y_{(t-1)}\right)$, if the following four conditions hold, then Theorem 1 in this paper holds:*

(i) $\frac{\partial g}{\partial \theta_i}, \frac{\partial^2 g}{\partial \theta_i \partial \theta_j}, \frac{\partial^3 g}{\partial \theta_i \partial \theta_j \partial \theta_k}, 1 \leq i,j,k \leq \ell+1$, exists and are continuous with respect to θ.

(ii) For $1 \leq i,j \leq \ell+1$, $\mathbb{E}\left|(Y_t - g)\frac{\partial g}{\partial \theta_i}\right| < \infty, \mathbb{E}\left|(Y_t - g)\frac{\partial^2 g}{\partial \theta_i \partial \theta_j}\right| < \infty, \mathbb{E}\left|\frac{\partial g}{\partial \theta_i}\frac{\partial g}{\partial \theta_j}\right| < \infty$.

(iii) For $1 \leq i,j,k \leq \ell+1$, there exist functions:

$$H^{(0)}(Y_{t-1},\ldots,Y_0), H_i^{(1)}(Y_{t-1},\ldots,Y_0), H_{ij}^{(2)}(Y_{t-1},\ldots,Y_0), H_{ijk}^{(3)}(Y_{t-1},\ldots,Y_0),$$

such that
$$|g| \leq H^{(0)}, \left|\frac{\partial g}{\partial \theta_i}\right| \leq H_i^{(1)}, \left|\frac{\partial^2 g}{\partial \theta_i \partial \theta_j}\right| \leq H_{ij}^{(2)}, \left|\frac{\partial^3 g}{\partial \theta_i \partial \theta_j \partial \theta_k}\right| \leq H_{ijk}^{(3)},$$
$$\mathbb{E}\left|Y_t \cdot H_{ijk}^{(3)}(Y_{t-1}, \ldots, Y_0)\right| < \infty,$$
$$H^{(0)}(Y_{t-1}, \ldots, Y_0) H_{ijk}^{(3)}(Y_{t-1}, \ldots, Y_0) < \infty,$$
$$\mathbb{E}\left|H_i^{(1)}(Y_{t-1}, \ldots, Y_0) H_{ij}^{(2)}(Y_{t-1}, \ldots, Y_0)\right| < \infty.$$

(iv) $\mathbb{E}(Y_t|Y_{t-1}, \ldots, Y_0) = \mathbb{E}\left(Y_t \middle| Y_{(t-1)}\right)$, a.e., $t \geq 1$,
$$\mathbb{E}\left(u_t^2(\theta) \left|\frac{\partial g}{\partial \theta_i} \frac{\partial g}{\partial \theta_j}\right|\right) < \infty,$$

where $u_t(\theta) = Y_t - \mathbb{E}(Y_t|Y_{t-1})$.

For model (4), $g(\theta) = \frac{\exp[\nu(Y_{t-1};\beta)]}{1+\exp[\nu(Y_{t-1};\beta)]} Y_{t-1} + \lambda$, for $1 \leq i, j, k \leq \ell$, we have:

$$|g(\theta)| < Y_{t-1} + \lambda, \left|\frac{\partial g}{\partial \theta_{\ell+1}}\right| = 1, \left|\frac{\partial g}{\partial \theta_i}\right| < \left|\frac{\partial \nu}{\partial \beta_i}\right| Y_{t-1}, \left|\frac{\partial^2 g}{\partial \theta_i \partial \theta_j}\right| < \left(\left|\frac{\partial g}{\partial \theta_i} \frac{\partial g}{\partial \theta_j}\right| + \left|\frac{\partial^2 \nu}{\partial \beta_i \partial \beta_j}\right|\right) Y_{t-1},$$

$$\left|\frac{\partial^3 g}{\partial \theta_i \partial \theta_j \partial \theta_k}\right| < \left(\left|\frac{\partial \nu}{\partial \beta_i} \frac{\partial \nu}{\partial \beta_j} \frac{\partial \nu}{\partial \beta_k}\right| + \left|\frac{\partial^2 \nu}{\partial \beta_i \partial \beta_k} \frac{\partial \nu}{\partial \beta_j}\right| + \left|\frac{\partial^2 \nu}{\partial \beta_j \partial \beta_k} \frac{\partial \nu}{\partial \beta_i}\right| + \left|\frac{\partial^2 \nu}{\partial \beta_i \partial \beta_j} \frac{\partial \nu}{\partial \beta_k}\right|\right) Y_{t-1} + \left|\frac{\partial^3 \nu}{\partial \beta_i \partial \beta_j \partial \beta_k}\right| Y_{t-1}.$$

Note that the second- and third-order partial derivatives of the function g with respect to λ are both 0. According to assumptions (A2) and (A3), $\frac{\partial g}{\partial \theta_i}, \frac{\partial^2 g}{\partial \theta_i \partial \theta_j}$, and $\frac{\partial^3 g}{\partial \theta_i \partial \theta_j \partial \theta_k}, 1 \leq i, j, k \leq \ell+1$, exist and are continuous with respect to θ. According to assumption (A5), $V(\theta_0)$ is non-singular. Based on assumptions (A1), (A4), and the Hölder inequality, all four conditions are satisfied. Thus, Theorem 1 holds.

Lemma A1. $\left\{M_t(\theta)M_t(\theta)'\right\}$ *is an integrable process.*

Note that $\frac{\exp[\nu(Y_{t-1};\beta)]}{1+\exp[\nu(Y_{t-1};\beta)]} < 1, \frac{1}{1+\exp[\nu(Y_{t-1};\beta)]} \leq 1$. According to assumption (A4), if $i \leq \ell$, then:

$$\mathbb{E}(m_{ti}m_{ti}) \leq \mathbb{E}\left\{\left[Y_t - \frac{\exp[\nu(Y_{t-1};\beta)]}{1+\exp[\nu(Y_{t-1};\beta)]} Y_{t-1} - \lambda\right]^2 \frac{\partial \nu(Y_{t-1};\beta)}{\partial \beta_i} \frac{\partial \nu(Y_{t-1};\beta)}{\partial \beta_i} Y_{t-1}^2\right\}$$

$$\leq \sqrt{\mathbb{E}\left[Y_t - \frac{\exp[\nu(Y_{t-1};\beta)]}{1+\exp[\nu(Y_{t-1};\beta)]} Y_{t-1} - \lambda\right]^4} \sqrt{\mathbb{E}(N^4(y) Y_{t-1}^4)}$$

$$\leq \sqrt{\mathbb{E}\left[Y_t - \frac{\exp[\nu(Y_{t-1};\beta)]}{1+\exp[\nu(Y_{t-1};\beta)]} Y_{t-1} - \lambda\right]^4} \sqrt{\sqrt{\mathbb{E}(N^8(y))} \sqrt{\mathbb{E} Y_{t-1}^8}} < \infty.$$

Similarly, we can derive that:
If $i, j \leq \ell, i \neq j$, then:

$$\mathbb{E}(m_{ti}m_{tj}) \leq \mathbb{E}\left\{\left[Y_t - \frac{\exp[\nu(Y_{t-1};\beta)]}{1+\exp[\nu(Y_{t-1};\beta)]} Y_{t-1} - \lambda\right]^2 \frac{\partial \nu(Y_{t-1};\beta)}{\partial \beta_i} \frac{\partial \nu(Y_{t-1};\beta)}{\partial \beta_j} Y_{t-1}^2\right\} < \infty.$$

If $i \leq \ell, j = \ell+1$, then:

$$\mathbb{E}(m_{ti}m_{tj}) \leq \mathbb{E}\left\{\left[Y_t - \frac{\exp[\nu(Y_{t-1};\beta)]}{1+\exp[\nu(Y_{t-1};\beta)]}Y_{t-1} - \lambda\right]^2 \frac{\partial \nu(Y_{t-1};\beta)}{\partial \beta_i}Y_{t-1}\right\} < \infty.$$

If $i = \ell + 1$, then:

$$\mathbb{E}(m_{ti}m_{ti}) \leq \mathbb{E}\left\{\left[Y_t - \frac{\exp[\nu(Y_{t-1};\beta)]}{1+\exp[\nu(Y_{t-1};\beta)]}Y_{t-1} - \lambda\right]^2\right\} < \infty.$$

Lemma A2. $max_{1 \leq t \leq T} \|M_t(\theta)\| = o_p\left(T^{\frac{1}{2}}\right)$.

Given assumption (A4) and Lemma A1, it follows that $\mathbb{E}(M_t(\theta)'M_t(\theta)) < \infty$, resulting in $\sum_{t=1}^{\infty} \mathbb{P}(M_t(\theta)'M_t(\theta)) < \infty$. As the $\{Y_t\}$ series is strictly stationary, the event $\left\{\|M_t(\theta)\| > t^{\frac{1}{2}}\right\}$ occurs only a finite number of times with probability 1.

By a similar reasoning, let $M_T^* = max_{1 \leq t \leq T}\|M_t(\theta)\|$, and for any $\varepsilon > 0$, with probability 1, there will be only a finite number of $T \in \mathbb{N}$ such that $M_T^* > \varepsilon\sqrt{T}$. Consequently:

$$limsup_T M_T^* T^{-\frac{1}{2}} \leq \varepsilon, \ a.s.$$

This result implies that $M_T^* = o_p$.

Lemma A3. $max_{1 \leq t \leq T} \frac{t}{\sum_{t=1}^{T}\mathbb{E}(m_{ti}m_{tj})} < \infty, \forall 1 \leq i,j \leq \ell+1$.

The ergodicity property of $\{Y_t\}$ and Lemma A1 lead to:

$$max_{1 \leq t \leq T}\frac{t}{\sum_{t=1}^{T}\mathbb{E}(m_{ti}m_{tj})} = max_{1 \leq t \leq T}\frac{t}{T}(\mathbb{E}(m_{ti}m_{tj}))^{-1} \leq (\mathbb{E}(m_{ti}m_{tj}))^{-1} = O(1).$$

Theorem A2. *Given the ergodicity property of $\{Y_t\}$ and Lemma A1, and applying Theorem 14.6 from Davidson [30], we have:*

$$\frac{1}{T}\sum_{t=2}^{T} M_t(\theta_0)M_t(\theta_0)' \xrightarrow{a.s.} \mathbb{E}\left(M_t(\theta_0)M_t(\theta_0)'\right).$$

Let $\mathcal{F}_n = \sigma(Y_1, Y_2, \ldots, Y_n)$, $\widetilde{M}_{ni} = \sum_{i=1}^{n} m_{ti}(\theta), 1 \leq i \leq \ell+1$. For $1 \leq i \leq \ell$, we have

$$\mathbb{E}\left(\widetilde{M}_{ni}\big|\mathcal{F}_{n-1}\right) = \widetilde{M}_{(n-1)i},$$

$$+\mathbb{E}\left(\left(Y_n - \frac{\exp[\nu(Y_{n-1};\beta)]}{1+\exp[\nu(Y_{n-1};\beta)]}Y_{n-1} - \lambda\right)\frac{\exp[\nu(Y_{n-1};\beta)]}{(1+\exp[\nu(Y_{n-1};\beta)])^2}\frac{\partial \nu(y;\beta)}{\partial \beta_i}Y_{n-1}\bigg|\mathcal{F}_{n-1}\right),$$

$$= \widetilde{M}_{(n-1)i}.$$

Similarly, $\mathbb{E}\left(\widetilde{M}_{n(\ell+1)}\big|\mathcal{F}_{n-1}\right) = \widetilde{M}_{(n-1)(\ell+1)}$. Thus, for $1 \leq i \leq \ell+1$, $\left\{\widetilde{M}_{ni}, \mathcal{F}_n, n \geq 0\right\}$ is a martingale. Based on this and the ergodicity property of $\{Y_t\}$, and using Lemmas 2 and 3, applying Theorem 25.4 from Davidson [30] establishes that the conditions of Theorem 25.3 in Davidson [30] are satisfied, resulting in:

$$\frac{1}{\sqrt{T}}\sum_{t=2}^{T} m_{ti}(\theta_0) \xrightarrow{d} N\left(0, \mathbb{E}\left(m_{ti}^2(\theta)\right)\right).$$

Furthermore, for any $(\ell+1)$-dimensional vector $c \neq 0$, we have:

$$\frac{1}{T}\sum_{t=2}^{T} c' M_t(\theta_0) \xrightarrow{d} N\left(0, \sigma^2\right).$$

Here, $\sigma^2 = \mathbb{E}\left(c' M_t(\theta_0) M_t(\theta_0)' c\right)$. Therefore, we have:

$$\frac{1}{\sqrt{T}}\sum_{t=2}^{T} M_t(\theta_0) \xrightarrow{d} N\left(0, \mathbb{E}\left(M_t(\theta_0) M_t(\theta_0)'\right)\right).$$

In summary, $H(\theta_0) \xrightarrow{d} \chi^2(\ell+1)$.

Lemma A4. *Let $\{Y_t\}$ be an ergodic stationary random variable sequence; for any $i \geq 2$, $\mathbb{E}(Y_t | Y_1, Y_2, \ldots, Y_{t-1}) = 0$, a.s., and $\mathbb{E}(Y_1^2) = 1$. Then:*

$$\limsup \frac{\sum_{t=1}^{T} Y_t}{\sqrt{2T \log \log T}} = 1.$$

The proof can be found in Stout [31].

Theorem A3. *Following steps similar to those in Yu, Wang, and Yang [18] and Qin and Lawless [20], we can show (by replacing the usage of the double logarithm law with Lemma A4):*

$$\gamma(\theta) = \left[\frac{1}{T}\sum_{t=2}^{T} M_t(\theta) M_t(\theta)'\right] \frac{1}{T}\sum_{t=}^{T} M_t(\theta) + o\left(T^{\frac{1}{3}}\right).$$

$$2\mathcal{L}_E(\theta_0) = \left[\sum_{t=}^{T} M_t(\theta_0)\right]' \left[\sum_{t=2}^{T} M_t(\theta_0) M_t(\theta_0)'\right]^{-1} \left[\sum_{t=}^{T} M_t(\theta_0)\right] + o(1).$$

Furthermore:

$$\sqrt{T}(\hat{\theta}_{EL} - \theta_0) = S_{22}^{-1} S_{21} S_{11}^{-1} \frac{1}{\sqrt{T}} \sum_{t=}^{T} M_t(\theta_0) + o_p(1) \xrightarrow{d} N\left(0, S_{22}^{-1}\right),$$

$$2\mathcal{L}_E(\hat{\theta}_{EL}) = \left[\sum_{t=}^{T} M_t(\theta_0)\right]' S_3 \left[\sum_{t=}^{T} M_t(\theta_0)\right] + o_p(1),$$

where:

$$S_3 = S_{11}^{-1}\left(I + S_{12} S_{22}^{-1} S_{21} S_{11}^{-1}\right),\ S_{22}^{-1} = \left[\mathbb{E}\left(\frac{\partial M_t(\theta_0)}{\partial \theta'}\right) \left(\mathbb{E}\left(M_t(\theta_0) M_t(\theta_0)'\right)\right)^{-1} \mathbb{E}\left(\frac{\partial M_t(\theta_0)}{\partial \theta}\right)\right]^{-1},$$

$$S_{11} = \mathbb{E}\left(M_t(\theta_0) M_t(\theta_0)'\right),\ S_{12} = \mathbb{E}\left(\frac{\partial M_t(\theta_0)}{\partial \theta'}\right),\ S_{21} = S_{12}'.$$

Based on this, we perform a Taylor expansion of $2\mathcal{L}_E\left(\theta_0^{(1)}, \widetilde{\theta}_{EL}^{(2)}\right) - 2\mathcal{L}_E\left(\hat{\theta}_{EL}^{(1)}, \hat{\theta}_{EL}^{(2)}\right)$ at $\theta = \theta_0, \gamma = 0$:

$$2\mathcal{L}_E\left(\theta_0^{(1)}, \widetilde{\theta}_{EL}^{(2)}\right) - 2\mathcal{L}_E\left(\hat{\theta}_{EL}^{(1)}, \hat{\theta}_{EL}^{(2)}\right) \xrightarrow{d}$$

$$\left[\mathbb{E}\left(M_t(\theta_0)M_t(\theta_0)'\right)^{-\frac{1}{2}} \frac{1}{\sqrt{T}} \sum_{t=1}^{T} M_t(\theta_0)\right]' \left(\mathbb{E}\left(M_t(\theta_0)M_t(\theta_0)'\right)\right)^{-\frac{1}{2}}$$

$$\times \left\{ \mathbb{E}\left(\frac{\partial M_t(\theta_0)}{\partial \theta}\right) \left[\mathbb{E}\left(\frac{\partial M_t(\theta_0)}{\partial \theta'}\right) \left(\mathbb{E}\left(M_t(\theta_0)M_t(\theta_0)'\right)\right)^{-1} \mathbb{E}\left(\frac{\partial M_t(\theta_0)}{\partial \theta}\right)\right]^{-1} \mathbb{E}\left(\frac{\partial M_t(\theta_0)}{\partial \theta'}\right) - \right.$$

$$\left. \left(\frac{\partial M_t(\theta_0)}{\partial \theta_{(1)}}\right) \left[\mathbb{E}\left(\frac{\partial M_t(\theta_0)}{\partial \theta'_{(1)}}\right) \left(\mathbb{E}\left(M_t(\theta_0)M_t(\theta_0)'\right)\right)^{-1} \mathbb{E}\left(\frac{\partial M_t(\theta_0)}{\partial \theta_{(1)}}\right)\right]^{-1} \mathbb{E}\left(\frac{\partial M_t(\theta_0)}{\partial \theta'_{(1)}}\right) \right\}$$

$$\times \left(\mathbb{E}\left(M_t(\theta_0)M_t(\theta_0)'\right)\right)^{-\frac{1}{2}} \left[\mathbb{E}\left(M_t(\theta_0)M_t(\theta_0)'\right)\right]^{-\frac{1}{2}} \frac{1}{\sqrt{T}} \sum_{t=2}^{T} M_t(\theta_0)\right] + o_p(1).$$

It is easy to see that

$$\mathbb{E}\left(\frac{\partial M_t(\theta_0)}{\partial \theta}\right) \left[\mathbb{E}\left(\frac{\partial M_t(\theta_0)}{\partial \theta'}\right) \left(\mathbb{E}\left(M_t(\theta_0)M_t(\theta_0)'\right)\right)^{-1} \mathbb{E}\left(\frac{\partial M_t(\theta_0)}{\partial \theta}\right)\right]^{-1} \mathbb{E}\left(\frac{\partial M_t(\theta_0)}{\partial \theta'}\right)$$

$$- \left(\frac{\partial M_t(\theta_0)}{\partial \theta_{(1)}}\right) \left[\mathbb{E}\left(\frac{\partial M_t(\theta_0)}{\partial \theta'_{(1)}}\right) (\mathbb{E}(M_t(\theta_0)M_t(\theta_0)'))^{-1} \mathbb{E}\left(\frac{\partial M_t(\theta_0)}{\partial \theta_{(1)}}\right)\right]^{-1} \mathbb{E}\left(\frac{\partial M_t(\theta_0)}{\partial \theta'_{(1)}}\right)$$

is a symmetric matrix; we will now show that this symmetric matrix is positive–semi definite:

$$\mathbb{E}\left(\frac{\partial M_t(\theta_0)}{\partial \theta}\right) \left[\mathbb{E}\left(\frac{\partial M_t(\theta_0)}{\partial \theta'}\right) \left(\mathbb{E}\left(M_t(\theta_0)M_t(\theta_0)'\right)\right)^{-1} \mathbb{E}\left(\frac{\partial M_t(\theta_0)}{\partial \theta}\right)\right]^{-1} \mathbb{E}\left(\frac{\partial M_t(\theta_0)}{\partial \theta'}\right)$$

$$\gtrsim \begin{bmatrix} \mathbb{E}\frac{\partial M_t(\theta_0)}{\partial \theta} & \mathbb{E}\left(\frac{\partial M_t(\theta_0)}{\partial \theta'_{(1)}}\right) \end{bmatrix} \begin{bmatrix} \mathbb{E}\left(\frac{\partial M_t(\theta_0)}{\partial \theta'_{(1)}}\right) (\mathbb{E}(M_t(\theta_0)M_t(\theta_0)'))^{-1} \mathbb{E}\left(\frac{\partial M_t(\theta_0)}{\partial \theta_{(1)}}\right) & 0 \\ 0 & 0 \end{bmatrix} \begin{bmatrix} \mathbb{E}\left(\frac{\partial M_t(\theta_0)}{\partial \theta_{(1)}}\right) \\ \mathbb{E}\left(\frac{\partial M_t(\theta_0)}{\partial \theta_{(2)}}\right) \end{bmatrix}$$

$$= \left(\mathbb{E}\frac{\partial M_t(\theta_0)}{\partial \theta_{(1)}}\right) \left[\mathbb{E}\left(\frac{\partial M_t(\theta_0)}{\partial \theta'_{(1)}}\right) \left(\mathbb{E}\left(M_t(\theta_0)M_t(\theta_0)'\right)\right)^{-1} \mathbb{E}\left(\frac{\partial M_t(\theta_0)}{\partial \theta_{(1)}}\right)\right]^{-1} \mathbb{E}\left(\frac{\partial M_t(\theta_0)}{\partial \theta'_{(1)}}\right),$$

here, $A \gtrsim B$ implies that $A - B$ is positive–semi definite. Therefore, by the result in Rao [32], we have:

$$\mathcal{L}_E\left(\theta_0^{(1)}, \widetilde{\theta}_{EL}^{(2)}\right) - \mathcal{L}_E\left(\hat{\theta}_{EL}^{(1)}, \hat{\theta}_{EL}^{(2)}\right) \xrightarrow{d} \chi^2(q).$$

Appendix A.2. Complementary Numerical Simulations

Table A1. Parameter Estimation Simulation Results.

Sample Size	$\beta_0^{(CLS)}$	$\beta_0^{(CML)}$	$\beta_1^{(CLS)}$	$\beta_1^{(CML)}$	$\lambda^{(CLS)}$	$\lambda^{(CML)}$
Parameter: $\beta_0 = 2$, $\beta_1 = -0.8$, $\lambda = 3.5$ $\phi_t\|y_{t-1}$ is fixed.						
T = 300						
BIAS	0.4431	0.3667	−0.1879	−0.1624	−0.0282	−0.0306
RMSE	2.5257	2.7556	1.4189	1.1704	0.2811	0.2811
MAPE	0.4738	0.4791	0.3841	0.3682	0.0637	0.0637
T = 500						
BIAS	0.2098	0.2051	−0.0759	−0.0741	−0.0074	−0.0085
RMSE	0.8623	0.8619	0.4246	0.4371	0.2066	0.2062
MAPE	0.3049	0.3041	0.2148	0.2141	0.0475	0.0474
T = 800						
BIAS	0.1109	0.1071	−0.0346	−0.0329	−0.0119	−0.0126
RMSE	0.5673	0.5609	0.1646	0.1619	0.1783	0.1775
MAPE	0.2229	0.2208	0.1509	0.1496	0.0404	0.0403
T = 1200						
BIAS	0.0862	0.0848	−0.0232	−0.0224	−0.0093	−0.0101
RMSE	0.4491	0.4477	0.1201	0.1193	0.1375	0.1371
MAPE	0.1773	0.1771	0.1169	0.1163	0.0313	0.0311
T = 2000						
BIAS	0.0278	0.0269	−0.0119	−0.0115	0.0007	0.0003
RMSE	0.3369	0.3359	0.0889	0.0889	0.1076	0.1074
MAPE	0.1339	0.1333	0.0864	0.0864	0.0246	0.0244
Parameter: $\beta_0 = 2$, $\beta_1 = -0.8$, $\lambda = 3.5$ $\phi_t\|y_{t-1}$ follows a uniform distribution.						
T = 300						
BIAS	0.5624	0.3983	−0.2331	−0.1424	−0.0404	−0.0203
RMSE	2.0828	1.1916	1.2894	0.4719	0.2807	0.2534
MAPE	0.4877	0.4146	0.4355	0.3232	0.0641	0.0581
T = 500						
BIAS	0.1717	0.1399	−0.0712	−0.0593	−0.0028	0.0079
RMSE	0.8577	0.7919	0.3289	0.2552	0.2153	0.1982
MAPE	0.3028	0.2852	0.2125	0.2012	0.0496	0.0543
T = 800						
BIAS	0.1036	0.0809	−0.0317	−0.0285	−0.0124	−0.0039
RMSE	0.5735	0.5538	0.1547	0.1499	0.1725	0.1563
MAPE	0.2212	0.2158	0.1458	0.1427	0.0405	0.0036

Table A1. Cont.

Sample Size	$\beta_0^{(CLS)}$	$\beta_0^{(CML)}$	$\beta_1^{(CLS)}$	$\beta_1^{(CML)}$	$\lambda^{(CLS)}$	$\lambda^{(CML)}$
T = 1200						
BIAS	0.0531	0.0367	−0.0152	−0.0131	−0.0114	−0.0067
RMSE	0.4479	0.4334	0.1217	0.1196	0.1445	0.1303
MAPE	0.1785	0.1723	0.1177	0.1167	0.0331	0.0301
T = 2000						
BIAS	0.0453	0.0385	−0.0143	−0.0129	−0.0048	−0.0029
RMSE	0.3493	0.3429	0.0912	0.0898	0.1091	0.0903
MAPE	0.1389	0.1354	0.0885	0.0871	0.0248	0.0231

Parameter : $\beta_0 = 2$, $\beta_1 = -0.8$, $\lambda = 3.5$
$\phi_t | y_{t-1}$ follows an exponential distribution.

Sample Size	$\beta_0^{(CLS)}$	$\beta_0^{(CML)}$	$\beta_1^{(CLS)}$	$\beta_1^{(CML)}$	$\lambda^{(CLS)}$	$\lambda^{(CML)}$
T = 300						
BIAS	0.5805	0.4213	−0.2463	−0.1944	−0.0092	0.0232
RMSE	2.2029	2.0433	1.0969	1.0533	0.2702	0.2058
MAPE	0.5443	0.4843	0.4557	0.3986	0.0614	0.0466
T = 500						
BIAS	0.1923	0.0879	−0.0723	−0.0451	−0.0131	−0.0071
RMSE	1.0283	0.8006	0.2859	0.2364	0.2127	0.1601
MAPE	0.3299	0.2888	0.2236	0.1963	0.0483	0.0359
T = 800						
BIAS	0.1439	0.0929	−0.0464	−0.0336	−0.0061	0.0047
RMSE	0.6386	0.5709	0.1855	0.1605	0.1724	0.1293
MAPE	0.2456	0.2238	0.1653	0.1497	0.0389	0.0291
T = 1200						
BIAS	0.0699	0.0416	−0.0201	−0.0167	−0.0095	0.0025
RMSE	0.4731	0.4405	0.1242	0.1169	0.1404	0.1049
MAPE	0.1869	0.1744	0.1172	0.1123	0.0322	0.0239
T = 2000						
BIAS	0.0519	0.0319	−0.0151	−0.0111	−0.0049	0.0007
RMSE	0.3669	0.3435	0.0976	0.9161	0.1106	0.0818
MAPE	0.1442	0.1369	0.0955	0.0908	0.0251	0.0185

Parameter : $\beta_0 = 2$, $\beta_1 = -0.8$, $\lambda = 3.5$
$\phi_t | y_{t-1}$ follows a chi−square distribution.

Sample Size	$\beta_0^{(CLS)}$	$\beta_0^{(CML)}$	$\beta_1^{(CLS)}$	$\beta_1^{(CML)}$	$\lambda^{(CLS)}$	$\lambda^{(CML)}$
T = 300						
BIAS	0.9824	0.4063	−0.5663	−0.1282	−0.0098	0.0078
RMSE	3.3564	2.2437	1.6833	0.6341	0.3081	0.1569
MAPE	0.8569	0.5793	0.7361	0.3699	0.0696	0.0361
T = 500						
BIAS	0.4831	0.2249	−0.1805	−0.0621	−0.0202	−0.0068
RMSE	1.4549	0.9875	0.8114	0.2533	0.2293	0.1187
MAPE	0.4856	0.3749	0.3716	0.2354	0.0514	0.0269

Table A1. *Cont.*

Sample Size	$\beta_0^{(CLS)}$	$\beta_0^{(CML)}$	$\beta_1^{(CLS)}$	$\beta_1^{(CML)}$	$\lambda^{(CLS)}$	$\lambda^{(CML)}$
T = 800						
BIAS	0.2344	0.0869	−0.092	−0.0305	−0.008	0.0036
RMSE	1.0181	0.7138	0.4998	0.1758	0.1916	0.0962
MAPE	0.3477	0.2792	0.2501	0.1712	0.0433	0.0221
T = 1200						
BIAS	0.1382	0.0428	−0.041	−0.015	−0.014	−0.0021
RMSE	0.6592	0.5481	0.1766	0.1351	0.1557	0.0782
MAPE	0.2531	0.2164	0.1649	0.1325	0.0353	0.0181
T = 2000						
BIAS	0.0751	0.0438	−0.0269	−0.0161	−0.0011	0.0019
RMSE	0.5081	0.4318	0.1322	0.1079	0.1211	0.0611
MAPE	0.2017	0.1713	0.1279	0.1061	0.0277	0.0141

Table A2. Simulation Results for Parameter Estimation under Model Misspecification. With likelihood function settled as $\phi_t|y_{t-1}$ it follows a chi-squared distribution.

Sample Size	$\beta_0^{(CLS)}$	$\beta_0^{(CML)}$	$\beta_1^{(CLS)}$	$\beta_1^{(CML)}$	$\lambda^{(CLS)}$	$\lambda^{(CML)}$
Parameter : $\beta_0 = 1, \beta_1 = -0.6, \lambda = 1.2$ $\phi_t|y_{t-1}$ follows a uniform distribution.						
T = 300						
BIAS	0.0865	−1.8661	−0.0456	−1.5741	0.0046	0.4508
RMSE	0.8065	4.647	0.2301	5.3142	0.1454	0.4777
MAPE	0.6267	2.7518	0.2865	3.1747	0.0964	0.3757
T = 500						
BIAS	0.0312	−2.0474	−0.0228	−0.788	0.0043	0.4587
RMSE	0.5636	5.0468	0.1567	5.2847	0.1052	0.4753
MAPE	0.4493	2.5831	0.2046	1.9182	0.0703	0.3823
T = 800						
BIAS	0.0292	−2.1058	−0.0165	−0.3596	0.0038	0.4548
RMSE	0.4503	3.2688	0.1244	3.0257	0.0852	0.4657
MAPE	0.3587	2.3491	0.1651	1.2312	0.0563	0.3789
T = 1200						
BIAS	0.0249	−2.1077	−0.0127	−0.0739	0.0003	0.4558
RMSE	0.3513	2.6833	0.0971	1.7031	0.0689	0.4641
MAPE	0.2815	2.1461	0.1289	0.7674	0.0464	0.3799
T = 2000						
BIAS	0.0062	−2.0216	−0.0041	0.0766	0.0016	0.4546
RMSE	0.2735	2.2846	0.0749	1.0373	0.0529	0.4591
MAPE	0.2165	2.0256	0.0983	0.5483	0.0353	0.3788

Table A3. Empirical Likelihood Test for the λ Parameter. The significance level is set at 0.05.

Parameter: $\beta_0 = 1$, $\beta_1 = -0.6$, $\lambda = 1.2$
$\phi_t | y_{t-1}$ is fixed.

T	300	500	800	1200	2000
$\mathbb{H}_0 : \lambda = 1.5$	0.537	0.705	0.865	0.995	1
$\mathbb{H}_0 : \lambda = 1.35$	0.235	0.263	0.375	0.542	0.757
$\mathbb{H}_0 : \lambda = 1.2$ (true)	0.038	0.045	0.043	0.052	0.055
$\mathbb{H}_0 : \lambda = 1.05$	0.176	0.33	0.415	0.593	0.823
$\mathbb{H}_0 : \lambda = 0.9$	0.554	0.806	0.96	1	1

Parameter: $\beta_0 = 1$, $\beta_1 = -0.6$, $\lambda = 1.2$
$\phi_t | y_{t-1}$ follows a uniform distribution.

T	300	500	800	1200	2000
$\mathbb{H}_0 : \lambda = 1.5$	0.461	0.754	0.905	0.984	1
$\mathbb{H}_0 : \lambda = 1.35$	0.212	0.304	0.417	0.54	0.786
$\mathbb{H}_0 : \lambda = 1.2$ (true)	0.059	0.06	0.062	0.059	0.05
$\mathbb{H}_0 : \lambda = 1.05$	0.167	0.321	0.407	0.588	0.845
$\mathbb{H}_0 : \lambda = 0.9$	0.645	0.845	0.975	1	1

Parameter: $\beta_0 = 1$, $\beta_1 = -0.6$, $\lambda = 1.2$
$\phi_t | y_{t-1}$ follows an exponential distribution.

T	300	500	800	1200	2000
$\mathbb{H}_0 : \lambda = 1.5$	0.495	0.722	0.943	0.991	1
$\mathbb{H}_0 : \lambda = 1.35$	0.171	0.31	0.505	0.593	0.844
$\mathbb{H}_0 : \lambda = 1.2$ (true)	0.049	0.046	0.055	0.058	0.047
$\mathbb{H}_0 : \lambda = 1.05$	0.235	0.286	0.442	0.605	0.884
$\mathbb{H}_0 : \lambda = 0.9$	0.57	0.815	0.972	1	1

Parameter: $\beta_0 = 1$, $\beta_1 = -0.6$, $\lambda = 1.2$
$\phi_t | y_{t-1}$ follows a chi-square distribution.

T	300	500	800	1200	2000
$\mathbb{H}_0 : \lambda = 1.5$	0.478	0.648	0.852	0.951	1
$\mathbb{H}_0 : \lambda = 1.35$	0.195	0.334	0.491	0.612	0.807
$\mathbb{H}_0 : \lambda = 1.2$ (true)	0.086	0.088	0.079	0.054	0.051
$\mathbb{H}_0 : \lambda = 1.05$	0.115	0.225	0.318	0.515	0.795
$\mathbb{H}_0 : \lambda = 0.9$	0.417	0.635	0.859	0.946	1

Table A4. Empirical Likelihood Test for β_1 with a True Value of 0.

Parameter : $\beta_0 = 1$, $\beta_1 = 0$, $\lambda = 1.2$
$\phi_t | y_{t-1}$ follows an exponential distribution, significance level 0.05.

T	300	500	800	1200	2000
$\mathbb{H}_0 : \beta_1 = 0$ (true)	0.437	0.446	0.416	0.51	0.427
$\mathbb{H}_0 : \beta_1 = -0.1$	0.71	0.787	0.863	0.954	0.982
$\mathbb{H}_0 : \beta_1 = -0.2$	0.813	0.933	0.989	0.997	1
$\mathbb{H}_0 : \beta_1 = -0.3$	0.912	0.983	1	1	1
$\mathbb{H}_0 : \beta_1 = -0.4$	0.945	0.982	1	1	1

Table A4. *Cont.*

Parameter : $\beta_0 = 1, \beta_1 = 0, \lambda = 1.2$					
$\phi_t\|y_{t-1}$ follows an exponential distribution, significance level 0.10.					
T	300	500	800	1200	2000
$\mathbb{H}_0 : \beta_1 = 0$ (true)	0.543	0.544	0.517	0.613	0.55
$\mathbb{H}_0 : \beta_1 = -0.1$	0.797	0.846	0.93	0.982	0.993
$\mathbb{H}_0 : \beta_1 = -0.2$	0.872	0.957	0.988	1	1
$\mathbb{H}_0 : \beta_1 = -0.3$	0.945	1	1	1	1
$\mathbb{H}_0 : \beta_1 = -0.4$	0.971	0.985	1	1	1

Appendix A.3. Complementary Figure

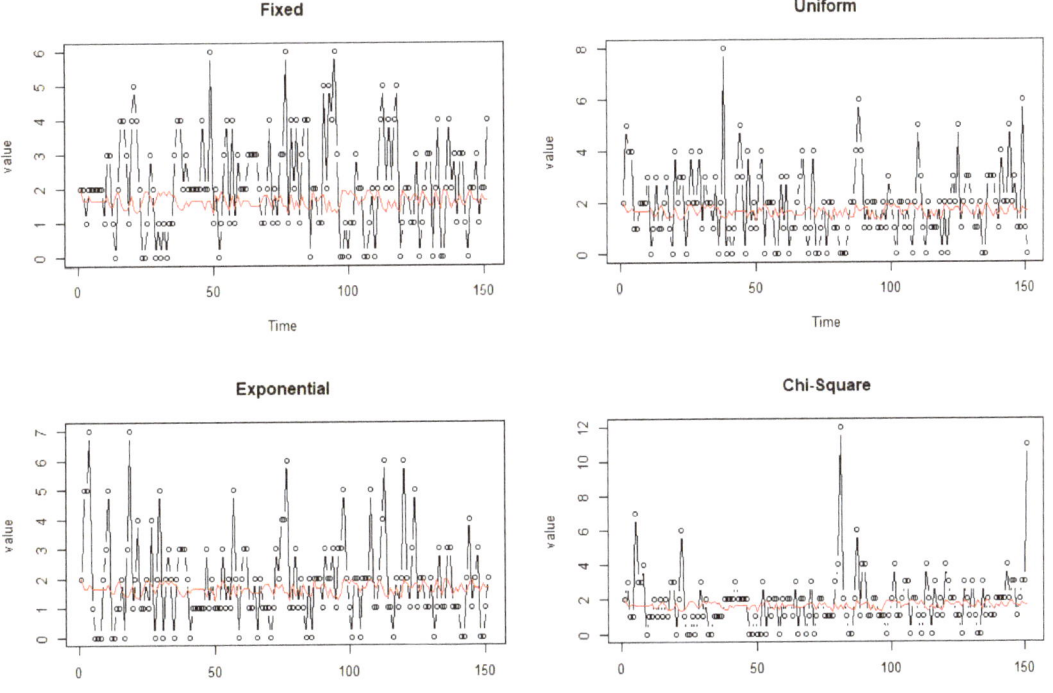

Figure A1. The black line represents the sample trajectory, and the red line denotes the one-step-ahead forecast trajectory. Parameter: $\beta_0 = 1, \beta_1 = -0.6, \lambda = 1.2$.

References

1. Steutel, F.W.; van Harn, K. Discrete analogues of self-decomposability and stability. *Ann. Probab.* **1979**, *7*, 893–899. [CrossRef]
2. Al-Osh, M.A.; Alzaid, A.A. First-order integer-valued autoregressive (INAR(1)) process. *J. Time Ser. Anal.* **1987**, *8*, 261–275. [CrossRef]
3. Latour, A. Existence and stochastic structure of a non-negative integer-valued autoregressive proces. *J. Time Ser. Anal.* **1998**, *719*, 439–455. [CrossRef]
4. Joe, H. Time series models with univariate margins in the convolution-closed infinitely divisible class. *J. Appl. Probab.* **1996**, *33*, 664–677. [CrossRef]
5. Zheng, H.T.; Basawa, I.V.; Datta, S. First-order random coefficient integer-valued autoregressive processes. *J. Stat. Plan. Inference* **2007**, *137*, 212–229. [CrossRef]
6. Gomes, D.; Castro, L.C. Generalized integer-valued random coefficient for a first order structure autoregressive (RCINAR) process. *J. Stat. Plan. Inference* **2009**, *139*, 4088–4097. [CrossRef]

7. Weiß, C.H.; Jentsch, C. Bootstrap-based bias corrections for INAR count time series. *J. Stat. Comput. Simul.* **2019**, *89*, 1248–1264. [CrossRef]
8. Kang, Y.; Wang, D.H.; Yang, K. A new INAR(1) process with bounded support for counts showing equidispersion, underdispersion and overdispersion. *Stat. Pap.* **2021**, *62*, 745–767. [CrossRef]
9. Pegram, G.G.S. An autoregressive model for multilag Markov chains. *J. Appl. Probab.* **1980**, *17*, 350–362. [CrossRef]
10. Salinas, D.; Flunkert, V.; Gasthaus, J.; Januschowski, T. DeepAR: Probabilistic forecasting with autoregressive recurrent networks. *Int. J. Forecast.* **2020**, *36*, 1181–1191. [CrossRef]
11. Huang, J.; Zhu, F.K.; Deng, D.L. A mixed generalized Poisson INAR model with applications. *J. Stat. Comput. Simul.* **2023**, 1–28. [CrossRef]
12. Mohammadi, Z.; Sajjadnia, Z.; Bakouch, H.S.; Sharafi, M. Zero-and-one inflated Poisson–Lindley INAR(1) process for mod-elling count time series with extra zeros and ones. *J. Stat. Comput. Simul.* **2022**, *92*, 2018–2040. [CrossRef]
13. Scotto, M.G.; Weiß, C.H.; Gouveia, S. Thinning-based models in the analysis of integer-valued time series: A review. *Stat. Model.* **2015**, *15*, 590–618. [CrossRef]
14. Zheng, H.T.; Basawa, I.V. First-order observation-driven integer-valued autoregressive processes. *Stat. Probab. Lett.* **2008**, *78*, 1–9. [CrossRef]
15. Triebsch, L.K. *New Integer-Valued Autoregressive and Regression Models with State-Dependent Parameters*; TU Kaiserslautern: Munich, Germany, 2008.
16. Monteiro, M.; Scotto, M.G.; Pereira, I. Integer-valued self-exciting threshold autoregressive processes. *Commun. Stat. Theory Methods* **2012**, *41*, 2717–2737. [CrossRef]
17. Ristić, M.M.; Bakouch, H.S.; Nastić, A.S. A new geometric first-order integer-valued autoregressive (NGINAR(1)) process. *J. Stat. Plan. Inference* **2009**, *139*, 2218–2226. [CrossRef]
18. Yu, M.J.; Wang, D.H.; Yang, K. A class of observation-driven random coefficient INAR(1) processes based on negative bi-nomial thinning. *J. Korean Stat. Soc.* **2018**, *48*, 248–264. [CrossRef]
19. Owen, A.B. Empirical likelihood ratio confidence intervals for a single functional. *Biometrika* **1988**, *75*, 237–249. [CrossRef]
20. Qin, J.; Lawless, J. Empirical likelihood and general estimating equations. *Ann. Stat.* **1994**, *22*, 300–325. [CrossRef]
21. Chen, S.X.; Keilegom, I.V. A review on empirical likelihood methods for regression. *Test* **2003**, *18*, 415–447. [CrossRef]
22. Billingsley, P. *Statistical Inference for Markov Processes*; The University of Chicago Press: Chicago, IL, USA, 1961.
23. Weiß, C.H. *An Introduction to Discrete-Valued Time Series*; John Wiley & Sons Ltd.: Hoboken, NJ, USA, 2018.
24. Du, J.G.; Li, Y. The Integer-Valued Autoregresive (INAR(p)) Model. *J. Time Ser. Anal.* **1989**, *12*, 129–142.
25. Fokianos, K.; Rahbek, R.; Tjøstheim, D. Poisson Autoregression. *J. Am. Stat. Assoc.* **2009**, *104*, 1430–1439. [CrossRef]
26. Freeland, R.K.; McCabe, B.P.M. Forecasting discrete valued low count time series. *Int. J. Forecast.* **2004**, *20*, 427–434. [CrossRef]
27. Tweedie, R.L. Sufficient conditions for ergodicity and recurrence of Markov chains on a general state space. *Stoch. Process. Appl.* **1975**, *3*, 385–403. [CrossRef]
28. Meyn, S.P.; Tweedie, R.L. *Markov Chains and Stochastic Stability*, 2nd ed.; Cambridge University Press: London, UK, 2009.
29. Klimko, L.A.; Nelson, P.I.; Datta, S. On conditional least squares estimation for stochastic processes. *Ann. Stat.* **1978**, *6*, 629–642. [CrossRef]
30. Davidson, J. *Stochastic Limit Theory—An Introduction for Econometricians*, 2nd ed.; Oxford University Press: Oxford, UK, 2021.
31. Stout, W.F. The Hartman-Wintner law of the iterated logarithm for martingaless. *Ann. Math. Stat.* **1970**, *41*, 2158–2160. [CrossRef]
32. Rao, C. *Linear Statistical Inference and Its Applications*; Wiley: New York, NY, USA, 1973.

Disclaimer/Publisher's Note: The statements, opinions and data contained in all publications are solely those of the individual author(s) and contributor(s) and not of MDPI and/or the editor(s). MDPI and/or the editor(s) disclaim responsibility for any injury to people or property resulting from any ideas, methods, instructions or products referred to in the content.

A Conway–Maxwell–Poisson-Binomial AR(1) Model for Bounded Time Series Data

Huaping Chen [1,*], Jiayue Zhang [2] and Xiufang Liu [3]

1. School of Mathematics and Statistics, Henan University, Kaifeng 475004, China
2. School of Mathematics, Jilin University, Changchun 130012, China
3. College of Mathematics, Taiyuan University of Technology, Taiyuan 030024, China
* Correspondence: chenhp0107@henu.edu.cn

Abstract: Binomial autoregressive models are frequently used for modeling bounded time series counts. However, they are not well developed for more complex bounded time series counts of the occurrence of n exchangeable and dependent units, which are becoming increasingly common in practice. To fill this gap, this paper first constructs an exchangeable Conway–Maxwell–Poisson-binomial (CMPB) thinning operator and then establishes the Conway–Maxwell–Poisson-binomial AR (CMPBAR) model. We establish its stationarity and ergodicity, discuss the conditional maximum likelihood (CML) estimate of the model's parameters, and establish the asymptotic normality of the CML estimator. In a simulation study, the boxplots illustrate that the CML estimator is consistent and the qqplots show the asymptotic normality of the CML estimator. In the real data example, our model takes a smaller AIC and BIC than its main competitors.

Keywords: CMPB thinning operator; bounded time series; CMPBAR model; under-dispersion; equi-dispersion; over-dispersion

1. Introduction

Bounded time series of counts are commonly observed in real-world applications. Its (binomial) index of dispersion (as a function of n, μ and σ^2) is defined by $\mathrm{BID}(X) = n\sigma^2/(\mu(n-\mu))$, where n is the predetermined upper limit of the range, $E(X) = \mu$ and $\mathrm{Var}(X) = \sigma^2$. If its $\mathrm{BID}(X) < 1$, then it is under-dispersed, if its $\mathrm{BID}(X) = 1$, then it is equi-dispersed, while if its $\mathrm{BID}(X) > 1$, then it is over-dispersed (or the extra-binomial variation).

A popular tool to establish a binomial autoregressive model (BAR) is the binomial thinning operator "∘" [1], which is introduced by

$$\alpha \circ X := \sum_{i=1}^{X} W_i, \qquad (1)$$

where X is a non-negative integer-valued random variable, $\{W_i, i = 1, 2, \cdots, n\}$ is an i.i.d. Bernoulli random variable sequence with $P(W_i = 1) = 1 - P(W_i = 0) = \alpha$ and independent of X. McKenzie [2] used the binomial thinning operator given in (1) to establish the binomial AR(1) model, which is a popular model for bounded time series and defined as follows

$$X_t = \alpha \circ X_{t-1} + \beta \circ (n - X_{t-1}), \qquad (2)$$

where $n \in \mathbb{N}$ is the predetermined upper limit of the range; X_0 follows the binomial distribution with $P(X_0 = k) = \binom{n}{k}\pi^k(1-\pi)^{n-k}$; $\alpha = \beta + \rho$ and $\beta = (1-\rho)\pi$ with $\rho \in (\max\{-\pi/(1-\pi), -(1-\pi)/\pi\}, 1)$ and $\pi \in (0,1)$; the counting series at time t are independent of the random variables $X_s, \forall s < t$; and all the counting series in "$\alpha \circ$" and "$\beta \circ$" are mutually independent sequences of independent Bernoulli random variables with parameters α and β, respectively. The binomial AR(1) process given in (2) is now well understood and it is an ergodic Markov chain with a stationary distribution Bin(n, π) with $\pi = \beta/(1-\rho)$ and $\rho = \alpha - \beta$. Hence, its BID$(X_t) = 1$, i.e., the BAR model given in (2), applies to equi-dispersed time series with finite range; see [3–7] for more discussion about the BAR(1) model.

Weiß and Pollett [8] extended the binomial AR(1) model as the density-dependent BAR(1) model (denoted as the DDBAR(1) model), whose thinning probabilities vary over time by assuming $\alpha_t = \alpha(X_{t-1}/n)$ and $\beta_t = \beta(X_{t-1}/n)$. In particular, for given n, if $\alpha_t = (1-\rho)(a+bX_{t-1}/n)$ and $\beta_t = (1-\rho)(a+bX_{t-1}/n) + \rho$, the DDBAR(1) model allows to analyze bounded integer-valued time series with under-dispersion, equi-dispersion and over-dispersion, see Section 4 in [8] for more details. To model extra-binomial variation for time series counts, Weiß and Kim [9] proposed the beta-binomial AR (BBAR) model based on the beta-binomial thinning operator "$\alpha \circ_\phi$", which is introduced by

$$\alpha_\phi \circ X = \sum_{i=1}^{X} B_i,$$

where X is a non-negative integer-valued random variable, $\{B_i, i = 1, 2, \cdots, n\}$ is an i.i.d. Bernoulli random variable sequence with $P(B_i = 1|\alpha_\phi) = 1 - P(B_i = 0|\alpha_\phi) = \alpha_\phi$ and $\alpha_\phi \sim \text{Beta}(\tau\alpha, \tau(1-\alpha))$, $\tau = (1-\phi)/\phi$, $\{B_i, i = 1, 2, \cdots, X\}$ is independent of X.

As discussed in Weiß [10], the BAR(1) model, DDBAR(1) model, and BBAR(1) model can be interpreted as a system with n mutually independent units and each unit being either in state "1" or state "0". Assume X_t is the number of units being in state "1" at time t. Then $\alpha \circ X_{t-1}$ ($\alpha_t \circ X_{t-1}$ or $\alpha_\phi \circ X_{t-1}$) is the number of units still in state "1" at time t with survival probability α (random survival probability α_t or α_ϕ), $\beta \circ (n - X_{t-1})$ ($\beta_t \circ (n - X_{t-1})$ or $\beta_\phi \circ (n - X_{t-1})$) is the number of units, which moved from state "0" to state "1" at time t with revival probability β (random revival probability β_t or β_ϕ). It is worth mentioning that all of BAR(1), DDBAR(1), and BBAR(1) models are aimed at a system with n independent units, but not a system with n dependent units, i.e., the counting series in "∘" is independent and identically distributed, but not dependent. To solve this dilemma, Kang et al. [11] proposed a generalized binomial AR (GBAR) model based on the generalized binomial thinning operator "$\alpha \circ_\theta$", which is proposed by Ristić et al. [12] and given as follows

$$\alpha_\theta \circ X = \sum_{i=1}^{X} U_i,$$

where $U_i = (1 - V_i)W_i + V_i Z$, $\{W_i\}$ and $\{V_i\}$ are two independent random sequences of iid random variables with Bernoulli(α) and Bernoulli(θ) distributions, Z is a Bernoulli(α) random variable and is responsible for the cross-dependence, $\forall i, j = 1, 2, ..., X$, $\{W_i\}$, $\{V_j\}$ and Z are mutually independent and each of them is independent of X.

Unfortunately, the GBAR model [11] can not use to analyze under-dispersed or equi-dispersed bounded data. To fill this gap, we are inspired by the Conway–Maxwell–Poisson-binomial (CMPB) distribution [13] and construct the Conway–Maxwell–Poisson-binomial thinning operator, whose counting series is exchangeablility. Furthermore, we propose a new Conway–Maxwell–Poisson-binomial autoregressive (CMPBAR) model, which not only allows us to analyze bounded data with over-dispersion but also allows us to model bounded data with equi-dispersion or under-dispersion. The second contribution of this paper is that we discuss the CML estimation of the parameters involved in the new model, and establish the asymptotic normality of the CML estimator. To illustrate that the new model is more flexible and superior, we apply the new model on the weekly rainy days at Hamburg–Neuwiedenthal in Germany.

The paper is organized as follows. Section 2 first gives a brief review of the Conway–Maxwell–Poisson-binomial distribution, then gives the definition of the exchangeable Conway–Maxwell–Poisson-binomial thinning operator and that of the Conway–Maxwell–Poisson-binomial AR model. The conditional maximum likelihood estimation and its asymptotic properties are established in Section 3. Section 4 gives a simulation study and

Section 5 gives real data to show the better performance of the new model. Conclusions are made in Section 6.

2. Model Formulation and Stability Properties

2.1. Conway–Maxwell–Poisson-Binomial Distribution

For readability, we first give a brief review of the CMPB distribution introduced by Shmueli et al. [13].

A random variable X taking values in $\{0, 1, 2, \ldots, n\}$ is said to follow the Conway–Maxwell–Poisson-binomial distribution with parameters (α, ν), if the probability mass function (pmf) of X takes the form $P(X = x | \alpha, \nu, n) = \binom{n}{x}^{\nu} \alpha^x (1-\alpha)^{n-x} / Z(\alpha, \nu)$, where $Z(\alpha, \nu) = \sum_{x=0}^{n} \binom{n}{x}^{\nu} \alpha^x (1-\alpha)^{n-x}$, $0 < \alpha < 1$, $\nu \in \mathbb{R}$ and $n \in \mathbb{N}$ is the predetermined upper limit of the range.

For simplicity, we write $X \sim \text{CMPB}(n, \alpha, \nu)$. Denote $\theta = \alpha/(1-\alpha)$, the pmf of X can be rewritten as

$$P(X = x | \theta, \nu, n) = \frac{1}{S(\theta, \nu)} \binom{n}{x}^{\nu} \theta^x, \quad (3)$$

where $S(\theta, \nu) = \sum_{x=0}^{n} \binom{n}{x}^{\nu} \theta^x$, $\theta > 0$ and $n \in \mathbb{N}$ is the predetermined upper limit of the range. Therefore, we obtain the moment-generating function of X as $M_X(s) = E(e^{sX}) = \frac{S(\theta e^s, \nu)}{S(\theta, \nu)}$. Furthermore,

$$E(X) = \theta \frac{S'(\theta, \nu)}{S(\theta, \nu)}, \quad \text{Var}(X) = \theta \frac{S'(\theta, \nu)}{S(\theta, \nu)} + \theta^2 \left(\frac{S''(\theta, \nu)}{S(\theta, \nu)} - \left(\frac{S'(\theta, \nu)}{S(\theta, \nu)} \right)^2 \right),$$

$$\text{BID} = \frac{n \text{Var}(X)}{E(X)(n - E(X))} = \frac{S(\theta, \nu) S'(\theta, \nu) + \theta S(\theta, \nu) S''(\theta, \nu) - \theta (S'(\theta, \nu))^2}{n S(\theta, \nu) S'(\theta, \nu) - \theta (S'(\theta, \nu))^2}, \quad (4)$$

where $S'(\theta, \nu) = \partial S(\theta, \nu) / \partial \theta$ and $S''(\theta, \nu) = \partial S'(\theta, \nu) / \partial \theta$ (see Shmueli et al. [13], Borges et al. [14], Daly and Gaunt [15], and Kadane [16] for more detailed discussion).

Unfortunately, the specific range of the BID for the CMPB distribution can not be obtained by (4). To solve this dilemma, we give an example in Figure 1 with $n = 7$, when α and ν are varying from $\{0.1, 0.2, 0.3, \cdots, 0.9\}$ and $\{-2, -1.5, -0.5, 0, 0.5, 1, 1.5, 2, 2.5\}$, respectively.

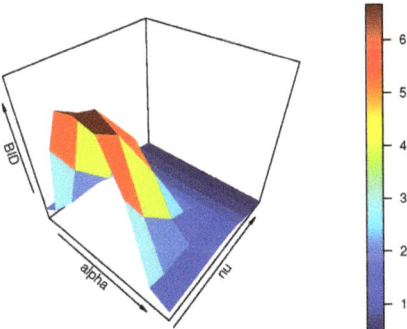

Figure 1. Plot of the BID of the CMPB distribution for different choices of α and ν.

From Figure 1, the BID of the CMPB distribution takes a value, which may be less than 1, equal to 1, or greater than 1 for different values α and ν. Additionally, it implies that the CMPB distribution allows us to analyze bounded time series counts with under-dispersion, equi-dispersion, and over-dispersion.

To further explore the dynamic change of the BID with α varying from $\{0.1, 0.2, \cdots, 0.9\}$ for given $n = 7$ and $\nu = -0.5, 0, 0.5, 1, 1.5,$ or 2, we present the plots of the BID in Figure 2.

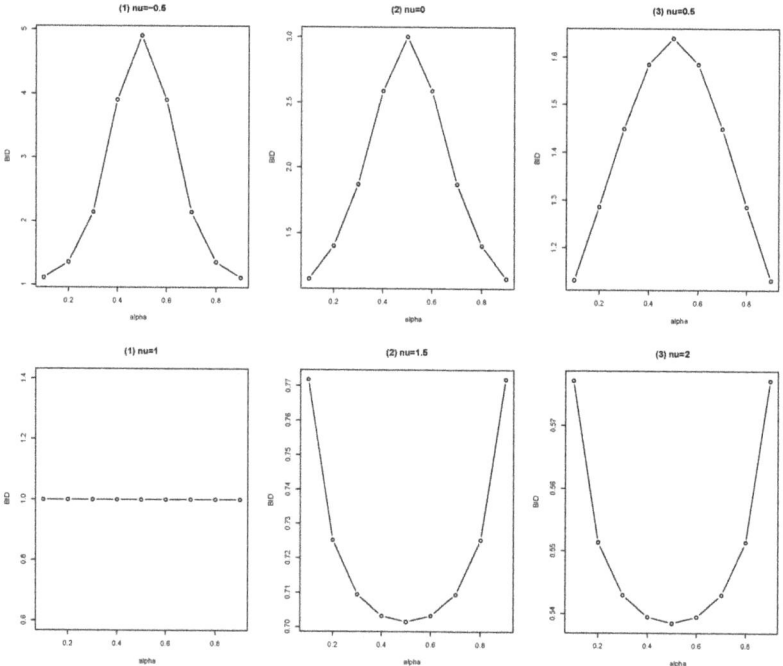

Figure 2. Plots of the BID of the CMPB distribution for different choices of α.

From Figure 2, we obtain the following observations. First, if $\nu < 1$, the BID is no less than 1. To be precise, its BID is increasing to maximum when α is varying from 0 to 0.5, and then decreasing to 1 when α is varying from 0.5 to 1. Second, if $\nu = 1$, its BID = 1, for all $\alpha \in (0, 1)$. Third, if $\nu > 1$, its BID is no more than 1. Precisely, its BID is decreasing to the minimum when α is varying from 0 to 0.5, and then increasing to 1 when α is varying from 0.5 to 1. To sum up, the Conway–Maxwell–Poisson-binomial distribution allows under-dispersion, equi-dispersion, and over-dispersion for bounded time series data.

Remark 1. *By (3), the pmf of the CMPB (n, α, ν) is expressed as that of the power series distribution and if $\nu = 0$, $P(X = x|\theta, \nu, n) = \theta^x / \sum_{x=0}^n \theta^x$, $\theta = \alpha/(1-\alpha)$, if $\nu = 1$, the CMPB(n, α, ν) reduces to binomial distribution with parameter α.*

2.2. Conway–Maxwell–Poisson-Binomial Thinning Operator

By Shmueli et al. [13], the CMPB distribution is a distribution on the sum of n dependent Bernoulli components without specifying anything else about the joint distribution of those components. Precisely, if $X \sim \text{CMPB}(n, \alpha, \nu)$, there exists a Bernoulli variable sequence $\{Z_i\}$ such that $X = \sum_{i=1}^n Z_i$, where

$$P_{z_1, \cdots, z_n} := P(Z_1 = z_1, \cdots, Z_n = z_n) = \frac{1}{\sum_{z_1=0}^1 \cdots \sum_{z_n=0}^1 \binom{n}{x}^{\nu-1} \theta^x} \binom{n}{x}^{\nu-1} \theta^x \quad (5)$$

with $\theta = \alpha/(1-\alpha)$, $x = \sum_{i=1}^n z_i$ and $(z_1, z_2, \cdots, z_n) \in \{0,1\}^n$.

Definition 1. *Let $\theta = \alpha/(1-\alpha)$. Then the exchangeable Conway–Maxwell–Poisson-binomial thinning operator is introduced by*

$$\alpha \diamond_\nu X := \sum_{i=1}^{X} Z_i, \qquad (6)$$

where X is a non-negative random variable, $\{Z_i, i = 1, 2, \cdots, X\}$ is an exchangeable Bernoulli variable sequence with its pmf taking the form (5) and independent of X.

To generate the random number of "$\alpha \diamond_\nu X$", we first let $X = n$, then $\alpha \diamond_\nu X | (X = n) \sim \text{CMPB}(n, \alpha, \nu)$. Therefore, $E(\alpha \diamond_\nu X | X = n) = \theta S'(\theta, \nu)/S(\theta, \nu)$, $\text{Var}(\alpha \diamond_\nu X | X = n) = \theta \frac{S'(\theta, \nu)}{S(\theta, \nu)} + \theta^2 \left(\frac{S''(\theta, \nu)}{S(\theta, \nu)} - \left(\frac{S'(\theta, \nu)}{S(\theta, \nu)} \right)^2 \right)$ and the conditional binomial index of dispersion (CBID) is $\text{CBID} = \frac{S(\theta,\nu)S'(\theta,\nu) + \theta S(\theta,\nu)S''(\theta,\nu) - \theta(S'(\theta,\nu))^2}{nS(\theta,\nu)S'(\theta,\nu) - \theta(S'(\theta,\nu))^2}$, where $S(\theta, \nu) = \sum_{x=0}^{n} \binom{n}{x}^\nu \theta^x$, $S'(\theta, \nu) = \partial S(\theta, \nu)/\partial \theta$, and $S''(\theta, \nu) = \partial S'(\theta, \nu)/\partial \theta$.

Second, we let $\theta = \alpha/(1-\alpha)$, then the pmf of $\alpha \diamond_\nu n$ takes the form (3). Third, we let $\theta = \lambda^\nu, \lambda > 0$. By (3), the pmf of the $\alpha \diamond_\nu n$ can be rewritten as

$$P(\alpha \diamond_\nu n = x) = \frac{1}{U(\lambda, \nu)} \left(\binom{n}{x} \lambda^x \right)^\nu \text{ with } U(\lambda, \nu) = \sum_{x=0}^{n} \left(\binom{n}{x} \lambda^x \right)^\nu.$$

Furthermore,

$$P(\alpha \diamond_\nu n = x + 1) = \left(\frac{n-x}{x+1} \lambda \right)^\nu P(\alpha \diamond_\nu n = x), \qquad (7)$$

by which an algorithm is used to generate a random number of $\alpha \diamond_\nu X$ with $X = n$ can be expressed as follows.

Remark 2. *By Kadane [16], the counting series $\{Z_i\}$ in Definition 1 is a dependent Bernoulli variable sequence with exchangeability of order 2. To account for the concept of exchangeability, we assume π is a permutation of $(z_1, z_2, \cdots, z_n)$. Then $P_{z_1, \cdots, z_n} = P_{\pi(1, \cdots, z_n)}$. By the definition of exchangeability in Section 6 in Kadane [16], $\sum_{i=1}^{n} Z_i$ is n-exchangeable. Kadane [16] stated that "de Finetti's Theorem shows that sums of exchangeable random variables are mixtures of Binomial random variables. Because the marginal distribution of each component is Bernoulli, interest centers on the joint distribution of pairs of such variables". By Theorem 4 in Kadane [16], n-exchangeability applies to every permutation of length n, it implies that n' is exchangeable for each $n' < n$. Hence, $\{Z_i\}$ is exchangeable with order 2 because every pair has the same distribution as every other pair, i.e., every pair of $\{Z_1, Z_2, \cdots, Z_n\}$ has the same distribution as every other pair and for any pair (Z_i, Z_j), $\forall i, j = 1, 2, \cdots, n$, and $i \neq j$, $P(Z_i = 0, Z_j = 1) = P(Z_i = 1, Z_j = 0) > 0$, $P(Z_i = 0, Z_j = 0) + 2P(Z_i = 0, Z_j = 1) + P(Z_i = 1, Z_j = 1) = 1$, $P(Z_i = 1, Z_j = 1) > 0$, and $P(Z_i = 0, Z_j = 0) > 0$; see [16] for more discussion.*

2.3. Binomial Autoregressive Model with the CMPB Operator

Now, we define the BAR(1) model with the CMPB operator by

$$X_t = \alpha \diamond_\nu X_{t-1} + \beta \diamond_\nu (n - X_{t-1}), \qquad (8)$$

where $0 < \alpha < 1$, $0 < \beta < 1$, both $\alpha \diamond_\nu X_{t-1} = \sum_{i=1}^{X_{t-1}} Z_i$ and $\beta \diamond_\nu (n - X_{t-1}) = \sum_{i=1}^{n-X_{t-1}} W_i$ are the CMPB thinning operators given in Definition 1, their counting series $\{Z_i\}$ and $\{W_i\}$ are the exchangeable Bernoulli variable sequence with their pmfs taking the form (5), $\{Z_i\}$ is independent of $\{W_j\}$, $\forall i = 1, 2, \ldots, X_{t-1}$, $j = 1, 2, \ldots, (n - X_{t-1})$, and all the thinnings at time t are independent of $\{X_s, s < t\}$, $n \in \mathbb{N}$, $\nu \in \mathbb{R}$.

For simplicity, we denote the new model as the CMPBAR(1) model. By (8), $\{X_t\}_\mathbb{N}$ is a Markov chain and its one-step transition probability takes the form

$$P_\eta(k|l) = P(X_t = k|X_{t-1} = l) = \frac{1}{S(\theta_1, \nu) S(\theta_2, \nu)} \sum_{i=0}^{\min\{k,l\}} \binom{l}{i}^\nu \binom{n-l}{k-i}^\nu \theta_1^i \theta_2^{k-i}, \quad (9)$$

where $S(\theta_1, \nu) = \sum_{i=0}^{l} \binom{l}{i}^\nu \theta_1^i$ and $S(\theta_2, \nu) = \sum_{i=0}^{n-l} \binom{n-l}{i}^\nu \theta_2^i$ with $\eta = (\theta_1, \theta_2, \nu)$ and $\theta_1 = \alpha/(1-\alpha)$ and $\theta_2 = \beta/(1-\beta)$.

Theorem 1. *If $\{X_t\}$ satisfies (8), then $\{X_t\}$ is ergodicity and strictly stationarity.*

Proof. Similar to that of Theorem 1 in Kang et al. [11], the state space of $\{X_t\}$ is $\{0, 1, \cdots, n\}$. Because $P(X_t = i | X_{t-1} = j) > 0, \forall i, j \in \{0, 1, \cdots, n\}$, so the state space of $\{X_t\}$ is an equivalence class. Furthermore, $\{X_t\}$ is an irreducible and aperiodic Markov chain; therefore, $\{X_t\}$ is ergodic with a unique stationary distribution by [17]. □

By Definition 1 and (8), for given X_{t-1}, $\{X_t\}$ given in (8) consists of two independent parts $\alpha \diamond_\nu X_{t-1}$ and $\beta \diamond_\nu (n - X_{t-1})$, where $\alpha \diamond_\nu X_{t-1} \sim \text{CMPB}(X_{t-1}, \alpha, \nu)$ and $\beta \diamond_\nu (n - X_{t-1}) \sim \text{CMPB}(n - X_{t-1}, \beta, \nu)$. Denote $\theta_1 = \alpha/(1-\alpha)$ and $\theta_2 = \beta/(1-\beta)$. Then

$$E(X_t|X_{t-1}) = \theta_1 S_1'/S_1 + \theta_2 S_2'/S_2,$$

$$\text{Var}(X_t|X_{t-1}) = \theta_1 \frac{S_1'}{S_1} + \theta_2 \frac{S_2'}{S_2} + \theta_1^2 \left(\frac{S_1''}{S_1} - \left(\frac{S_1'}{S_1}\right)^2\right) + \theta_2^2 \left(\frac{S_2''}{S_2} - \left(\frac{S_2'}{S_2}\right)^2\right),$$

and the conditional binomial index of dispersion (CBID) is

$$\text{CBID} = \frac{\theta_1^2 S_2^2 (S_1 S_1'' - (S_1')^2) + \theta_2^2 S_1^2 (S_2 S_2'' - (S_2')^2) + \theta_1 S_1 S_1' S_2^2 + \theta_2 S_2 S_2' S_1^2}{(n S_1 S_2 - \theta_1 S_1' S_2 - \theta_2 S_1 S_2')(\theta_1 S_2 S_1' + \theta_2 S_1 S_2')}$$

where $S_1 := S_1(\theta_1, \nu) = \sum_{x=0}^{X_{t-1}} \binom{X_{t-1}}{x}^\nu \theta_1^x$, $S_1' := S_1'(\theta_1, \nu) = \partial S_1(\theta_1, \nu)/\partial \theta_1$, $S_1'' := S_1''(\theta_1, \nu) = \partial S_1'(\theta_1, \nu)/\partial \theta_1$, $S_2 := S_2(\theta_2, \nu) = \sum_{x=0}^{n-X_{t-1}} \binom{n-X_{t-1}}{x}^\nu \theta_2^x$, $S_2' := S_2'(\theta_2, \nu) = \partial S_2(\theta_2, \nu)/\partial \theta_2$, $S_2'' := S_2''(\theta_2, \nu) = \partial S_2'(\theta_2, \nu)/\partial \theta_2$.

Unfortunally, because of the complexity of $S_1(\theta_1, \nu)$ and $S_2(\theta_2, \nu)$, we can not obtain the marginal distribution of $\{X_t\}$ and its the autocorrelation structure, including the $E(X_t)$, $\text{Var}(X_t)$, and BID. To resolve this dilemma, for given $n = 10$, we create some plots of the BID (in Figure 3) by generating some samples from the CMPBAR(1) model with $\nu \in \{-5, -4.5, -4, \cdots, 4.5, 5\}$ and sample size $T = 500$, when $(\alpha, \beta) = (0.2, 0.2), (0.2, 0.5), (0.2, 0.6), (0.5, 0.6)$, i.e., $(\theta_1, \theta_2) = (0.25, 0.25), (0.25, 1), (0.25, 1.5), (1, 1.5)$.

From Figure 3, we have the following observations. First, if $\nu < 1$, the BID of the CMPBAR(1) model is greater than 1, i.e., the CMPBAR(1) model allows us to analyze bounded integer-valued time series with overdispersion. Second, if $\nu > 1$, the BID of the CMPBAR(1) model is less than 1, i.e., the CMPBAR(1) model allows us to analyze bounded integer-valued time series with underdispersion. Third, if $\nu = 1$, the CMPBAR(1) model becomes to the BAR(1) given in (2) and its BID is equal to 1, i.e., equi-dispersed bounded integer-valued time series is allowed.

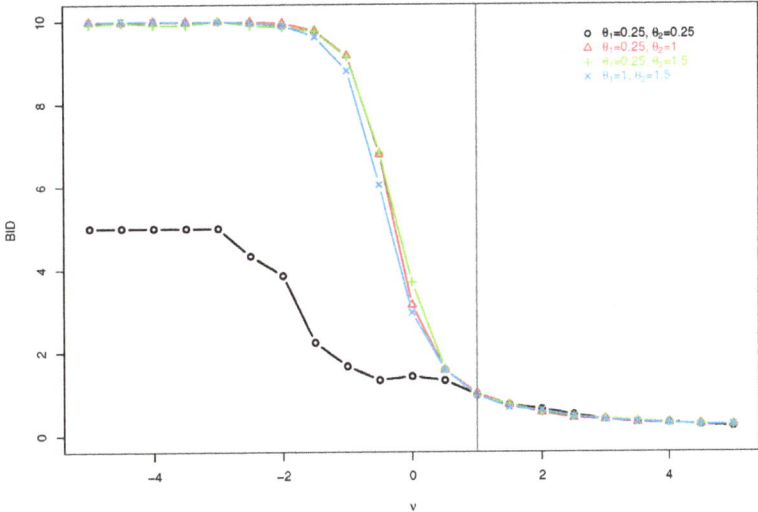

Figure 3. Plots of BID of the CMPBAR model.

3. Parameter Estimation

In this section, we use the conditional maximum likelihood method to estimate the parameters (denoted as $\eta = (\theta_1, \theta_2, \nu)^\top$) involving in the CMPBAR(1) model. Let $\{X_0, X_1, \ldots, X_T\}$ be a realization of $\{X_t\}$, and generate by the CMPBAR(1) process based on Algorithm 1, where $T \in \mathbb{N}$ represents the size of sample.

Algorithm 1: Random number generation algorithm for the CMPB distribution

---

Step 1. generate a random number u, $u \sim \text{Uniform}(0,1)$;
Step 2. $x = 0$, $p = P(\alpha \diamond_\nu n = 0|\theta, \nu, n)$, $F = p$, where $P(\alpha \diamond_\nu n = 0|\theta, \nu, n)$ is given in (3);
Step 3. if $u < F$, set $\alpha \diamond_\nu n = x$ and stop;
Step 4. else $p = p \times \left(\frac{n-x}{x+1}\lambda\right)^\nu$ by (7), $F = F + p$, $x = x + 1$;
Step 5. go to Step 3.

By using (9), the conditional log-likelihood function can be written as:

$$\ell(\eta) = \sum_{t=1}^{T} \log P_\eta(X_t|X_{t-1})$$
$$= \sum_{t=1}^{T} \log \left(\sum_{i=0}^{m} \binom{X_{t-1}}{i}^\nu \binom{n-X_{t-1}}{X_t - i}^\nu \theta_1^i \theta_2^{X_t - i} \right) - \log(S(\theta_1, \nu)) - \log(S(\theta_2, \nu)), \quad (10)$$

where $S(\theta_1, \nu) = \sum_{i=0}^{X_{t-1}} \binom{i}{X_{t-1}}^\nu \theta_1^i$ and $S(\theta_2, \nu) = \sum_{i=0}^{n-X_{t-1}} \binom{i}{n-X_{t-1}}^\nu \theta_2^i$ with $m = \min\{X_t, X_{t-1}\}$, $\theta_1 > 0$, $\theta_2 > 0$, and $\nu \in \mathbb{R}$. Then the CML estimate $\hat{\eta}^{cml}$ is obtained by minimizing (10).

Assumption 1. *If there exists a $t \geq 1$, such that $X_t(\eta) = X_t(\eta_0)$, P_{η_0} a.s., then $\eta = \eta_0$, where P_{η_0} is the probability measure under the true parameter η_0 with $\eta_0 = \{\theta_1^0, \theta_2^0, \nu^0\}$.*

Theorem 2. *Let $\{X_t\}$ be generalized by the CMPBAR(1) model. If Assumption 1 holds, there exists an estimator $\hat{\eta}^{cml}$ such that*

$$\hat{\eta}^{cml} \xrightarrow{a.s.} \eta_0 \text{ and } \sqrt{T}(\hat{\eta}^{cml} - \eta_0) \xrightarrow{d} \mathcal{N}\left(0, J^{-1}(\eta_0) I(\eta_0) J^{-1}(\eta_0)\right), T \to \infty,$$

where $I(\eta_0) = E\left[\frac{\partial \log P_{\eta_0}(X_t|X_{t-1})}{\partial \eta}\frac{\partial \log P_{\eta_0}(X_t|X_{t-1})}{\partial \eta^\top}\right]$ and $J(\eta_0) = E\left(\frac{\partial^2 \ell(\eta_0)}{\partial \eta \partial \eta^\top}\right)$.

Proof. To prove the consistence of $\hat{\eta}^{cml}$, we denote $\ell_t(\eta) = \log P_\eta(X_t|X_{t-1})$. Hence, $\ell(\eta) = \sum_{t=1}^T \ell_t(\eta)$. Similar to the first item of Theorem 4 in Chen et al. [18], we can verify that the assumptions of Theorem 4.1.2 in Amemiya [19] hold under Assumption 1, i.e., $E\ell_t(\eta)$ attains a strict local maximum at η_0; therefore, there exists an estimator $\hat{\eta}^{cml}$ such that $\hat{\eta}^{cml} \xrightarrow{a.s.} \eta_0$.

In the following, we prove the asymptotic normality of $\hat{\eta}^{cml}$. It is easy to see $\partial \ell_t(\eta)/\partial \theta_1$, $\partial \ell_t(\eta)/\partial \theta_2$, and $\partial \ell_t(\eta)/\partial \nu$ exist and are three times continuous differentiable in Θ. Thus, there exist a $N(\eta_0)$ such that $\partial^2 \ell_t(\eta)/(\partial \eta \partial \eta^\top)$ attains the maximum value at $\tilde{\eta} \in N(\eta_0)$. Therefore,

$$E\|\sup_{\eta \in N(\eta_0)} \frac{\partial^2 \ell_t(\eta)}{\partial \eta \partial \eta^\top}\| = E\|\frac{\partial^2 \ell_t(\tilde{\eta})}{\partial \eta_i \partial \eta_j}\| < \infty.$$

Similar to the second item of Theorem 4 in [18], we can prove that

$$T^{-1}\sum_{t=1}^T \frac{\partial^2 \ell_t(\eta)}{\partial \eta \partial \eta^\top} \xrightarrow{p} E\left(\frac{\partial^2 \ell_t(\eta_0)}{\partial \eta \partial \eta^\top}\right)$$

by Theorem 4.1.3 in Amemiya [19]. Furthermore,

$$T^{-1}\sum_{t=1}^T \partial \ell_t(\eta_0)/\partial \eta \xrightarrow{p} E(\partial \ell_t(\eta_0)/\partial \eta)$$

by using ergodic theorem. Using the Martingale central limit theorem and the Cramér device, it is direct to show that

$$T^{-1/2}\partial \ell(\eta_0)/\partial \eta \xrightarrow{d} N(0, I(\eta_0)).$$

Then the asymptotic normal distribution of $\hat{\eta}^{cml}$ is obtained based on the Taylor series expansion of $\partial \ell(\hat{\eta}^{cml})/\partial \eta$ around η_0. □

4. Simulation

In this section, we conduct a simulation study to illustrate the large sample property of the CMPBAR(1) model.

In the simulation, we fix $n = 10$, let sample size $T = 100, 300, 500$, and use the optim function in R to optimize $\ell(\eta)$ in (10). To check the finite sample performance, we use the following parameter combinations of $(\theta_1, \theta_2, \nu)$ as

(A1) = (0.25, 0.25, 0.5), (A2) = (0.25, 1, 0.5), (A3) = (0.25, 1.5, 0.5), (A4) = (1, 1.5, 0.5),
(B1) = (0.25, 0.25, 1), (B2) = (0.25, 1, 1), (B3) = (0.25, 1.5, 1), (B4) = (1, 1.5, 1),
(C1) = (0.25, 0.25, 1.5), (C2) = (0.25, 1, 1.5), (C3) = (0.25, 1.5, 1.5), (C4) = (1, 1.5, 1.5),

where $\nu = 0.5, 1$ and 1.5 to reflect overdispersion, equidispersion, and underdispersion, respectively.

For the simulated sample, performances of mean and standard deviation (sd) are given. For a scale parameter φ, sd $= \sqrt{\frac{1}{m-1}\sum_{i=1}^m (\hat{\varphi}_i - \varphi)^2}$, where $\hat{\varphi}_i$ is the estimator of φ in the ith replication and $m = 10,000$. Summaries of the simulation results are given in Tables 1–3.

To illustrate the consistency and the asymptotic normality of the CML estimators, we present the boxplots of the CML estimates for (A1), (B1), and (C1) in Figures 4, 5, and 6, and their qqplots with $T = 500$ in Figure 7, 8, and 9, respectively. Others are similar and we omit them.

Table 1. Mean and sd in parentheses of estimates for (A1)–(A4).

	100	300	500
		(A1) = (0.25, 0.25, 0.5)	
θ_1	0.2336 (0.1425)	0.2435 (0.0881)	0.2471 (0.0683)
θ_2	0.2408 (0.0829)	0.2467 (0.0498)	0.2479 (0.0391)
ν	0.5682 (0.2484)	0.5231 (0.1371)	0.5135 (0.1065)
		(A2) = (0.25, 1, 0.5)	
θ_1	0.2420 (0.0847)	0.2471 (0.0477)	0.2483 (0.0369)
θ_2	1.0058 (0.0935)	1.0022 (0.0510)	1.0010 (0.0390)
ν	0.5236 (0.1353)	0.5070 (0.0742)	0.5044 (0.0567)
		(A3) = (0.25, 1.5, 0.5)	
θ_1	0.2450 (0.0644)	0.2483 (0.0374)	0.2490 (0.0288)
θ_2	1.5283 (0.1677)	1.5092 (0.0936)	1.5053 (0.0710)
ν	0.5269 (0.1505)	0.5072 (0.0821)	0.5046 (0.0628)
		(A4) = (1, 1.5, 0.5)	
θ_1	1.0032 (0.1132)	1.0002 (0.0622)	1.0005 (0.0481)
θ_2	1.5446 (0.2335)	1.5176 (0.1389)	1.5097 (0.1066)
ν	0.5246 (0.1336)	0.5087 (0.0755)	0.5052 (0.0585)

Table 2. Mean and sd in parentheses of estimates for (B1)–(B4).

	100	200	500
		(B1) = (0.25, 0.25, 1)	
θ_1	0.2442 (0.1286)	0.2475 (0.0755)	0.2487 (0.0586)
θ_2	0.2484 (0.0693)	0.2497 (0.0402)	0.2496 (0.0313)
ν	1.0484 (0.2317)	1.0152 (0.1288)	1.0094 (0.0997)
		(B2) = (0.25, 1, 1)	
θ_1	0.2483 (0.0906)	0.2491 (0.0508)	0.2496 (0.0393)
θ_2	1.0114 (0.1667)	1.0033 (0.0906)	1.0016 (0.0692)
ν	1.0390 (0.2070)	1.0130 (0.1140)	1.0083 (0.0873)
		(B3) = (0.25, 1.5, 1)	
θ_1	0.2507 (0.0770)	0.2497 (0.0440)	0.2499 (0.0339)
θ_2	1.5215 (0.2412)	1.5097 (0.1417)	1.5053 (0.1082)
ν	1.0409 (0.2167)	1.0128 (0.1201)	1.0084 (0.0922)
		(B4) = (1, 1.5, 1)	
θ_1	1.0219 (0.1985)	1.0042 (0.1127)	1.0028 (0.0876)
θ_2	1.5420 (0.3113)	1.5251 (0.2070)	1.5151 (0.1632)
ν	1.0318 (0.1883)	1.0114 (0.1057)	1.0067 (0.0820)

Table 3. Mean and sd in parentheses of estimates for (C1)–(C4).

	100	200	500
		(C1) = (0.25, 0.25, 1.5)	
θ_1	0.2563 (0.1402)	0.2517 (0.0784)	0.2514 (0.0611)
θ_2	0.2550 (0.0732)	0.2513 (0.0435)	0.2506 (0.0336)
ν	1.5431 (0.2529)	1.5169 (0.1553)	1.5103 (0.1191)
		(C2) = (0.25, 1, 1.5)	
θ_1	0.2586 (0.1141)	0.2524 (0.0620)	0.2515 (0.0479)
θ_2	1.0332 (0.2637)	1.0094 (0.1449)	1.0052 (0.1120)
ν	1.5408 (0.2482)	1.5157 (0.1497)	1.5100 (0.1153)

Table 3. Cont.

		100	200	500
		(C3) = (0.25, 1.5, 1.5)		
	θ_1	0.2625 (0.1000)	0.2523 (0.0559)	0.2515 (0.0433)
	θ_2	1.5186 (0.3340)	1.5169 (0.2200)	1.5100 (0.1730)
	ν	1.5383 (0.2512)	1.5167 (0.1531)	1.5103 (0.1180)
		(C4) = (1, 1.5, 1.5)		
	θ_1	1.0528 (0.2914)	1.0134 (0.1701)	1.0075 (0.1329)
	θ_2	1.5339 (0.3820)	1.5310 (0.2724)	1.5221 (0.2243)
	ν	1.5398 (0.2350)	1.5161 (0.1396)	1.5100 (0.1082)

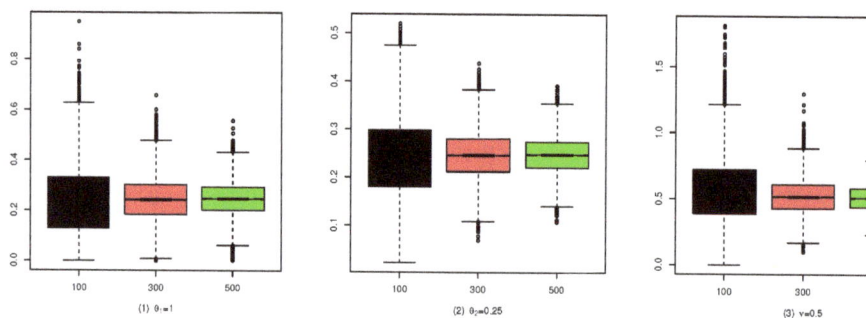

Figure 4. Boxplots of the CML estimates for (A1).

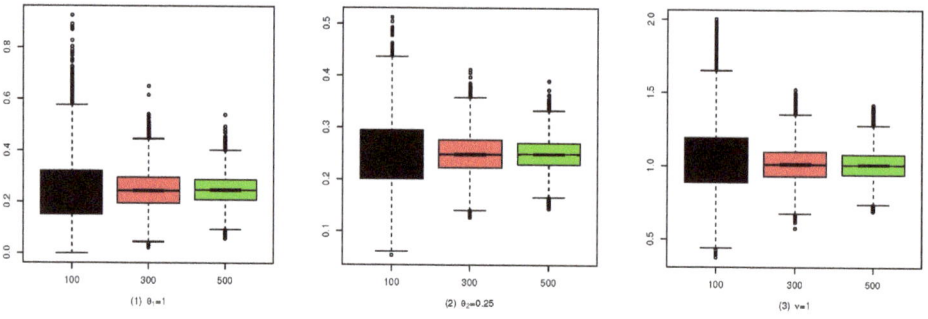

Figure 5. Boxplots of the CML estimates for (B1).

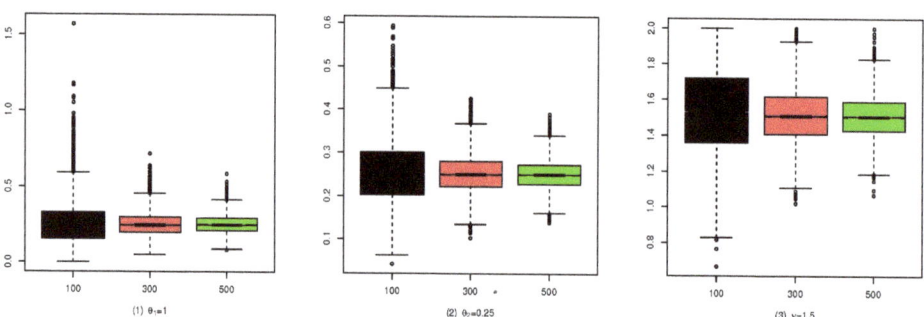

Figure 6. Boxplots of the CML estimates for (C1).

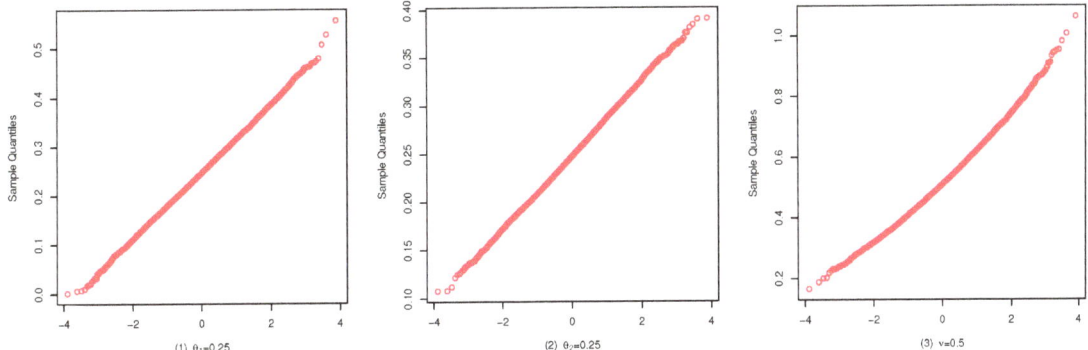

Figure 7. qqplots of the CML estimates for (A1) with $T = 500$.

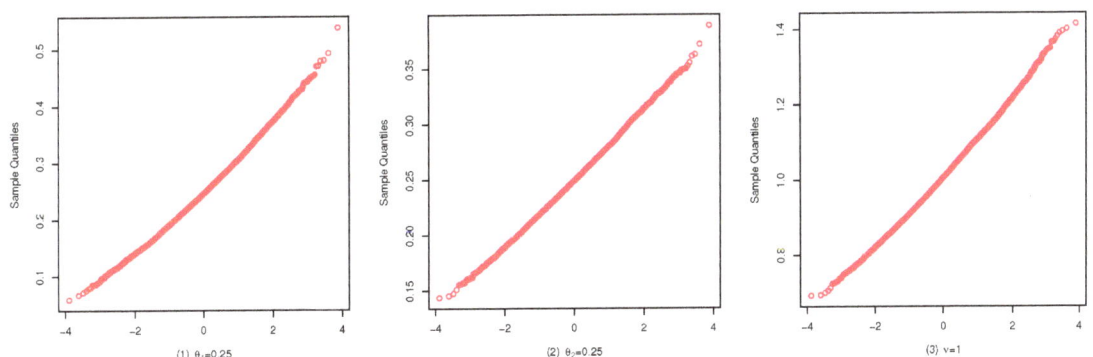

Figure 8. qqplots of the CML estimates for (B1) with $T = 500$.

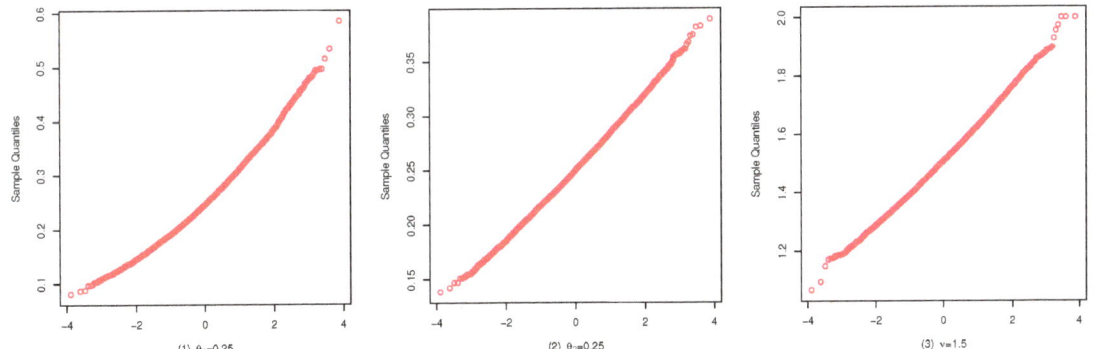

Figure 9. qqplots of the CML estimates for (C1) with $T = 500$.

These studies indicate that the CML method seems to perform reasonably well. First, Tables 1–3 show that the standard deviation of the CML estimator is decreasing with the sample size increase and the mean of the CML estimator is closer to the true parameter value in general cases. Second, Figures 4–6 account for the location and dispersion of the estimates, all of which indicate the consistency of the estimators. Third, Figures 7–9 indicate the asymptotic normality of the CML estimator.

5. Real Data Example

In this section, we consider the number of weekly rainy days for the period from 1 January 2005 to 31 December 2010 at Hamburg–Neuwiedenthal in Germany, where a week is defined as being from Saturday to Friday and $n = 7$. The data were collected from the German Weather Service (http://www.dwd.de/, accessed on 12 December 2018). The sample path and the ACF and PACF plots of the observations are given in Figures 10 and 11, respectively.

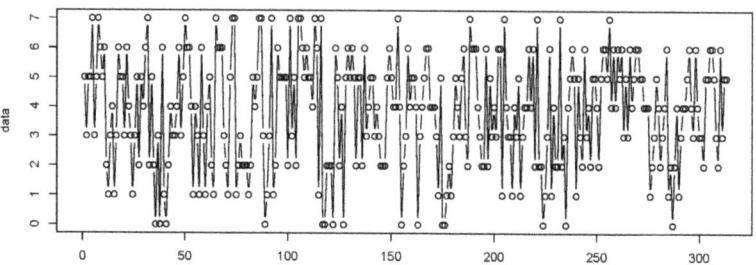

Figure 10. Path of the weekly rainy days.

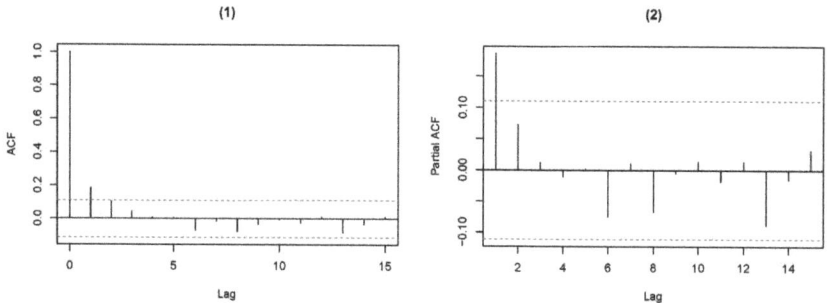

Figure 11. ACF and PACF plots of the weekly rainy days. (**1**) shows that the ACF exhibits significant value for lag 1, and (**2**) presents that the PACF indicates an AR(1)-like autocorrelation structure.

By computation, the sample mean and variance are 3.8371 and 3.6753, and the BID of the data is 1.2371, which implies the data exhibits extra-binomial variation. Hence, we use the CMPBAR(1) model, BAR(1) model [2], BBAR(1) model [9], and GBAR(1) model [11] to fit data by the CML method. We compare the estimated standard error (SE), $-$log-likelihood ($-$log-lik), Akaike's information criterion (AIC) and Bayesian information criterion (BIC), which are summarized in Table 4, including the fitted results of the CML estimate.

From Table 4, the CMPBAR(1) model takes the smallest values of the $-$log-lik, AIC, and BIC. Hence, the CMPBAR(1) model might be more appropriate for the weekly rainy days.

To illustrate the adequacy of the CMPBAR(1) model, we consider the fitted Pearson residual analysis of the CMPBAR(1) model. By computation, the mean and variance of the fitted Pearson residual are 0.0760 and 1.0500, respectively. The residual analysis in Figure 12 shows that this model performs rather well.

Table 4. Estimates for the weekly rainy days and SE are shown in parentheses.

Model	Estimates			−log-lik	AIC	BIC
BAR(1)	$\hat{\pi}$ 0.5476 (0.0122)	$\hat{\rho}$ 0.1323 (0.0325)		691.5400	1387.0800	1394.5720
BBAR(1)	$\hat{\pi}$ 0.5475 (0.0177)	$\hat{\rho}$ 0.1408 (0.0507)	$\hat{\phi}$ 0.2827 (0.0320)	623.6617	1253.33233	1264.5619
GBAR(1)	$\hat{\pi}$ 0.5493 (0.0169)	$\hat{\rho}$ 0.1396 (0.0492)	$\hat{\phi}$ 0.5209 (0.0279)	625.4958	1256.9916	1268.2303
CMPBAR(1)	$\hat{\theta}_1$ 1.2313 (0.0627)	$\hat{\theta}_2$ 0.9547 (0.0532)	$\hat{\upsilon}$ 0.0995 (0.0681)	622.6669	1251.3337	1262.5723

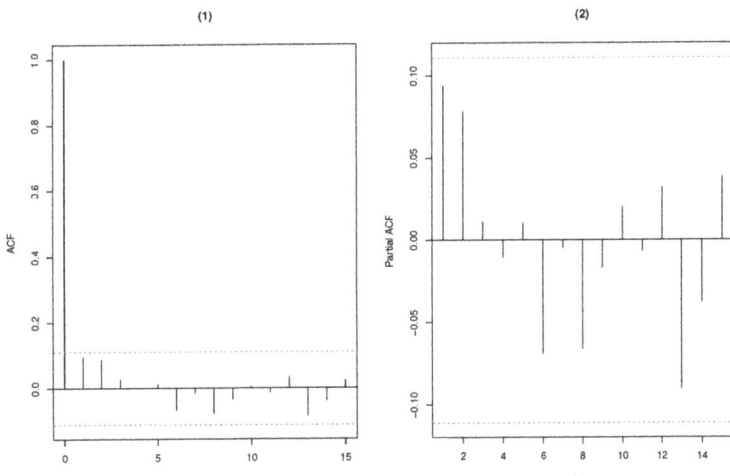

Figure 12. Pearson residual analysis of the weekly rainy days. (1) ACF (2) PACF.

In addition, to further check the adequacy of the CMPBAR(1) model, we present the probability integral transform (PIT) (if the fitted model is adequate, its PIT histogram looks like that of a uniform distribution, see [10] for more discussion) in Figure 13 based on the fitted CMPBAR(1) model.

As can be seen in Figure 13, the PIT histogram of the CMPBAR(1) model is close to uniformity, i.e., the PIT histogram confirms that the fitted CMPBAR(1) model works reasonably well for the weekly rainy days.

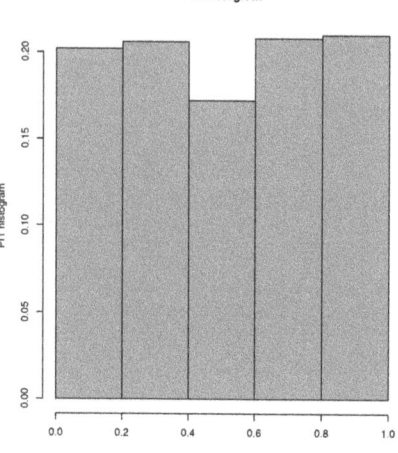

Figure 13. PIT histogram based on the fitted CMPBAR(1) model.

6. Concluding Remarks

This paper considers a new CMPB thinning operator and proposes a new CMPBAR(1) model, which provides an available method to model bounded data with under-dispersion, equi-dispersion, and over-dispersion. We discuss some properties of the new model, the estimate of the parameters, and its large-sample properties. Simulations are conducted to examine the finite sample performance of estimators. A real data example is provided to illustrate the applicability of the CMPBAR(1) model.

There are several directions in which we plan to take this work forward. First, the random coefficient CMPBAR(1) model can be introduced by

$$X_t = \alpha_t \circ_\nu X_{t-1} + \beta_t \circ_\nu (n - X_{t-1}),$$

where $\alpha_t = \alpha(X_{t-1}/n)$ and $\beta_t = \beta(X_{t-1}/n)$, "$\circ_\nu$" is the CMPB thinning operator and the counting series in "$\alpha_t \circ_\nu$", and that in "$\beta_t \circ_\nu$" is independent and all of the counting series at time t is independent of $\{X_s, s < t\}$; see Weiß and Pollett [8] for the random coefficient BAR(1) model. Second, a correlated sign-thinning operator can be established by

$$\alpha \diamond_\nu X = \text{sign}(\alpha)\text{sign}(X) \sum_{i=1}^{X} Z_i,$$

where $\text{sign}(x) = 1$ if $x \geq 0$ and $\text{sign}(x) = -1$ if $x < 0$, $\{Z_i, i = 1, 2, \cdots, X\}$ is an exchangeable Bernoulli variable sequence with its pmf taking the form (5). Based on the correlated sign thinning operator, one can construct a $\mathbb{Z}$-valued autoregressive model to analyze data with a range $\mathbb{Z}$ and under-dispersed, equi-dispersed, and over-dispersed. Third, a class of Conway–Maxwell–Poisson-binomial generalized autoregressive conditional heteroskedasticity models can be considered by

$$Z_t | \mathcal{F}_{t-1} \sim \text{CMPB}(n, \alpha_t, \nu), \ \alpha_t = g_\eta(Z_{t-1}/n, \alpha_{t-1}),$$

where η is the parameter vector involving in the model (see Ristić et al. [20] and Chen et al. [18] for ARCH-type models, Lee and Lee [21] and Chen et al. [22] for GARCH-type models for bounded data). In addition, a semi-parameter version can be considered by

$$Z_t | \mathcal{F}_{t-1} \sim \text{CMPB}(n, \alpha_t, \nu), \ \alpha_t = g_\eta(Z_{t-1}/n, \alpha_{t-1}) + f_\gamma(X_t),$$

where η is the parameter vector involved in the model, $\{X_t\}$ is the covariate process imposed in the observe process $\{Z_t\}$, and γ is the parameter vector involving in $f(\cdot)$.

Supplementary Materials: The following supporting information can be downloaded at: https://www.mdpi.com/article/10.3390/e25010126/s1.

Author Contributions: Conceptualization, H.C.; methodology, H.C. and J.Z.; software, H.C. and J.Z.; validation, H.C., J.Z. and X.L.; formal analysis, H.C.; investigation, H.C.; resources, H.C.; data curation, H.C. and J.Z.; writing—original draft preparation, H.C.; writing—review and editing, H.C.; visualization, H.C., J.Z. and X.L.; supervision, H.C., J.Z. and X.L. All authors have read and agreed to the published version of the manuscript.

Funding: Chen's work is supported by the Natural Science Foundation of Henan Province No. 222300420127 and Postdoctoral research in Henan Province No. 202103051. Liu's work is supported by the Basic Research Programs of Shanxi Province No. 202103021223084.

Data Availability Statement: The weekly rainy days for the period from 1st January 2005 to 31st December 2010 at Hamburg–Neuwiedenthal in Germany is collected from the German Weather Service (http://www.dwd.de/ accessed on 12 December 2018), where a week is defined as being from Saturday to Friday and $n = 7$ and the data can be found in supplementary materials.

Acknowledgments: The authors thank the Editor-in-Chief and the anonymous referees for the valuable comments and suggestions that result in a substantial improvement of this paper.

Conflicts of Interest: The authors declare no conflict of interest.

References

1. Steutel, F.W.; van Harn, K. Discrete analogues of self-decomposability and stability. *Ann. Probab.* **1979**, *7*, 893–899. [CrossRef]
2. McKenzie, E. Some simple models for discrete variate time series. *Water Resour Bull.* **1985**, *21*, 645–650. [CrossRef]
3. Weiß, C.H. Monitoring correlated processes with binomial marginals. *J. Appl. Stat.* **2009**, *36*, 399–414. [CrossRef]
4. Weiß, C.H. Jumps in binomial AR(1) processes. *Stat. Probab. Lett.* **2009**, *79*, 2012–2019. [CrossRef]
5. Weiß, C.H. A new class of autoregressive models for time series of binomial counts. *Commun. Stat.-Theory Methods* **2009**, *38*, 447–460. [CrossRef]
6. Weiß, C.H.; Kim, H.Y. Binomial AR(1) processes: Moments, cumulants, and estimation. *Statistics* **2013**, *47*, 494–510. [CrossRef]
7. Weiß, C.H.; Kim, H.Y. Parameter estimation for binomial AR(1) models with applications in finance and industry. *Stat. Pap.* **2013**, *54*, 563–590. [CrossRef]
8. Weiß, C.H.; Pollett, P.K. Binomial autoregressive processes with density dependent thinning. *J. Time Ser. Anal.* **2014**, *35*, 115–132. [CrossRef]
9. Weiß, C.H.; Kim, H.Y. Diagnosing and modeling extra-binomial variation for time-dependent counts. *Appl. Stoch. Model. Bus. Ind.* **2014**, *30*, 588–608. [CrossRef]
10. Weiß, C.H. *An Introduction to Discrete-Valued Time Series*; John Wiley & Sons: Chichester, UK, 2018.
11. Kang, Y.; Wang, D.H.; Yang, K. Extended binomial AR(1) processes with generalized binomial thinning operator. *Commun. Stat.-Theory Methods* **2020**, *49*, 3498–3520. [CrossRef]
12. Ristić, M.M.; Nastić, A.S.; Ilić, A.V.M. A geometric time series model with dependent Bernoulli counting series. *J. Time Ser. Anal.* **2013**, *34*, 466–476. [CrossRef]
13. Shmueli, G.; Minka, T.P.; Kadane, J.B.; Borle, S.; Boatwright, P. A useful distribution for fitting discrete data: Revival of the Conway-Maxwell-Poisson distribution. *Appl. Stat.* **2005**, *54*, 127–142. [CrossRef]
14. Borges, P.; Rodrigues, J.; Balakrishnan, N.; Bazn, J. A COM-Poisson type generalization of the binomial distribution and its properties and applications. *Stat. Probab. Lett.* **2014**, *87*, 158–166. [CrossRef]
15. Daly, F.; Gaunt, R.E. The Conway-Maxwell-Poisson distribution: Distributional theory and approximation. *ALEA Lat. Am. J. Probability Math. Stat.* **2016**, *13*, 635–658. [CrossRef]
16. Kadane, J.B. Sums of possibly associated bernoulli variables: The Conway-Maxwell-Binomial distribution. *Bayesian Anal.* **2016**, *11*, 363–374. [CrossRef]
17. Seneta, E. *Non-Negative Matrices and Markov Chains*, 2nd ed.; Springer: New York, NY, USA, 1983.
18. Chen, H.; Li, Q.; Zhu, F. Two classes of dynamic binomial integer-valued ARCH models. *Braz. J. Probab. Stat.* **2020**, *34*, 685–711. [CrossRef]
19. Amemiya, T. *Advanced Econometrics*; Harvard University Press: Cambridge, UK, 1985; pp. 110–112.
20. Ristić, M.M.; Weiß, C.H.; Janjić, A.D. A binomial integer-valued ARCH model. *Int. J. Biostat.* **2016**, *12*, 20150051. [CrossRef] [PubMed]

21. Lee, Y.; Lee, S. CUSUM test for general nonlinear integer–valued GARCH models: Comparison study. *Ann. Inst. Stat. Math.* **2019**, *71*, 1033–1057. [CrossRef]
22. Chen, H.; Li, Q.; Zhu, F. A new class of integer-valued GARCH models for time series of bounded counts with extra-binomial variation. *AStA Adv. Stat. Anal. Vol.* **2022**, *106*, 243–270. [CrossRef]

Disclaimer/Publisher's Note: The statements, opinions and data contained in all publications are solely those of the individual author(s) and contributor(s) and not of MDPI and/or the editor(s). MDPI and/or the editor(s) disclaim responsibility for any injury to people or property resulting from any ideas, methods, instructions or products referred to in the content.

Article

Time Series of Counts under Censoring: A Bayesian Approach

Isabel Silva [1], Maria Eduarda Silva [2], Isabel Pereira [3,*] and Brendan McCabe [4]

[1] Faculdade de Engenharia, Universidade do Porto, CIDMA, 4200-465 Porto, Portugal
[2] Faculdade de Economia, Universidade do Porto, LIADD-INESC TEC, 4200-464 Porto, Portugal
[3] Departamento de Matemática, Universidade de Aveiro, CIDMA, 3810-193 Aveiro, Portugal
[4] School of Management, University of Liverpool, Liverpool L69 3BX, UK
* Correspondence: isabel.pereira@ua.pt

Abstract: Censored data are frequently found in diverse fields including environmental monitoring, medicine, economics and social sciences. Censoring occurs when observations are available only for a restricted range, e.g., due to a detection limit. Ignoring censoring produces biased estimates and unreliable statistical inference. The aim of this work is to contribute to the modelling of time series of counts under censoring using convolution closed infinitely divisible (CCID) models. The emphasis is on estimation and inference problems, using Bayesian approaches with Approximate Bayesian Computation (ABC) and Gibbs sampler with Data Augmentation (GDA) algorithms.

Keywords: Bayesian estimation; censored time series; convolution closed infinitely divisible; Poisson INAR(1) model

1. Introduction

Observations collected over time or space are usually correlated rather than independent. Time series are often observed with data irregularities such as missing values or detection limits. For instance, a monitoring device may have a technical detection limit and it records the limit value when the true value exceeds/precedes the detection limit. Such data is called censored (type 1) data and are common in environmental monitoring, physical sciences, business and economics. In particular, in the context of time series of counts, censored data arise in call centers. In fact, the demand measured by the number of calls is limited by the number of operators. When the number of calls is higher than the number of operators the data is right censored and the call center incurs under-staffing and poor service to the costumers.

The main consequence of neglecting censoring in the time series analysis is the loss of information that is reflected in biased and inconsistent estimators and altered serial correlation. These consequences can be summarized as problems in inference that lead to model misspecification, biased parameter estimation, and poor forecasts.

These problems have been solved in regression settings (i.i.d.) and partially solved for Gaussian time series (see for instance [1–7]). However, the problem of modelling time series under censoring in the context of time series of counts has, as yet, received little attention in the literature even though its relevance for inference. Count time series occur in many areas such as telecommunications, actuarial science, epidemiology, hydrology and environmental studies where the modelling of censored data may be invaluable in risk assessment.

In the context of time series of counts, Ref. [8] deal with correlated under-reported data through INAR(1)-hidden Markov chain models. A naïve method of parameter estimation was proposed, jointly with the maximum likelihood method based on a revised version of the forward algorithm. Additionally, Ref. [9] propose a random-censoring Poisson model for under-reported data, which accounts for the uncertainty about both the count and the data reporting processes.

Here, the problem of modelling count data under censoring is considered under a Bayesian perspective. In this paper, we consider a general class of convolution closed infinitely divisible (CCID) models as proposed by [10].

We investigate two natural approaches to analyse censored convolution closed infinitely divisible models of first order, CCID(1), using the Bayesian framework: the Approximate Bayesian Computation (ABC) methodology and the Gibbs sampler with Data Augmentation (GDA).

Since the CCID(1) under censoring presents an intractable likelihood, we resort to the Approximate Bayesian Computation methodology for estimating the model parameters. The presupposed model is simulated by using sample parameters taken from the prior distribution, then a distance between the simulated dataset and the observations is computed and when the simulated dataset is *very close* to the observed, the corresponding parameter samples are accepted as part of the posterior.

In addition, a widely used strategy to deal with censored data is to fill in censored data in order to create a data-augmented (complete) dataset. When the data-augmented posterior and the conditional pdf of the latent process are both available in a tractable form, the Gibbs sampler allows us to sample from the posterior distribution of the parameters of the complete dataset. This methodology is called Gibbs sampler with Data Augmentation (GDA). Here, a modified GDA, in which the data augmentation is achieved by multiple sampling of the latent variables from the truncated conditional distributions (GDA-MMS), is adopted.

The Poisson integer-valued autoregressive models of first-order, PoINAR(1), is one of the most popular classes of CCID models. It was proposed by [11,12] and extensively studied in the literature and applied to many real-world problems because of its ease of interpretation. To motivate the proposed approaches, we present in Figure 1 a synthetic dataset with $n = 350$ observations generated from a PoINAR(1) process with parameters $\alpha = 0.5$ and $\lambda = 5$ (X_t, blue line) and the respective right-censored dataset (Y_t, red line), at $L = 11$, corresponding to 30% of censoring. If we disregard the censoring, the estimates for the parameters (assuming an PoINAR(1) model without censoring) present a strong bias. For instance, in the frequentist framework, the conditional maximum likelihood estimates are $\hat{\alpha}_{CML} = 0.6174$ and $\hat{\lambda}_{CML} = 3.4078$, while in the Bayesian framework, the Gibbs sampler gives $\hat{\alpha}_{Bayes} = 0.6242$ and $\hat{\lambda}_{Bayes} = 3.3297$. On the other hand, if we assume a PoINAR(1) model under censoring, the parameter estimates given by the proposed approaches described in this work are, respectively, $\hat{\alpha}_{ABC} = 0.4623$ and $\hat{\lambda}_{ABC} = 5.2259$, and $\hat{\alpha}_{GDA} = 0.4834$ and $\hat{\lambda}_{GDA} = 4.9073$. Therefore, it is important to consider the censoring in data in order to avoid some inference issues that lead to a poor time series analysis.

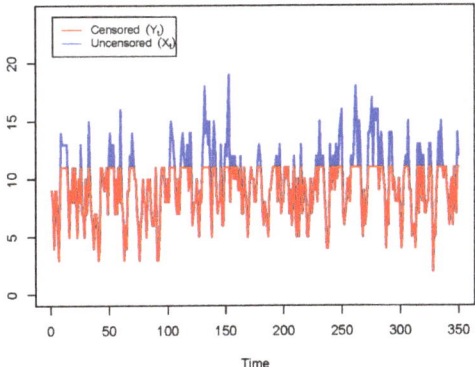

Figure 1. Synthetic dataset with $n = 350$ observations generated from a PoINAR(1) process with parameters $\alpha = 0.5$ and $\lambda = 5$ (X_t, blue line) and the respective right censored dataset (Y_t, red line), at $L = 11$.

The remainder of this work is organized as follows. Section 2 presents a general class of convolution closed infinitely divisible (CCID) models under censoring. Two Bayesian approaches proposed to estimate the parameters of the censored CCID(1) model are described in Section 3. The proposed methodologies are illustrated and compared with synthetic data in Section 4. Finally, Section 5 concludes the paper.

2. A Model for Time Series of Counts under Censoring

This section introduces a class of models adequate for censored time series of counts based on the convolution closed infinitely divisible (CCID) models as proposed by [10].

2.1. Convolution Closed Models for Count Time Series

First we introduce some notation. Consider a random variable X with a distribution F_μ, $\mu > 0$, belonging to the convolution closed infinitely divisible (CCID) parametric family [10]. This means, in particular, that the distribution F_μ is closed under convolution, $F_{\mu_1} * F_{\mu_2} = F_{\mu_1+\mu_2}$, where $*$ is the convolution operator. Let $R(\cdot)$ denote a random operator on X such that $R(X) \sim F_{\alpha\mu}$, $0 < \alpha < 1$ and the conditional distribution of $R(X)$ given $X = x$ is $G_{\alpha\mu,(1-\alpha)\mu,x}$, $R(X)|X = x \sim G_{\alpha\mu,(1-\alpha)\mu,x}$. As an example, consider a Poisson random variable, $X \sim \text{Po}(\mu)$ and a binomial thinning operation, $R(X) = \alpha \circ X = \sum_{i=1}^{X} \xi_i$, $\xi_i \sim^{iid} \text{Ber}(\alpha)$. Then F_μ is the Poisson distribution with parameter μ, $R(X) \sim \text{Po}(\alpha\mu)$ and $R(X)|X = x \sim \text{Bi}(x, \alpha)$, $G_{\alpha\mu,(1-\alpha)\mu,x}$ is the Binomial distribution with parameters x and α.

A stationary time series, $\{X_t; t = 0, \pm 1, \pm 2, \ldots\}$ with margin F_μ, $X_t \sim F_\mu$, is called a convolution closed infinitely divisible process of order 1, CCID(1), if it satisfies the following equation

$$X_t = R_t(X_{t-1}) + e_t, \qquad (1)$$

where the innovations e_t are independently and identically distributed (i.i.d.) with distribution $F_{(1-\alpha)\mu}$ and $\{R_t(\cdot) : t = 0, \pm 1, \pm 2, \ldots\}$ are independent replications of the random operator $R(\cdot)$ [10]. Note that the above construction leads to time series with the same marginal distribution as that of the innovations.

Model (1) encompasses many AR(1) models proposed in the literature for integer valued time series. In particular, the Poisson INAR(1), PoINAR(1), the negative binomial INAR(1), NBINAR(1), and the generalised Poisson INAR(1), GPINAR(1) [13], summarized in Table 1 (marginal distribution, random operation and its pmf $g(\cdot|\cdot)$, set of parameters θ), have been widely used in the literature to model time series of counts, see *inter alia* [14,15], among others.

Table 1. Methods for constructing integer valued AR(1) models with specified marginals F_μ and innovations $e_t \stackrel{iid}{=} F_\lambda$, $\lambda = \mu(1-\alpha)$. $B(\cdot,\cdot)$ denotes the beta function.

Marginal Distribution	Random Operator	$g(s\|X_{t-1};\alpha)$	Innovations	θ
Poison Po(μ)	binomial thinning	$\binom{X_{t-1}}{s}\alpha^s(1-\alpha)^{X_{t-1}-s}$	Po(λ)	(μ,α)
Negative binomial NB(μ,ξ)	beta binomial thinning	$\binom{X_{t-1}}{s}\alpha^s(1-\alpha)^{X_{t-1}-s}$	NB(λ,ξ)	(μ,α,ξ)
Generalised Poisson GP(μ,ξ)	quasi binomial thinning	$\binom{X_{t-1}}{s}\alpha(\alpha+s(\xi/\mu))^{s-1}(1-\alpha-s(\xi/\mu))^{X_{t-1}-s}$	GP(λ,ξ)	(μ,α,ξ)

If one chooses $F_{(1-\alpha)\mu}$ as Poisson$((1-\alpha)\mu)$, and the random operation as the usual binomial thinning operation (based on underlying Bernoulli random variables) $R_t(X_{t-1}) = \alpha \circ X_{t-1} = \sum_{i=1}^{X_{t-1}} \xi_{ti}$, $\xi_{ti} \stackrel{iid}{\sim} Ber(\alpha)$, then F_μ is Poisson(μ) and the Poisson integer-valued autoregressive model, PoINAR(1), as proposed by [11,12], is recovered with the familiar representation

$$X_t = \alpha \circ X_{t-1} + \epsilon_t. \quad (2)$$

Since model (1) is Markovian [10], given a time series $x = (X_1, \ldots, X_n)$, the conditional likelihood is as follows

$$L(\theta) = \prod_{t=2}^{n} f_{X_t|X_{t-1}}(x_t|x_{t-1}), \quad (3)$$

with

$$f_{X_t|X_{t-1}}(k|l) = P(X_t = k|X_{t-1} = l) = \sum_{j=0}^{\min\{k,l\}} g(j|l)P(e_t = k - j). \quad (4)$$

2.2. Modelling Censoring in CCID(1) Time Series

Given a model as in (1), a basic question is whether it properly describes all the observations of a given time series, or whether some observations have been affected by censoring. Here, we describe a model for dealing with censored observations in CCID(1) processes and study some of its properties.

Exogenous censoring can be modelled assuming (1) as a latent process and $Y_t = \min\{L, X_t\}$ as the observed process, where L is a constant that is assumed to be known. For simplicity of exposition we assume exogenous right censoring but all the results are easily extended to left-censoring or interval censoring. Hence, for right exogenous censoring

$$\begin{aligned} Y_t &= \min\{X_t, L\} = \begin{cases} X_t, & \text{if } X_t < L, \\ L, & \text{if } X_t \geq L, \end{cases} \\ X_t &= R_t(X_{t-1}) + e_t. \end{aligned} \quad (5)$$

Although X_t, a CCID(1) process is Markovian, the exogenous censoring implies that Y_t is not Markovian because Y_t depends on X_t and L. Furthermore, Y_t is not CLAR (Conditionally Linear AutoRegressive). In fact,

$$\begin{aligned} E[Y_t|Y_{t-1} = y_{t-1}] &= E[Y_t|Y_{t-1} = y_{t-1}]\mathbb{I}_{\{y_{t-1}<L\}} + E[Y_t|Y_{t-1} = L]\mathbb{I}_{\{y_{t-1}=L\}} \\ &= \left(E[X_t|X_{t-1} = y_{t-1}] - \sum_{j=0}^{+\infty} jP[X_t = L+j|X_{t-1} = y_{t-1}]\right)\mathbb{I}_{\{y_{t-1}<L\}} \\ &+ \left(E[X_t|X_{t-1} \geq L] - \sum_{j=0}^{+\infty} jP[X_t = L+j|X_{t-1} \geq L]\right)\mathbb{I}_{\{y_{t-1}=L\}} \end{aligned}$$

The authors Zeger and Brookmeyer [1] established a procedure to obtain the likelihood of an observed time series under censoring, $y = (Y_1, \ldots, Y_n)$, which becomes infeasible when the proportion of censoring is large. To overcome this issue, this work considers a Bayesian approach.

3. Bayesian Modelling

The Bayesian approach to the inference of an unknown parameter vector θ is based on the posterior distribution $\pi(\theta|y)$, defined as

$$\pi(\theta|y) \propto L(y|\theta)\pi(\theta),$$

where $L(y|\theta)$ is the likelihood function of the observed data y and $\pi(\theta)$ is the prior distribution of the model parameters.

When the likelihood is computationally prohibitive or even impossible to handle, but it is feasible to simulate samples from the model (bypass the likelihood evaluation), as is the case of censored CCID(1) processes, Approximate Bayesian Computation (ABC) algorithms are an alternative. This methodology accepts the parameter draws that produce a match between the observed and the simulated sample, depending on a set of summary statistics, a chosen distance and a selected tolerance. The accepted parameters are then used to estimate (an approximation of) the posterior distribution (conditioned on the summary statistics that afforded the match).

On the other hand, the idea of imputation arises naturally in the context of censored data. The Gibbs sampler with Data Augmentation (GDA) allows us to obtain an augmented dataset from the censored data by using a modified version of the Gibbs sampler, which samples not only the parameters of the model from its complete conditional but also the censored observations. The usual inference procedures may then be applied to the augmented data set.

3.1. Approximate Bayesian Computation

Approximate Bayesian Computation (ABC) is based on an acceptance–rejection algorithm. ABC is used to compute a draw from an approximation of the posterior distributions, based on simulated data obtained from the assumed model in situations where its likelihood function is intractable or numerically difficult to handle. Summary statistics from the synthetic data are compared with the corresponding statistics from the observed data and a parameter draw is retained when there is a *match* (in some sense) between the simulated sample and the observed time series observation.

Recently, Ref. [16] provided the asymptotic results pertaining to the ABC posterior, such as Bayesian consistency and asymptotic distribution of the posterior mean.

Let $y^0 = (Y_1^0, \ldots, Y_n^0)$ be the fixed (observed) data and $\eta(\cdot)$ the model from which the data is generated. The most basic *approximate acceptance/rejection algorithm*, based on the works of [17,18], is as follows:

1. draw a value θ from the prior distribution, $\pi(\theta)$,
2. simulate a sample $y = (Y_1, \ldots, Y_n)$ from the model $\eta(.|\theta)$,
3. accept θ if $d(S(y), S_0)) \leq \delta$ for some distance measure $d(.,.)$ and some non-negative tolerance value δ, where $S(\cdot)$ is a summary statistic and $S_0 = S(y^0)$ is a fixed value.

It is well known that, if we use a proper distance measure, then as δ tends to zero, the distribution of the accepted values tends to the posterior distribution of the parameter given the data. When the summary statistics are sufficient for the parameter, then the distribution of the accepted values tends to the true posterior as δ tends to zero, assuming a proper distance measure on the space of sufficient statistics. The latent structure of the thinning operator means that the reduction to a sufficient set of statistics of dimension smaller than the sample size is not feasible and, therefore, informative summary statistics are often used [19].

In this work, given the characteristics of the data under study to compare the observed data (y^0) and the synthetic (simulated) data (y), we consider two distinctive characteristics of CCID(1) time series which are affected by the censoring: (i) the empirical marginal distribution and (ii) lag 1 auto-correlation.

To measure the similarity between the empirical marginal distributions the Kullback-Leibler (Note that Kullback-Leibler distance measures the dissimilarity between two probability distributions.) distance is calculated as

$$S_1(y) = d_{KL}(\hat{p}^0, \hat{p}) = \sum_j \ln\left(\frac{\hat{p}_j^0}{\hat{p}_j}\right) \hat{p}_j^0, \tag{6}$$

where $\hat{p}_j^0$ and $\hat{p}_j$ denote the empirical marginal distribution of the observed time series and that of the simulated time series, respectively, estimated by the corresponding sample proportions, $\left(\hat{p}_j^0 = \frac{1}{n}\sum_{t=1}^n I_{\{Y_t^0=j\}} \text{ and } \hat{p}_j = \frac{1}{n}\sum_{t=1}^n I_{\{Y_t=j\}}\right)$. Whenever $\hat{p}_j^0$ is zero, the contribution of the jth term is interpreted as zero because $\lim_{p\to 0} p \ln(p) = 0$.

On the other hand, lag 1 sample autocorrelations, $S_2(y^0) = \hat{\rho}_{Y^0}(1)$ and $S_2(y) = \hat{\rho}_Y(1)$, are compared by their squared difference.

Additionally, we estimated the censoring rates, $S_3(y^0) = \frac{1}{n}\sum_{t=1}^n I_{\{y_t^0=L\}}$ and $S_3(y) = \frac{1}{n}\sum_{t=1}^n I_{\{Y_t=L\}}$, which are also compared by their squared difference.

Thus, for each set of parameters, $\theta^{(k)}$, a time series $x^{(k)}$ is generated from the model CCID(1) and right censored at L, yielding $y^{(k)} = (Y_1^{(k)}, \ldots, Y_n^{(k)})$ and the above statistics, $S_1(y^{(k)})$, $S_2(y^{(k)})$ and $S_3(y^{(k)})$ are computed. Combining these statistics in a metric leading to the choice of the parameters θ requires scaling. Thus, we propose the following metric

$$d_S^{(k)} = \frac{S_1(y^{(k)})^2}{V(S_1(y))} + \sum_{i=2}^3 \frac{[S_i(y^0) - S_i(y^{(k)})]^2}{V(S_i(y^0) - S_i(y))} \tag{7}$$

where $S_i(y^0)$ and $S_i(y^{(k)})$ are the ith statistics obtained respectively from the observed and kth simulated data and $V(S_1(y))$ and $V(S_i(y^0) - S_i(y))$ are the corresponding sample variances across the replications.

In summary, we propose Algorithm 1 for ABC approach based on [20]:

Algorithm 1 ABC for censored CCID(1)

For $k = 1, \ldots, N$
  Sample $\theta^{(k)}$ from the prior distribution $\pi(\theta)$
  Generate a time series with n observations, $x^{(k)}$ from the CCID(1) model
  Right truncate at L $x^{(k)}$ to obtain the simulated data $y^{(k)}$
  Compute $S_1(y^{(k)})$, $S_2(y^{(k)})$ and $S_3(y^{(k)})$
Compute $d_S^{(k)} = \frac{S_1(y^{(k)})^2}{V(S_1(y))} + \sum_{i=2}^3 \frac{[S_i(y^0) - S_i(y^{(k)})]^2}{V(S_i(y^0) - S_i(y))}, k = 1, \ldots, N$
Select the values $\theta^{(k)}$ corresponding to the 0.1% quantile of $d_S^{(k)}, k = 1, \ldots, N$

Implementation issues regarding the prior distributions and the number of draws N for the CensPoINAR(1) model are addressed in Sections 3.3 and 4.1.

3.2. Gibbs Sampler with Data Augmentation

Gibbs sampling is a Markov chain Monte Carlo (MCMC) algorithm that can generate samples of the posterior distribution from their full conditional distributions [21]. When the data are under censoring or there are missing values, both cases leading to an incomplete data set, Ref. [22] proposed to combine the Gibbs sampler with data augmentation. This methodology implies imputing the censored (or missing) data, thus obtaining a complete dataset, and then dealing with the posterior of the complete data through the iterative

Gibbs sampler. Therefore, the Gibbs sampler is modified in order to sample not only the parameters of the model from their complete conditionals but also the censored observations, obtaining an augmented (complete) dataset $\mathbf{z} = (z_1, \ldots, z_n)$ where

$$z_t = \begin{cases} Y_t, & \text{if } Y_t < L \\ z_t \sim F_\mu(x|x \geq L), & \text{if } Y_t = L \end{cases} \quad (8)$$

where $F_\mu(x|x \geq L)$ is the truncated marginal distribution of the CCID(1) model with support in $[L, +\infty[$. Furthermore, we consider a modified sampling procedure for the imputation, designated as Mean of Multiple Simulation (MMS), proposed by [23] consisting in sampling from $F_\mu(x|x \geq L)$ multiple times, say m, and then imputing with the (nearest integer value) median of the m samples. This procedure is designated by **GDA-MMS**.

The augmented dataset can be considered as a CCID(1) time series and with a conditional likelihood function given by Equation (3). The posterior distribution of $\boldsymbol{\theta}$ is given by

$$p(\boldsymbol{\theta}|\mathbf{z}) \propto L(\mathbf{z}|\boldsymbol{\theta}) \pi(\boldsymbol{\theta}) \quad (9)$$

where $\pi(\boldsymbol{\theta})$ is the prior distribution of the parameters. In CCID(1) models the complexity of $p(\boldsymbol{\theta}|\mathbf{z})$ requires resorting to Markov Chain Monte Carlo (MCMC) techniques for sampling from the full conditional distributions. The procedure is summarized in Algorithm 2 and detailed for the CensPoINAR(1) case in Sections 3.3 and 4.1.

Algorithm 2 GDA-MMS for censored CCID(1)

Initialize with $\mathbf{y} = (Y_1, \ldots, Y_n)$, $\boldsymbol{\theta}^{(0)} = (\theta_1^{(0)}, \ldots, \theta_p^{(0)})$, $L \in \mathbb{R}$, and n, m, $N \in \mathbb{N}$
Set $\mathbf{z}^{(0)} = \mathbf{y}$
For $k = 1, \ldots, N$
    Sample $\theta_i^{(k)} \sim \pi(\theta_i|\boldsymbol{\theta}_{(-i)}^{(k-1)}, \mathbf{z}^{(k-1)})$ ($\mathbf{x}_{(-i)}$ represents the vector $\mathbf{x}$ with the ith element removed.), $i = 1, \ldots, p$
    For $t = 1, \ldots, n$
        If $Y_t = L$
            For $j = 1, \ldots, m$
                Sample $z_t^{(j)} \sim F(x|\boldsymbol{\theta}^{(k)}, x \geq L)$
            $z_t^{(k)} := \frac{1}{m} \sum_{j=1}^m z_t^{(j)}$
        Else
            $z_t^{(k)} := Y_t$
Return $\boldsymbol{\theta} = [\boldsymbol{\theta}^{(1)}, \ldots, \boldsymbol{\theta}^{(N)}]'$ and $\mathbf{z}^{(N)}$.

3.3. The Particular Case of CensPoINAR(1)

This section details the ABC and GDA-MMS procedures to estimate a censored CCID(1) with the binomial thinning operation and Poisson marginal distribution, the censored Poisson INAR(1), CensPoINAR(1), model.

Consider the censored observations $\mathbf{y} = (Y_1, \ldots, Y_n)$ from a PoINAR(1) time series $\mathbf{x} = (X_1, \ldots, X_n)$ defined as

$$\begin{aligned} Y_t &= \min\{X_t, L\} = \begin{cases} X_t, & \text{if } X_t < L \\ L, & \text{if } X_t \geq L \end{cases} \\ X_t &= \alpha \circ X_{t-1} + e_t, \end{aligned} \quad (10)$$

with $\alpha \circ X_{t-1} = \sum_{i=1}^{X_{t-1}} \xi_{ti}$, $\xi_{ti} \stackrel{iid}{\sim} Ber(\alpha)$, $e_t \sim Po(\lambda)$ and $X_t \sim Po(\frac{\lambda}{1-\alpha})$. Then $\boldsymbol{\theta} = (\alpha, \lambda)$ and given $\mathbf{x}$, the conditional likelihood function is given by

$$L(\theta) = \prod_{t=2}^{n} f_{X_t|X_{t-1}}(x_t|x_{t-1}) \quad (11)$$

$$= \prod_{t=2}^{n} \sum_{j=0}^{\min\{x_t,x_{t-1}\}} \binom{x_{t-1}}{j} \alpha^j (1-\alpha)^{(x_{t-1}-j)} \frac{e^{-\lambda} \lambda^{x_t-j}}{(x_t-j)!}.$$

Under a Bayesian approach, we need a prior distribution for θ. In the absence of prior information, we use weakly informative prior distributions for θ detailed below.

3.3.1. ABC for Censored PoINAR(1)

The ABC procedure described in Algorithm 1 is now implemented for the censored PoINAR(1). For the parameter $0 < \alpha < 1$, we choose a non-informative prior $U(0, 1)$, while for the positive parameter λ, we choose a non-informative $U(0, 10)$. The former allows us to explore all the support space for α. The choice of $U(0, 10)$ as a prior for $\lambda > 0$ allows us to explore a restricted support for the parameter that is in accordance with small counts.

3.3.2. GDA-MMS for Censored PoINAR(1)

Under the GDA-MMS approach, we first need to obtain a complete data set $z = (z_1, \ldots, z_n)$ by imputing the censored observations, see (8). In this work, we draw $m = 10$ replicates of the right truncated at L Poisson distribution with parameter $\frac{\lambda}{1-\alpha}$, $w_i \sim \text{Po}\left(\frac{\lambda}{1-\alpha}\right) \times I_{(w_i \geq L)}$ and set $z_t = \lceil median(w) \rceil$ ($\lceil c \rceil$ represents the integer ceiling of c), $w = (w_1, \ldots, w_m)$, if $Y_t = L$. Figure 2 shows an augmented dataset (Z_t, black line) from the synthetic data presented in Figure 1.

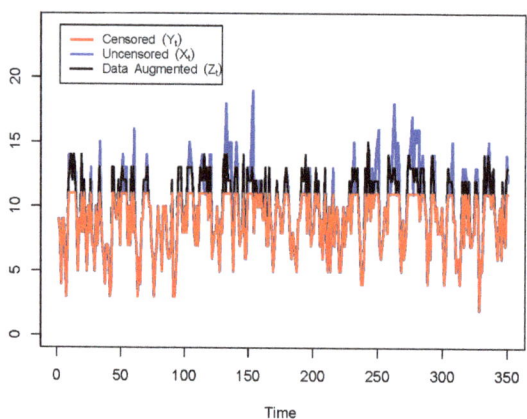

Figure 2. Synthetic dataset with $n = 350$ observations generated from a PoINAR(1) process with parameters $\alpha = 0.5$ and $\lambda = 5$ (X_t, blue line), the respective right-censored dataset (Y_t, red line), at $L = 11$, and an example of data augmentation (Z_t, black line).

As remarked above, given the complexity of the posterior distribution, Markov Chain Monte Carlo techniques are required for sampling from the full conditional distributions. Thus, the Adaptive Rejection Metropolis Sampling (ARMS) is used inside the Gibbs sampler [24]. Also in this approach, in the absence of prior information, we use weakly informative prior distributions for (α, λ). Thus, for the parameter $0 < \alpha < 1$, we choose a non-informative beta prior, conjugate of the binomial distribution, with parameters (a, b), while for the positive parameter μ, we choose a non-informative Gamma (shape, rate) prior, conjugate of the Poisson distribution, with parameters (c,d). The full conditional of λ is given by

$$p(\lambda|\alpha,z) = \frac{p(\lambda,\alpha|z)}{p(\alpha|z)} \propto \exp[-(d+(n-1))\lambda]\lambda^{c-1}\prod_{t=2}^{n}\sum_{i=0}^{\min\{z_t,z_{t-1}\}}C(t,i)\lambda^{(z_t)-i}, \quad (12)$$

where

$$C(t,i) = \frac{1}{((z_t)-i)!}\binom{z_{t-1}}{i}\alpha^i(1-\alpha)^{(z_{t-1})-i} \quad \text{and} \quad \lambda > 0.$$

The full conditional distribution of α is given by

$$p(\alpha|\lambda,z) = \frac{p(\lambda,\alpha|z)}{p(\lambda|z)} \propto \alpha^{a-1}(1-\alpha)^{b-1}\prod_{t=2}^{n}\sum_{i=0}^{\min\{z_t,z_{t-1}\}}K(t,i)\alpha^i(1-\alpha)^{(z_{t-1})-i}, \quad (13)$$

where

$$K(t,i) = \frac{\lambda^{(z_t)-i}}{((z_t)-i)!}\binom{X_{t-1}}{i} \quad 0 < \alpha < 1.$$

The parameters α and λ are computed as the posterior mean.

The GDA-MMS procedure to estimate a censored PoINAR(1) process is detailed in Algorithm 3.

Algorithm 3 GDA-MMS for CensPoINAR(1)

Initialize with y, $\theta^{(0)} = (\alpha^{(0)}, \lambda^{(0)})$, $L \in \mathbb{R}$, and N, $m \in \mathbb{N}$
Set $z^{(0)} = y$
For $k = 1, ..., N$
    Using ARMS
        Sample $\lambda^{(k)} \sim p(\lambda|\alpha^{(k-1)}, z^{(k-1)})$
        Sample $\alpha^{(k)} \sim p(\alpha|\lambda^{(k)}, z^{(k-1)})$
    For $t = 1, ..., n$
        If $Y_t = L$
            For $j = 1, ..., m$
                Sample $w^{(j)} \sim \text{Po}\left(\frac{\lambda^{(k)}}{1-\alpha^{(k)}}\right) \times I_{(w^{(j)}\geq L)}$
            $z_t^{(k)} := \lceil median(w) \rceil, w = (w^{(1)}, ..., w^{(m)})$,
        Else
            $z_t^{(k)} := Y_t$
Return $\theta = [\theta^{(1)}, ..., \theta^{(N)}]'$ and $z^{(N)}$.

---

4. Illustration

This section illustrates the procedures proposed above to model CCID(1) right-censored time series in the particular case of Poisson distribution and binomial thinning operation.

4.1. Illustration with CensPoINAR(1)

In this section, the performance of the Bayesian approaches previously proposed is illustrated via synthetic data. Thus, realizations with $n = 100, 350, 1000$ observations of CensPoINAR(1) models were simulated, with parameters $\theta = (0.2, 3)$ and $\theta = (0.5, 5)$, considering for each case two levels of censorship, namely 30% and 5%.

For the ABC estimates, we run $N = 10^6$ replications and choose the pairs (α, λ) corresponding to the 0.1% lower quantile of $d_S^{(k)}$, Equation (7), in total of 1000 values from which the estimates are computed as the mean value. Software R [25] was used to implement the ABC algorithm.

To implement GDA-MMS algorithm we consider the initial values $\theta^{(0)} = (\alpha^{(0)}, \lambda^{(0)})$ given by the Conditional Least squares estimates of α and λ [24]. The hyper-parameters for the prior distributions of α and λ are the following $\alpha \sim Beta(a = 2, b = 2)$ and

$\lambda \sim Gamma(c = 0.1, d = 0.1)$. In this work, the function *armspp* was used from the package *armspp* [26] in R to sample from the full conditional distributions. Several experiments were carried out to analyse the size that the chain should have in order to be stable and, thus, the number of Gibbs sampler iterations used in this work is N = 15,000. Among these, we ignored the first 5000 simulations as burn-in time and, to reduce autocorrelation between MCMC observations, we considered only simulations from every 30 iterations. Therefore, we use a simulated sample with size 323 to obtain the Bayesian estimates. A convergence analysis with the usual diagnostic tests was performed with the package *coda* [27] in R [25].

Tables 2 and 3 summarize ABC and GDA-MMS results for the several scenarios described above: point estimates, $\hat{\alpha}$ and $\hat{\lambda}$, obtained as sample means, the corresponding bias, standard deviation and the coefficient of variation. The results indicate that the bias tends to decrease for large sample sizes and small censoring rates. The results also indicate that overall ABC presents estimates with smaller bias but larger variability when compared with GDA-MMS.

Table 2. ABC and GDA-MMS results for parameter α (sample mean, and the corresponding bias, standard deviation and coefficient of variation) for synthetic data generated from CensPoINAR(1) models.

				ABC				GDA-MMS			
α	λ	L	n	$\bar{\hat{\alpha}}$	Bias($\hat{\alpha}$)	s.d.($\hat{\alpha}$)	CV($\hat{\alpha}$)	$\bar{\hat{\alpha}}$	Bias($\hat{\alpha}$)	s.d.($\hat{\alpha}$)	CV($\hat{\alpha}$)
0.2	3	4 (30%)	100	0.2571	0.0571	0.0911	0.3544	0.3155	0.1155	0.0323	0.1024
			350	0.2067	0.0067	0.0579	0.2803	0.2274	0.0274	0.0178	0.0783
			1000	0.1793	−0.0207	0.0398	0.2217	0.2025	0.0025	0.0157	0.0775
0.2	3	6 (5%)	100	0.2268	0.0268	0.0760	0.3350	0.2738	0.0738	0.0270	0.0986
			350	0.2302	0.0302	0.0511	0.2221	0.2309	0.0309	0.0140	0.0606
			1000	0.1931	−0.0069	0.0327	0.1692	0.1915	−0.0085	0.0112	0.0585
0.5	5	11 (30%)	100	0.5304	0.0304	0.0800	0.1508	0.5596	0.0596	0.0170	0.0304
			350	0.4637	−0.0363	0.0535	0.1153	0.4834	−0.0166	0.0124	0.0257
			1000	0.5115	0.0115	0.0320	0.0626	0.5050	0.0050	0.0072	0.0143
0.5	5	14 (5%)	100	0.5230	0.0230	0.0815	0.1559	0.5363	0.0363	0.0175	0.0326
			350	0.4671	−0.0329	0.0461	0.0987	0.4796	−0.0204	0.0107	0.0223
			1000	0.4992	−0.0008	0.0291	0.0584	0.5008	0.0008	0.0070	0.0140

Table 3. ABC and GDA-MMS results for the parameter λ (sample mean, and the corresponding bias, standard deviation and coefficient of variation) for synthetic data generated from CensPoINAR(1) models.

				ABC				GDA-MMS			
α	λ	L	n	$\bar{\hat{\lambda}}$	Bias($\hat{\lambda}$)	s.d.($\hat{\lambda}$)	CV($\hat{\lambda}$)	$\bar{\hat{\lambda}}$	Bias($\hat{\lambda}$)	s.d.($\hat{\lambda}$)	CV($\hat{\lambda}$)
0.2	3	4 (30%)	100	2.6623	−0.3377	0.3699	0.1389	2.3265	−0.6735	0.1144	0.0492
			350	2.8530	−0.1470	0.2353	0.0825	2.6639	−0.3361	0.0672	0.0252
			1000	3.1398	0.1398	0.1668	0.0531	2.9757	−0.0243	0.0603	0.0203
0.2	3	6 (5%)	100	2.7918	−0.2082	0.3203	0.1147	2.5719	−0.4281	0.1007	0.0392
			350	2.9507	−0.0493	0.2173	0.0736	2.7846	−0.2154	0.0579	0.0208
			1000	3.1342	0.1342	0.1448	0.0462	3.0417	0.0417	0.0460	0.0151
0.5	5	11 (30%)	100	4.3432	−0.6568	0.7504	0.1728	3.9528	−1.0472	0.1600	0.0405
			350	5.2315	0.2315	0.5265	0.1006	4.9073	−0.0927	0.1177	0.0240
			1000	4.9102	−0.0898	0.3247	0.0661	4.8974	−0.1026	0.0720	0.0147
0.5	5	14 (5%)	100	4.4488	−0.5512	0.7828	0.1760	4.2574	−0.7426	0.1682	0.0395
			350	5.1333	0.1333	0.4286	0.0835	4.9877	-0.0123	0.1088	0.0218
			1000	5.0613	0.0613	0.2826	0.0558	4.9964	−0.0036	0.0708	0.0142

Additionally, Figures 3 and 4 represent the corresponding posterior densities. The plots show unimodal and approximately symmetric distributions, with a dispersion that clearly decreases with increasing sample size and smaller censoring rate. The posterior densities indicate that the ABC approach produces posteriors that are flatter but with modes very close to the true value, while the corresponding GDA-MMS approach, despite producing posteriors which are more concentrated also evidence higher bias. However, the behaviour of GDA-MMS estimates varies with the parameters and even the sample sizes. These results are representative of the properties of GDA-MMS estimates across a large number of experiments, not reported here for conciseness.

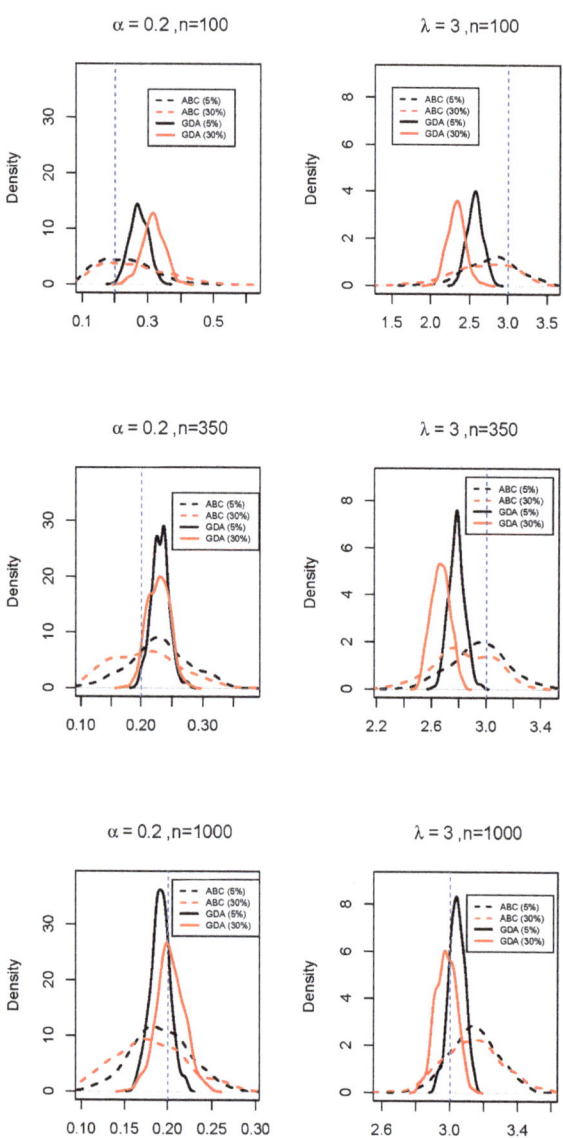

Figure 3. ABC and GDA-MMS posterior densities of the parameters for a realization of 100, 350 and 1000 observations of a CensPoINAR(1) model with $\theta = (0.2, 3)$, considering two levels of censoring. Note that the scale of x-axis of the six plots are different.

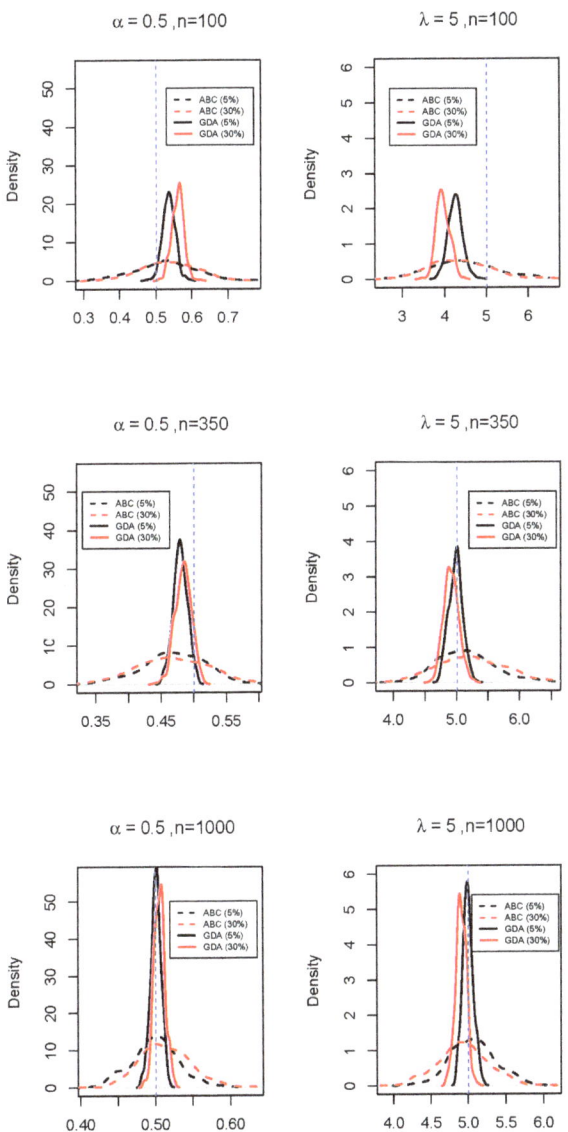

Figure 4. ABC and GDA-MMS posterior densities of the parameters for a realization of 100, 350 and 1000 observations of a CensPoINAR(1) model with $\theta = (0.5, 5)$, considering two levels of censoring. Note that the scale of x-axis of the six plots are different.

4.2. Simulation Study for GDA-MMS

This section presents the results of a simulation study designed to further analyse the sample properties of GDA-MMS, in particular the bias of the resulting Bayesian estimates.

For that purpose, realizations with sample sizes $n = 100$ and $n = 350$ of CensPoINAR(1) models with parameters $\theta = (0.2, 3)$ and $\theta = (0.5, 5)$, are generated, considering two levels of censorship, namely 30% and 5%. To analyse the performance of the procedure, the sample posterior mean, standard deviation and mean squared error were calculated over 50 repetitions.

Boxplots of the sample bias for the 50 repetitions of GDA-MMS methodology are presented in Figures 5 and 6. The bias increases with the rate of censoring and the variability decreases with the sample size. Furthermore, in general, the estimates for α presents positive sample mean biases, indicating that α is overestimated, whilst the estimates for λ shows negative sample biases, indicating underestimation for λ. Both bias and dispersion seem larger for λ.

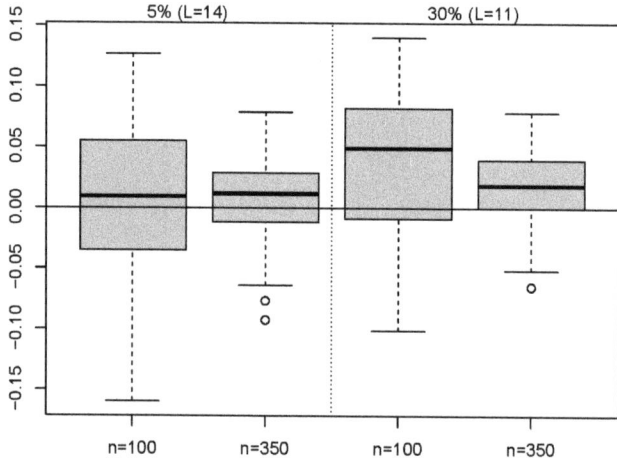

Figure 5. Boxplots of bias for GDA-MMS estimates of α, when $\theta = (0.5, 5)$.

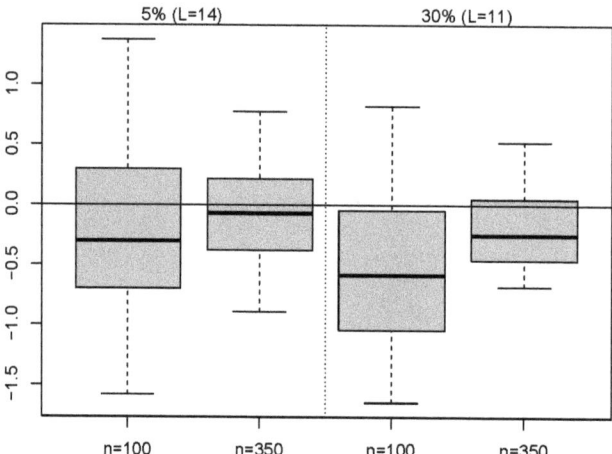

Figure 6. Boxplots of bias for GDA-MMS estimates of λ, when $\theta = (0.5, 5)$.

Tables 4 and 5 present the sample posterior measures for $\hat{\alpha}$ and $\hat{\lambda}$, respectively. We can see improvement of the estimation methods performance when the sample size increases. Additionally, the higher the censoring percentage, the worse the behavior of the proposed methods.

Table 4. Sample posterior mean, standard errors (in brackets) and root mean square error for GDA-MMS estimates of α.

α	λ	L	n	$\bar{\hat{\alpha}}$ (s.e.($\hat{\alpha}$))	RMSE($\hat{\alpha}$)
0.2	3	4 (30%)	100	0.2918 (0.0977)	0.1341
			350	0.2385 (0.0698)	0.0797
0.2	3	6 (5%)	100	0.2739 (0.0680)	0.1004
			350	0.2229 (0.0487)	0.0538
0.5	5	11 (30%)	100	0.5404 (0.0632)	0.0750
			350	0.5156 (0.0344)	0.0378
0.5	5	14 (5%)	100	0.5142 (0.0626)	0.0642
			350	0.5066 (0.0386)	0.0392

Table 5. Sample posterior mean, standard errors (in brackets) and root mean square error for GDA-MMS estimates of λ.

α	λ	L	n	$\bar{\hat{\lambda}}$ (s.e.($\hat{\lambda}$))	RMSE($\hat{\lambda}$)
0.2	3	4 (30%)	100	2.5283 (0.3077)	0.5632
			350	2.7842 (0.2462)	0.3274
0.2	3	6 (5%)	100	2.6814 (0.2984)	0.4365
			350	2.8934 (0.1843)	0.2129
0.5	5	11 (30%)	100	4.4861 (0.6710)	0.8452
			350	4.7976 (0.3357)	0.3920
0.5	5	14 (5%)	100	4.7593 (0.6391)	0.6829
			350	4.9229 (0.4177)	0.4248

5. Final Comments

This work approaches the problem of estimating CCID(1) models for time series of counts under censoring from a Bayesian perspective. Two algorithms are proposed: one is based on ABC methodology and the second a Gibbs Data Augmentation modified with multiple sampling. Experiments with synthetic data allow us to conclude that both approaches lead to estimates that present less bias than those obtained neglecting the censoring. Moreover, the GDA-MMS approach allows us to obtain a complete data set, making it a valuable method in other situations such as missing data.

In this study, we focus on the most popular CCID(1) model, the Poisson INAR(1). However, if the data under study present over- or under-dispersion, other CCID(1) models with appropriate distributions for the innovations, such as Generalised Poisson or Negative Binomial, can easily be entertained. Furthermore, one can consider different models for time series of counts under censoring, based on INGARCH models, ([28,29] using a switching mechanism) if they are more suitable to the data set to be modeled. These issues are beyond the scope of this paper and are a topic for a future research project.

Author Contributions: The authors contributed equally to this work. All authors have read and agreed to the published version of the manuscript.

Funding: The first and third authors were partially supported by The Center for Research and Development in Mathematics and Applications (CIDMA) through the Portuguese Foundation for Science and Technology (FCT—Fundação para a Ciência e a Tecnologia), reference UIDB/04106/2020. The second author is partially financed by National Funds through the Portuguese funding agency, FCT, within project UIDB/50014/2020.

Conflicts of Interest: The authors declare no conflict of interest.

References

1. Zeger, S.; Brookmeyer, R. Regression Analysis with Censored Autocorrelated Data. *J. Am. Stat. Assoc.* **1986**, *81*, 722–729. [CrossRef]
2. Hopke, P.K.; Liu, C.; Rubin, D.R. Multiple imputation for multivariate data with missing and below-threshold measurements: Time-series concentrations of pollutants in the arctic. *Biometrics* **2001**, *57*, 22–33. [CrossRef] [PubMed]
3. Park, J.W.; Genton, M.G.; Ghosh, S.K. Censored time series analysis with autoregressive moving average models. *Can. J. Stat.* **2007**, *35*, 151–168. [CrossRef]
4. Mohammad, N.M. Censored Time Series Analysis. Ph.D. Thesis, The University of Western Ontario, London, ON, Canada, 2014.
5. Schumacher, F.L.; Lachos, V.H.; Dey, D.K. Censored regression models with autoregressive errors: A likelihood-based perspective. *Can. J. Stat.* **2017**, *45*, 375–392. [CrossRef]
6. Wang, C.; Chan, K.S. Carx: An R Package to Estimate Censored Autoregressive Time Series with Exogenous Covariates. *R J.* **2017**, *9*, 213–231. [CrossRef]
7. Wang, C.; Chan, K.S. Quasi-Likelihood Estimation of a Censored Autoregressive Model with Exogenous Variables. *J. Am. Stat. Assoc.* **2018**, *113*, 1135–1145. [CrossRef]
8. Fernández-Fontelo, A.; Cabaña, A.; Puig, P.; Moriña, D. Under-reported data analysis with INAR-hidden Markov chains. *Stat. Med.* **2016**, *35*, 4875–4890. [CrossRef]
9. De Oliveira, G.L.; Loschi, R.H.; Assunção, R.M. A random-censoring Poisson model for underreported data. *Stat. Med.* **2017**, *36*, 4873–4892. [CrossRef]
10. Joe, H. Time series model with univariate margins in the convolution-closed infinitely divisible class. *Appl. Probab.* **1996**, *33*, 664–677. [CrossRef]
11. Al-Osh, M.A.; Alzaid, A.A. First-order integer-valued autoregressive (INAR(1)) process. *J. Time Ser. Anal.* **1987**, *8*, 261–275. [CrossRef]
12. McKenzie, E. Some ARMA models for dependent sequences of Poisson counts. *Adv. Appl. Probab.* **1988**, *20*, 822–835. [CrossRef]
13. Alzaid, A.A.; Al-Osh, M.A. Some autoregressive moving average processes with generalized Poisson marginal distributions. *Ann. Inst. Stat. Math.* **1993**, *45*, 223–232. [CrossRef]
14. Han, L.; McCabe, B. Testing for parameter constancy in non-Gaussian time series. *J. Time Ser. Anal.* **2013**, *34*, 17–29. [CrossRef]
15. Jung, R.C.; Tremayne, A.R. Useful models for time series of counts or simply wrong ones? *AStA Adv. Stat. Anal.* **2011**, *95*, 59–91. [CrossRef]
16. Frazier, D.T.; Martin, G.M.; Robert, C.P.; Rousseau, J. Asymptotic properties of approximate Bayesian computation. *Biometrika* **2018**, *105*, 593–607. [CrossRef]
17. Plagnol, V.; Tavaré, S. Approximate Bayesian computation and MCMC. In *Monte Carlo and Quasi-Monte Carlo Methods*; Niederreiter, H., Ed.; Springer: Berlin/Heidelberg, Germany, 2003; pp. 99–113.
18. Wilkinson, R.D. Approximate Bayesian computation (ABC) gives exact results under the assumption of model error. *Stat. Appl. Genet. Mol. Biol.* **2013**, *12*, 129–141. [CrossRef]
19. Frazier, D.T.; Maneesoonthorn, W.; Martin, G.M.; McCabe, B. Approximate Bayesian forecasting. *Int. J. Forecast.* **2019**, *35*, 521–539. [CrossRef]
20. Biau, G.; Cérou, F.; Guyader, A. New Insights into Approximate Bayesian Computation. *Ann. Henri Poincaré (B) Probab. Stat.* **2015**, *51*, 376–403. [CrossRef]
21. Gelfand, A.; Smith, A. Sampling-based approaches to calculating marginal densities. *J. R. Stat. Soc. Ser. B* **1990**, *85*, 398–409. [CrossRef]
22. Chib, S. Bayes Inference in the Tobit Censored Regression Model. *J. Econom.* **1992**, *51*, 79–99. [CrossRef]
23. Sousa, R.; Pereira, I.; Silva, M.E.; McCabe, B. Censored Regression with Serially Correlated Errors: A Bayesian approach. *arXiv* **2023**, arXiv:2301.01852.
24. Silva, I.; Silva, M.E.; Pereira, I.; Silva, N. Replicated INAR(1) Processes. *Methodol. Comput. Appl. Probab.* **2005**, *7*, 517–542. [CrossRef]
25. R Core Team. *R: A Language and Environment for Statistical Computing*; R Foundation for Statistical Computing: Vienna, Austria, 2022. Available online: https://www.R-project.org/ (accessed on 20 March 2023).
26. Bertolacci, M. Armspp: Adaptive Rejection Metropolis Sampling (ARMS) via 'Rcpp'. R Package Version 0.0.2. 2019. Available online: https://CRAN.R-project.org/package=armspp (accessed on 20 March 2023).
27. Plummer, M.; Best, N.; Cowles, K.; Vines, K. CODA: Convergence Diagnosis and Output Analysis for MCMC. *R News* **2006**, *6*, 7–11.
28. Chen, C.W.S.; Lee, S. Generalized Poisson autoregressive models for time series of counts. *Comput. Stat. Data Anal.* **2016**, *99*, 51–67. [CrossRef]
29. Ferland, R.; Latour, A.; Oraichi, D. Integer-Valued GARCH Process. *J. Time Ser. Anal.* **2006**, *27*, 923–942. [CrossRef]

Disclaimer/Publisher's Note: The statements, opinions and data contained in all publications are solely those of the individual author(s) and contributor(s) and not of MDPI and/or the editor(s). MDPI and/or the editor(s) disclaim responsibility for any injury to people or property resulting from any ideas, methods, instructions or products referred to in the content.

Review

A Systematic Review of INGARCH Models for Integer-Valued Time Series

Mengya Liu [1], Fukang Zhu [2,*], Jianfeng Li [1] and Chuning Sun [3]

[1] School of Mathematics and Statistics, Central China Normal University, Wuhan 430079, China
[2] School of Mathematics, Jilin University, Changchun 130012, China
[3] School of Business, Zhengzhou University, Zhengzhou 450001, China
* Correspondence: fzhu@jlu.edu.cn

Abstract: Count time series are widely available in fields such as epidemiology, finance, meteorology, and sports, and thus there is a growing demand for both methodological and application-oriented research on such data. This paper reviews recent developments in integer-valued generalized autoregressive conditional heteroscedasticity (INGARCH) models over the past five years, focusing on data types including unbounded non-negative counts, bounded non-negative counts, $\mathcal{Z}$-valued time series and multivariate counts. For each type of data, our review follows the three main lines of model innovation, methodological development, and expansion of application areas. We attempt to summarize the recent methodological developments of INGARCH models for each data type for the integration of the whole INGARCH modeling field and suggest some potential research topics.

Keywords: INGARCH; count time series; conditional distribution; dynamic structure; robust estimation

1. Introduction

This paper reviews the development of modeling and inference for four types of integer time series, including the unbounded $\mathcal{N}$-valued counts, $\mathcal{Z}$-valued time series, bounded $\mathcal{N}$-valued counts and multivariate counts. Firstly, the unbounded $\mathcal{N}$-valued counts, also known as count time series, refer to discrete time series taking values in the range $\mathcal{N} = \{0, 1, 2, \cdots\}$. For example, during an influenza outbreak, the number of new confirmed cases is reported daily, even down to each community. The analysis of such data is one of the fundamental tasks in epidemic forecasting and policy implementation (see Agosto and Giudici [1], Agosto et al. [2] and Giudici et al. [3] among others). Secondly, $\mathcal{Z}$-valued count time series taking values in the range $\mathcal{Z} = \{\cdots, -1, 0, 1 \cdots\}$ are the appropriate tool to employ when attention is turned to, for example, the changes in athletic performance by the difference between the number of goals scored in each game and that in the previous one. Thirdly, the study of bounded count time series with a range of $\{0, \cdots, n\}$ for a given $n \in \mathcal{N}$ is also a concern, such as the rigorously recorded data set of water quality in the estuary. Finally, with the development of data acquisition technology and the expansion of storage space, multivariate count time series have emerged in various fields, and the related research is also developing.

For the above four types of integer time series, this review focuses on the the relevant research fields of the integer-valued generalized autoregressive conditional heteroscedasticity (INGARCH) models, which assume that the observations follow a discrete distribution and are conditioned on an accompanying intensity process that drives the dynamics, which are classified as observation-driven models. For instance, given the intensity process, the observations of a count time series follow a Poisson distribution and the intensity process is a linear combination of its lagged values and lagged observations (Fokianos et al. [4]). The diversity of count time series urges the INGARCH-type models to evolve to a broader domain. This is accompanied by the development of many subdivision directions, including the selection and even the creation of suitable conditional distributions, the exploration

Citation: Liu, M.; Zhu, F.; Li, J.; Sun, C. A Systematic Review of INGARCH Models for Integer-Valued Time Series. *Entropy* **2023**, *25*, 922. https://doi.org/10.3390/e25060922

Academic Editor: José María Amigó

Received: 10 May 2023
Revised: 5 June 2023
Accepted: 9 June 2023
Published: 11 June 2023

Copyright: © 2023 by the authors. Licensee MDPI, Basel, Switzerland. This article is an open access article distributed under the terms and conditions of the Creative Commons Attribution (CC BY) license (https://creativecommons.org/licenses/by/4.0/).

of nonlinear dynamic structures, the proof of stationarity and ergodicity, the relaxation of restrictions, the updating of methodologies, and so on. The purpose of this paper is to review recent developments, including the above subdivision directions, in INGARCH models over the past five years, as earlier work has been summarized. For the generally methodological development of count time series, not only INGARCH-type models, please refer to Weiß [5] and Davis et al. [6].

The INGARCH-type models for count time series are the most well-developed of these four types of integer-valued time series modeling, and they are reviewed in Section 2. On the one hand, to model some specific features of count time series, researchers have innovated the conditional distribution assumption. Indeed, Gorgi [7] and Qian and Zhu [8] focus on heavy-tailed count time series, while Silva and Souza [9] is more concerned with the generality of distribution and Souza et al. [10] further extends Silva and Souza [9] to the time-varying version. Accordingly, they all propose new conditional distribution assumptions that can model the features of interest. On the other hand, there are some new developments in the dynamic structure of INGARCH. For example, Weiß et al. [11] introduces the softplus function in the linear dynamic structure to implement the possibility of negative ACF; similar to Souza et al. [10], Roy [12] also focuses on the time-varying feature, but proceeds from the construction of a semi-parametric dynamic structure based on a Bayesian framework; some advances are made in the study of nonlinear dynamic structures with the help of threshold structures and the hysteresis model (Liu et al. [13], Chen and Khamthong [14], Liu et al. [15] and Chen et al. [16]). Moreover, when it comes to proving the existence and uniqueness of stationary distributions, which has been a challenging work in this field, the common approaches recently are based on approximation techniques, weak dependence or theoretical frameworks using the Feller property, e-chain and Lyapunov methods. In contrast, the absolute regularity for nonlinear GARCH and INGARCH models under a mild assumption are considered by Doukhan and Neumann [17], and a series of methodological studies are specifically reviewed in Section 2.

In contrast, the INGARCH-type models for the other three integer-valued time series are all still areas of ongoing research. The development patterns of all three areas are similar to that of the above-mentioned INGARCH-type models for count time series, including innovation in conditional distribution assumptions, flexibility in dynamic structure and methodological establishment, etc., but the challenges faced by each are different. For $\mathcal{Z}$-valued time series, the difficulty of finding a suitable conditional distribution can be overcome, since it can be handled directly with the help of a sequence of binary or ternary random variables (Hu [18], Hu and Andrews [19] and Xu and Zhu [20]). However, the INGARCH-type models for $\mathcal{Z}$-valued time series require more complex and subtle dynamic structures to implement, and the corresponding theoretical proofs are of higher difficulty. For bounded counts with range $\{0, \cdots, n\}$ for given $n \in \mathcal{N}$, two research frameworks are currently divided into those based on vector form (Fokianos and Truquet [21]) and those based on scalar form (Weiß and Jahn [22], Chen et al. [23], Chen et al. [24] among others). The challenge encountered in the former is the computational stress of the estimation of matrix-type unknown parameters, while the difficulty in the latter lies in the rarity of suitable distributions, which are required to be discrete distributions with a range of values, and the accompanying proof of stationarity. Finally, the development of INGARCH-type models for multivariate numerical time series is still in its nascent stage. Lee et al. [25], Kim et al. [26] and Cui and Zhu [27] focus on the INGARCH-type models for bivariate count times series, but, subject to the development of multivariate Poisson distribution, more multivariate models related to these studies have not been extended. Another emerging area is the INGARCH-type models for time-varying network data (Armillotta and Fokianos [28], Armillotta et al. [29], Armillotta and Fokianos [30] and Tao et al. [31]), and many of the attempts that come with the challenge are worthwhile, such as optimization of conditional distributions, nonlinear dynamic structures and time-varying network data with upper bounds on the number of edges.

The rest of the paper attempts to review the INGARCH-type models, which are presented below. Sections 2–5 present the INGARCH-type models for unbounded $\mathcal{N}$-valued counts, $\mathcal{Z}$-valued time series, bounded $\mathcal{N}$-valued counts and multivariate counts, respectively (Figure 1). Since there is much content related to unbounded $\mathcal{N}$-valued counts, Section 2 is divided into three subsections to elaborate on the latest advances in models, methodologies and applications, but the other sections are also arranged in this way although they are not split again. Finally, we present some potential research topics from a personal perspective in Section 6.

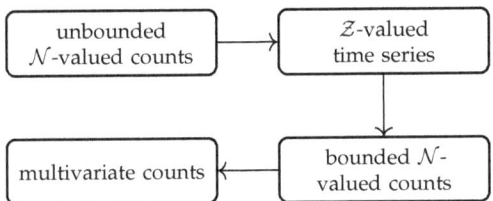

Figure 1. Flowchart of the types of time series reviewed.

2. Count Time Series

We consider count time series in this section, the possible outcomes of which are contained in $\mathcal{N}$. Let $\{X_t\}_{t\in\mathcal{Z}}$ be a count process, $x_1, \cdots, x_T$ be a finite set of observations, and F_t be defined as σ-field generated by observations up to and including time t.

The INGARCH(p,q)-type models with $p \geq 1$ and $q \geq 0$ require two types of assumptions: first, a distributional assumption for X_t conditioned on $X_{t-1}, X_{t-2}, \cdots$ is needed to guarantee the discrete range of X_t; secondly, the conditional mean $M_t = E(X_t|X_{t-1}, \cdots)$ is required to be a linear or nonlinear expression in the last p observations and the last q conditional means, thus constructing a dynamic structure. For example, the Poisson INGARCH model assumes that X_t, conditioned on $X_{t-1}, X_{t-2}, \cdots$, is Poisson distributed with intensity parameters M_t and

$$M_t = \omega + \sum_{i=1}^{p} \alpha_i X_{t-i} + \sum_{j=1}^{q} \beta_j M_{t-j}. \tag{1}$$

This section begins with a review of recent modeling work on improving these two types of assumptions separately.

2.1. Recent Advances in Assumptions of Conditional Distribution

This subsection begins with the progress of research on count time series with outliers. For the INGARCH models, the common approach is to assume a heavy-tailed conditional distribution. Gorgi [7] introduced a heavy-tailed mixture of negative binomial distributions, known as the beta-negative binomial (BNB) distribution, as an distribution of X_t conditioned on $X_{t-1}, X_{t-2}, \cdots$. See Table 1 for the specific definition of BNB. Relatedly, Qian and Zhu [8] was also concerned with heavy-tailed count time series and thus uses the generalized Conway–Maxwell–Poisson (GCOMP) distribution (in Table 1), which has one more parameter than the Conway–Maxwell–Poisson distribution, but provides a unified framework to handle over- or under-dispersed, zero-inflated, and heavy-tailed count data.

Table 1. Summary of basic properties of distributions related to INGARCH.

Distribution	Definition
BNB (Gorgi [7])	The probability mass function (PMF) of BNB is $$P(X=x) = \frac{\Gamma(x+r)}{\Gamma(x+1)\Gamma(r)} \frac{B(\alpha+r,\beta+x)}{B(\alpha,\beta)},$$ where $\Gamma(\cdot)$ and $B(\cdot,\cdot)$ denote the gamma function and beta function, respectively. And $E(X) = \frac{\beta r}{\alpha - 1}$.
GCOMP (Qian and Zhu [8])	The PMF of GCOMP is is $$P(X=x) = \frac{\Gamma(\nu+x)^r \zeta^x}{x! C(r,\nu,\zeta)},$$ where the normalizing constant $C(r,\nu,\zeta) = \sum_{k=0}^{\infty} \frac{\Gamma(\nu+k)^r \zeta^k}{k!}$, for $r < 1$, $\nu > 0$ and $\zeta > 0$ or $r = 1, \nu > 0$ and $0 < \zeta < 1$. $E(X) = \frac{C(r,\nu+1,\zeta)\zeta}{C(r,\nu,\zeta)}$ and $Var(X) = \frac{C(r,\nu+2,\zeta)\zeta^2}{C(r,\nu,\zeta)} + \frac{C(r,\nu+1,\zeta)\zeta}{C(r,\nu,\zeta)} - \frac{C^2(r,\nu+1,\zeta)\zeta^2}{C^2(r,\nu,\zeta)}$
MP (Silva and Souza [9])	Let Z be a continuous positive random variable belonging to the exponential family with density function given by $$f(z) = \exp\{\phi[z\zeta_0 - b(\zeta_0)] + c(z;\phi)\},$$ where $b(\cdot)$ is a continuous three-times differentiable function and ζ_0 is such that $b'(\zeta_0) = 1$ and $c(\cdot;\cdot)$ is a function that maps $\mathcal{R}^+ \times \mathcal{R}^+$ into R. Thus, $E(Z) = b'(\zeta_0) = 1$ and $Var(Z) = \phi^{-1} b''(\zeta_0)$. We denote this as $Z \sim EF(\phi)$. Let $X\|Z = z \sim Poisson(\mu z)$, with $\mu > 0$ and $Z \sim EF(\phi)$. Then, X belongs to the class of mixed Poisson distributions. If Z follows a gamma distribution with mean 1 and dispersion parameter ϕ, we find that X follows a negative binomial distribution with parameters μ and ϕ. Its probability function is given by $$p(x;\mu,\phi) = \frac{\Gamma(x+\phi)}{x!\Gamma(\phi)} \left(\frac{\mu}{\mu+\phi}\right)^x \left(\frac{\phi}{\mu+\phi}\right)^{\phi}.$$

Definition 1. *The following model is denoted by GCOMP-INGARCH(p,q),*

$$X_t | F_{t-1} \sim GCOMP(\lambda_t, \nu, r), \quad \lambda_t = \alpha_0 + \sum_{i=1}^{p} \alpha_i X_{t-i} + \sum_{j=1}^{q} \beta_j \lambda_{t-1}, \tag{2}$$

where $\alpha_0 > 0, \alpha_i \geq 0, \beta_i \geq 0, i = 1, \cdots, p, j = 1, \cdots, q, p \geq 1, q \geq 0, \nu > 0$ and $r < 1$. Specifically, the model (1) is GCOMP-INARCH(p) when $q = 0$.

According to the properties of the GCOMP distribution, the approximate conditional expectation and variance are given by

$$E(X_t|F_{t-1}) \approx \lambda_t + \frac{(2\nu-1)r}{2(1-r)}, \quad Var(X_t|F_{t-1}) \approx \frac{\lambda_t}{1-r},$$

and these are fundamentally the keys to the flexibility of the GCOMP-INGARCH(p,q) model. Qian and Zhu [8] also established some properties by assuming that model (2) is approximately stationary.

Additionally, Silva and Souza [9] proposes a general class of INGARCH models by introducing the mixed Poisson (MP) distribution proposed by Barreto-Souza and Simas [32]. This not only enriches the distribution types of INGARCH models, but also extends the negative binomial INGARCH process proposed by Zhu [33], and can even further evolves new models such as Poisson inverse Gaussian and Poisson generalized hyperbolic slope processes. Related works include Manaa and Bentarzi [34], Almohaimeed [35], Almohaimeed [36], and Cui and Wang [37], where Manaa and Bentarzi [34] focused on the expansion of time-varying parameters, which extends the time-invariant negative binomial INGARCH(1,1) studied by Zhu [33] to the periodic negative binomial INGARCH(1,1) model. Besides the flexibility of distributions, Silva and Souza [9] construct an expectation-maximization algorithm for estimating the parameters, in particular, the dispersion parameter. This provides a new framework for parameter estimation of the INGARCH-type models, which we believe also leaves a wide scope for further research.

Definition 2 (MP-INGARCH model). *The process $\{X_t\}_{t\in\mathcal{Z}}$ follows a MP-INGARCH(p,q) model if it satisfies*

$$\begin{cases} X_t | F_{t-1} \sim MP(\mu_t, \phi), & \forall t \in \mathcal{Z}, \\ \mu_t = \alpha_0 + \sum_{i=1}^{p} \alpha_i X_{t-i} + \sum_{j=1}^{q} \beta_j \mu_{t-j}, \end{cases} \quad (3)$$

where ϕ is the dispersion parameter, $\alpha_i \geq 0$, for $i = 0, 1, \cdots, p$ and $\beta_j \geq 0$, for $j = 0, 1, \cdots, q$, with $p, q \in \mathcal{N}$.

Note that the dispersion parameter ϕ in Definition 2 is independent of time t, and thus Souza et al. [10] is further extended for this setting. Indeed, the dispersion parameter is made time-varying, ϕ_t, by creating a dynamic structure for it similar to that for the intensity parameter. The resulting model in Definition 3 is called a linear time-varying dispersion INGARCH (tv-DINGARCH) model (nonlinear tv-DINGARCH(p_1, p_2, q_1, q_2) is omitted here). An interesting feature of the linear tv-DINGARCH processes is that to some extent it is more analogous to the original GARCH model than other INGARCH models. In particular, the mean of this model can be constant, while the variance depends on time as in an ordinary GARCH model. This feature is not possible to be accommodated by the standard INGARCH. Hence, they refer to the model as a pure INGARCH process to highlight the degree of association with GARCH that distinguishes it from other INGARCH models.

Definition 3 (Linear tv-DINGARCH model). *A linear tv-DINGARCH(1,1,1,1) model $\{X_t\}$ is defined by $X_t | F_{t-1} \sim MP(\mu_t, \phi_t)$, $\forall t \in \mathcal{Z}$, with*

$$\lambda_t = \beta_0 + \beta_1 Y_{t-1} + \beta_2 \lambda_{t-1}, \quad \phi_t = \alpha_0 + \alpha_1 Y_{t-1} + \alpha_2 \phi_{t-1}, \quad (4)$$

where $\beta_0, \alpha_0 > 0$ and $\beta_1, \beta_2, \alpha_1, \alpha_2 \geq 0$.

In addition, zero- or zero-one-inflated INGARCH models are valuable in applications such as insurance. Lee and Kim [38] proposes more general multiple values-inflated INGARCH models in Definition 4. The conditional distribution $q(\cdot|\eta)$ containing only one unknown parameter is indeed a streamlined setup, but it cannot be ignored that the multiple inflated values will be accompanied by a simultaneous increase in the number of parameters ρ_m and the resulting complexity can hardly be offset. From our personal point of view, a possible reason for such a dilemma is that the idea of a multi-inflated model constructed by superimposing indicator functions remains fundamentally indistinguishable from that of the zero-inflated model.

Definition 4 (multiple values-inflated INGARCH). *Let $\{Y_t, t \geq 1\}$ be a time series of counts taking values $0, 1, \ldots$ following the multiple values-inflated INGARCH model with the conditional distribution of the one-parameter exponential family:*

$$Y_t | F_{t-1} \sim p(\cdot|\eta_t), \quad X_t := E(Y_t|F_{t-1}) = f_\theta(X_{t-1}, Y_{t-1}), \quad (5)$$

where F_{t-1} is the σ-field generated by $\eta_1, Y_1, \cdots, Y_t$, and $f_\theta(x,y)$ is a non-negative bivariate function defined on $[0, \infty) \times \mathcal{N}$, depending on the parameter $\theta \in \Theta \subset \mathcal{R}^d$, and $p(\cdot|\cdot)$ is a probability mass function given by

$$p(z|\eta) = \sum_{m=0}^{M-1} \left\{ \rho_m + \left(1 - \sum_{m=0}^{M-1} \rho_m\right) q(m|\eta) \right\} I(z=m) \\ + \left(1 - \sum_{m=0}^{M-1} \rho_m\right) q(z|\eta) I(z \geq M), \quad (6)$$

$$q(z|\eta) = \exp\{\eta z - A(\eta)\} h(z), \ z \geq 0, \ \text{and} \ 0 \leq \rho_m < 1.$$

Here, η is the natural parameter, $A(\cdot)$ and $h(\cdot)$ are known functions, and $B(\cdot) := A'(\cdot)$ exists and is strictly increasing. In Model (5), $Y_t | F_{t-1} \sim p(\cdot|\eta_t)$ implicates $Y_t = F_{X_t}^{-1}(U_t)$, where $F_x(y) = \sum_{m=0}^{y} p(m|\eta)$ with $\eta = B^{-1}(x)$ and U_t are i.i.d. uniform random variables over (0,1).

An alternative approach to model generalization is to weaken the assumptions of the structural model allowing for more generalized link functions, and Wechsung and Neumann [39] make some contributions to the estimation of linkage functions. Wechsung and Neumann [39] considered a nonparametric version of the INGARCH(1,1) model, where the link function in the recursion for the variances was not specified by finite-dimensional parameters. This work completed the asymptotic analysis based on the mixed property, benefiting from the application of powerful exponential tail bounds in connection with a chaining argument. This work is highly theoretical, and shows in principle that a sufficiently regular link function of an INGARCH(1,1) process with hidden intensities can be estimated using a nonparametric least squares estimator, where the estimate is chosen from a truly nonparametric class of candidate functions. Further, the consistency rate of the estimator was shown to be nearly optimal. For practical purposes, Wechsung and Neumann [39] also reported an approximate version of the theoretical estimator.

The assumption about the distribution has also been approached from another perspective. Instead of specifying the conditional distribution, Aknouche and Demmouche [40] presents a double mixed Poisson autoregression whose conditional distribution is a superposition of two mixtures of Poisson distributions. It is more flexible compared to Poisson mixtures at the cost of just a few additional parameters. Further, Doukhan et al. [41] considered the mixture of nonlinear Poisson autoregressions, and Mao et al. [42] proposed a more general mixture INGARCH model, which includes s negative binomial mixture INGARCH of Diop et al. [43] and generalized Poisson mixture INGARCH models.

Definition 5. *A generalized mixture INGARCH model is defined as follows:*

$$\begin{cases} X_t = \sum_{k=1}^K \mathbf{1}(\eta_t = k) Y_{kt}, \\ E(Y_{kt}|F_{t-1}) = \lambda_{kt}, \\ Var(Y_{kt}|F_{t-1}) = \nu_{k0} \lambda_{kt} + \nu_{k1} \lambda_{kt}^2, \\ \lambda_{kt} = \alpha_{k0} + \sum_{i=1}^{p_k} \alpha_{ki} X_{t-i} + \sum_{j=1}^{q_k} \beta_{kj} \lambda_{k(t-j)}, \end{cases} \quad (7)$$

where $\nu_{k0} \geq 0, \nu_{k1} \geq 0$, but not simultaneously equal to zero, $\alpha_{k0} > 0, \alpha_{ki} \geq 0, \beta_{kj} \geq 0$ for $i = 1, \cdots, p_k, j = 1, \cdots, q_k$, $\mathbf{1}(\cdot)$ denotes the indicator function, F_{t-1} indicates the information given up to time $t-1$, η_t is a sequence of i.i.d. random variables with $\mathcal{P}(\eta_t = k) = \alpha_k$, $\alpha_1 + \alpha_2 + \cdots + \alpha_K = 1, \alpha_k \geq 0$ and $k = 1, \cdots, K$. Furthermore, it is assumed that X_{t-j} and η_t are independent for all t and $j > 0$, the variables Y_{kt} and η_t are conditionally independent given F_{t-1} and $\alpha_1 \geq \alpha_2 \geq \cdots \geq \alpha_K$ for identifiability. If $\beta_{kj} = 0$ for all $j = 1, \cdots, q_k$ and $k = 1, \cdots, K$, it reduces to the generalized mixture integer-valued ARCH model.

2.2. Recent Advances in Dynamic Structures

This subsection reviews the latest advances in dynamic structures of INGARCH. Many recent works have made attempts in this area, but due to space limitations, we mainly focus on the essential dynamic structural innovations, some of which will not be mentioned one by one; for example, Souza et al. [10] also considered the inclusion of covariate/exogenous time series in tv-DINGARCH model, which enhances the applicability.

First, the INGARCH model with linear dynamic structure has been limited in its setting. Specifically, since the mean of a count random variable is a positive real number, the constraints $a_0 > 0$ and $a_1, \cdots, a_p, b_1, \cdots, b_q \geq 0$ have to both hold in (1). This severely dampens the possibility of a negative ACF. The existing log-linear INGARCH model is an implementable solution to achieve negative ACF values, but at the cost of losing the linear conditional mean and the ARMA-like autocorrelation structure. Resolving the dilemma between the linear dynamic structure and a wider range of achievable ACF values is one of the key steps in the further development of the INGARCH-type models. The contribution of Weiß et al. [11] is to resolve this dilemma with the help of the softplus function. The definition of softplus INGARCH model is given as follows.

Definition 6 (softplus Poisson INGARCH model). *X_t follows from the softplus Poisson INGARCH model if satisfying*

$$X_t | F_{t-1} : Poi(M_t), \quad M_t = sp_c\left(\alpha_0 + \sum_{i=1}^{p} \alpha_i X_{t-i} + \sum_{j=1}^{q} \beta_j M_{t-j}\right), \tag{8}$$

where the softplus function $sp_c(x) = c \ln(1 + \exp(x/c))$ with $c > 0$, and $\alpha_0, \cdots, \alpha_p, \beta_1, \cdots, \beta_q \in \mathcal{R}$. The default choice for c is $c = 1$.

The softplus function $sp_c(x)$ avoids the drawback of not being differentiable in zero, while being approximately linear for $x > 0$. The excellent properties of the softplus functions play a key role as they coincide with the breakthrough of the dilemma of INGARCH mentioned above.

What follows is a progression of time-varying features in INGARCH, which has been mentioned in the previous introduction of linear tv-DINGARCH processes in Definition 3, but Souza et al. [10] approached it from the distribution. Roy [12] proceeded from the construction of a semi-parametric dynamic structure, and proposed a time-varying autoregressive models for count time-series based on a Bayesian framework (Definition 7). This is the first attempt to model possibly autoregressive count time series with time-varying coefficients, and the success of this attempt can be attributed in part to the Bayesian framework.

Definition 7 (time-varying Bayesian INGARCH model). *If the conditional distribution for X_t given F_{t-1} is $Poi(\lambda_t)$ and*

$$\lambda_t = \mu(t/T) + \sum_{i=1}^{p} a_i(t/T) X_{t-i} + \sum_{j=1}^{q} b_j(t/T) \lambda_{t-j}, \tag{9}$$

where the hierarchical prior on unknown functions $\mu(\cdot)$, $a_i(\cdot)$ and $b_i(\cdot)$ are based on B-spline bases, $\{X_t\}$ is said to follow from the time-varying Bayesian INGARCH model.

Take (1) with $p = q = 1$ for example, the essential role of the original dynamic structure is to drive the time-varying nature of the conditional mean M_t, but where the parameters ω, α_1 and β_1 do not vary with time. This means that the dynamic structure is allowed to take into account the heterogeneity changes at $t-1$ moments and the effects over the entire time period reflected by the time-constant parameters. In contrast, the semi-parametric time-varying dynamic structure (9) corresponds to further strengthening the time-varying property while weakening the average effect over the whole period. This modeling idea is better suited for rapidly changing count time series. In conclusion, Roy [12] contributes a new idea to the study of INGARCH from a Bayesian viewpoint. Its framework for the study of time-varying Bayesian INGARCH models is adapted to general non-stationary time series.

Next are some advances in the study of nonlinear dynamic structures with the help of threshold structures. The classical two-regime threshold autoregressive model allows for many properties associated with nonlinearity, and thus, both Liu et al. [13] and Chen and Khamthong [14] draw on this classical model to implement a nonlinear study of the dynamic structure of INGARCH, with the difference that the latter is based on the Markov switching approach and introduces covariates. It is worth mentioning that Lee and Hwang [44] proposed a generalized regime-switching INGARCH(1,1) model in a fashion similar to, yet different from, Chen and Khamthong [14], which also employs Markov chains with time-varying dependent transition probabilities. The difference is that Lee and Hwang [44] derives recursive formulas for the conditional probabilities of regimes in Markov chains given past information, starting from the Poisson parameters of the INGARCH(1,1) process.

Definition 8. *The Markov switching INGARCHX model with state variables is defined by*

$$p(X_t|F_{t-1}) = \binom{X_t + r - 1}{r - 1} \left(\frac{1}{1+\lambda_t}\right)^r \left(\frac{\lambda_t}{1+\lambda_t}\right), \quad X_t \geq 0 \tag{10}$$

$$\lambda_t = \begin{cases} \alpha_0^{(1)} + \alpha_1^{(1)} X_{t-1} + \beta_1^{(1)} \lambda_{t-1}, & \text{if } s_t = 1 \\ \alpha_0^{(2)} + \alpha_1^{(2)} Y_{t-1} + \beta_1^{(2)} \lambda_{t-1}, & \text{if } s_t = 2, \end{cases} \tag{11}$$

where s_t follows a first-order Markov chain with the following transition matrix:

$$\vartheta = \begin{pmatrix} Pr(s_t = 1|s_{t-1} = 1) & Pr(s_t = 2|s_{t-1} = 1) \\ Pr(s_t = 1|s_{t-1} = 2) & Pr(s_t = 2|s_{t-1} = 2) \end{pmatrix} = \begin{pmatrix} p_{11} & p_{12} \\ p_{21} & p_{22} \end{pmatrix}$$

and $\alpha_i = (\alpha_0^{(i)}, \alpha_1^{(i)}, \beta_1^{(i)})'$ is a non-negative parameter vector in state i. Naturally, changing the form of λ_t yields the threshold INGARCHX as follows:

$$\lambda_t = \begin{cases} \alpha_0^{(1)} + \alpha_1^{(1)} Y_{t-1} + \beta_1^{(1)} \lambda_{t-1}, & \text{if } Y_{t-d} \leq c \\ \alpha_0^{(2)} + \alpha_1^{(2)} Y_{t-1} + \beta_1^{(2)} \lambda_{t-1}, & \text{if } Y_{t-d} > c, \end{cases}$$

where Y_{t-d} is the threshold variable determining the dynamic switching mechanism of the model, d is a delay lag and c is the threshold value.

The segmented dynamic structure may lead to sudden changes in the probability of the INGARCH process, which is a crux that needs to be improved. Li et al. [45] considered a hysteretic process with the hysteresis variable $\{S_t\}$ and the hysteresis zone $(r_L, r_U]$, which makes the regime-switching mechanism more flexible. On this basis, Liu et al. [15] improved the self-excited threshold negative binomial autoregression (TNBAR) of Liu et al. [13] and proposed the self-excited hysteretic negative binomial autoregression (SEHNBAR) with the hysteresis variable $S_t = X_t$.

Definition 9. *Let $\{N_t, t \in \mathbb{Z}\}$ be a sequence of the i.i.d. negative binomial process given in Liu et al. [13], then $\{X_t\}$ is said to follow the SEHNBAR model, if*

$$X_t = N(0, \lambda_t]$$

with

$$\lambda_t = \begin{cases} d_1 + a_1 \lambda_{t-1} + b_1 X_{t-1}, & R_t = 1, \\ d_2 + a_2 \lambda_{t-1} + b_2 X_{t-1}, & R_t = 0, \end{cases} \tag{12}$$

where $d_i, a_i, b_i > 0, i = 1, 2,$

$$R_t = \begin{cases} 1, & X_{t-1} \leq r_L, \\ 0, & X_{t-1} > r_U, \\ R_{t-1}, & otherwise. \end{cases} \quad (13)$$

R_t is the regime indicator with the hysteresis variable Y_{t-1}, and (r_L, r_U) are boundary parameters of the hysteresis zone satisfying $0 \leq r_L \leq r_U < \infty$.

λ_t is at the lower regime when $X_{t-1} < r_L$, while at the upper regime when $X_{t-1} > r_U$. If X_{t-1} falls within the hysteresis zone, the regime indicator remains unchanged, which means that the regime indicator at time t is the same with that at $t-1$. Formally, the hysteresis model with piecewise linear structure enjoys a more flexible regime-switching mechanism. Another less intuitive but proven advantage is that the lagged observations on which λ_t relies are infinitely far away, which is the key difference between hysteresis models and traditional threshold models. Chen et al. [16] also proposed a similar INGARCH model based on the Bayesian framework.

Aknouche and Scotto [46] proposes a multiplicative INGARCH model, which is defined as the product of a unit-mean integer-valued i.i.d. sequence, and an integer-valued dependent process defined as a binomial thinning operation of its own past and of the past of the observed process. This model combines some features of the INGARCH, the autoregressive conditional duration, and the integer autoregression processes, so it can be used to model high overdispersion, persistence, and heavy-tailedness. Furthermore, Weiß and Zhu [47] propose an integer-valued analog of multiplicative error models based on a multiplicative operator, and the resulting models are closely related to the class of INGARCH models.

Last but not least, some migratory research works that have received attention in other fields but are poorly known in the INGARCH model are also of interest. Sim et al. [48] established the overall framework for the study of general-order INGARCH(p, q) models without the restriction $p = q = 1$. Similarly, the purpose of Tsamtsakiri and Karlis [49] was to select the most appropriate order of INGARCH(p, q) using a trans-dimensional Bayesian approach, and Tian et al. [50] focused on order shrinkage and selection for the INGARCH(p, q) model. Furthermore, the temporal aggregation and systematic sampling, which were widely studied in continuous-valued time series, have received the attention of Su and Zhu [51] in integer-valued time series.

For clarity, we have sorted out the main relationships of the models reviewed in this section in Figure 2.

2.3. Methodologies

There are many theory-oriented advances. The proofs for the existence and uniqueness of stationary distributions in the above-mentioned literature or in the earlier INGARCH literature are based on approximation techniques, the weak dependence or the theoretical framework using the Feller property, e-chain and Lyapunov's method. In contrast, to prove the existence and uniqueness of a stationary distribution and absolute regularity for nonlinear GARCH and INGARCH models under a mild assumption, Doukhan and Neumann [17] treated $Z_t = (X_t, \cdots, X_{t-p+1}, \lambda_t, \cdots, \lambda_{t-q+1})$ as a time-homogeneous Markov chain where $\{\lambda_t\}$ is the accompanying process of random intensities, and compensated for missing Feller properties with coupling results. Specially, besides a geometric drift condition, only a semi-contractive condition was imposed, which means a subgeometric, rather than the more usual geometric, decay rate of the mixing coefficients. This result not only enriches the theoretical proof technique, but also broadens the application area of the INGARCH model.

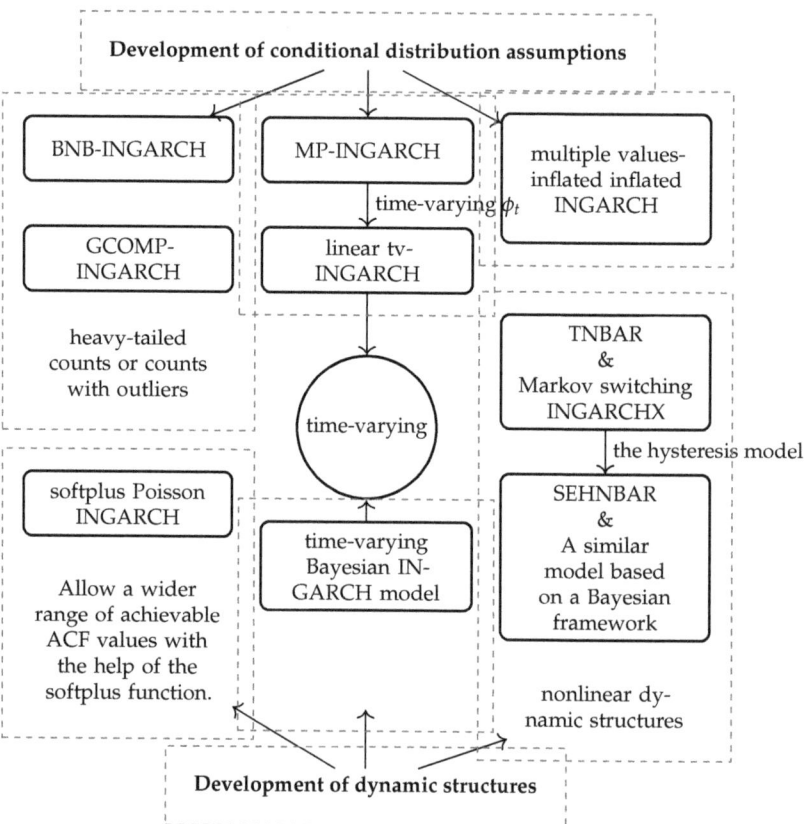

Figure 2. Flowchart of some major INGARCH-type models for unbounded counts.

Further, supposed that $\{X_t\}$ follow from Poisson INGARCH, Neumann [52] use the contraction property twice: first, under the contraction condition of the intensity process, Neumann [52] obtained the contraction property of the Markov kernel connected with Z_t in terms of a suitable Wasserstein metric, and then the existence and uniqueness of a stationary distribution follows via the Banach fixed point theorem; next, the almost effortlessly absolute regularity of the count process was established by using the contraction property once more; finally, Neumann [52] constructed a coupling of the original and the bootstrap process, and proved the existence and uniqueness of a stationary version of this joint process as well as absolute regularity of the joint count processes. Notably, the last item implies that the model-based bootstrap method proposed by Neumann [52] is more general than most of the existing papers on the consistency of bootstrap. Specifically, the bootstrap process mimics the random behavior of the original counting process, rather than being limited to the plausibility of certain specific statistics. More broadly, Doukhan et al. [53] derived the absolute regularity at a geometric rate not only for stationary Poisson GARCH processes, but also for models with an explosive trend. Recently, Aknouche and Francq [54] considered the existence of a stationary and ergodic solution of a general Markov-Switching autoregressive conditional mean model, of which the INGARCH model is one of the variants.

The contribution of Douc et al. [55] is establishing the necessary and sufficient conditions for the identifiability of observation-driven models including the pure INGARCH model and its numerous extensions, such as the pure INGARCH model with thresholds or exogenous covariates.

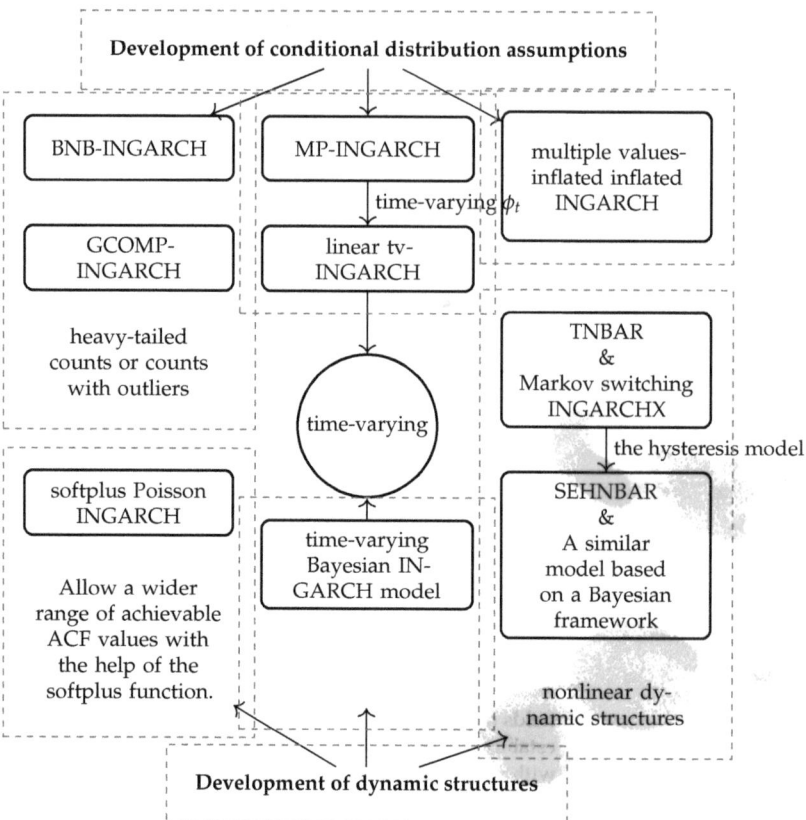

Recent advances in estimation methods regarding the INGARCH model are reviewed as follows. One of the main reasons for the utility of negative binomial models is that Poisson INGARCH is less flexible than models based on overdispersed conditional distributions when modeling overdispersed series. However, parameter estimation for a negative binomial INGARCH model is usually implemented based on Poisson quasi-maximum likelihood estimation (quasi-MLE or QMLE), where the pitfall is that Poisson-QMLE is likely to fail to achieve its full asymptotic efficiency with the presence of overdispersion. To clear this hurdle, Aknouche et al. [56] proposed two negative binomial QMLEs (NB-QMLEs), including the profile NB-QMLE calculated while arbitrarily fixing the dispersion parameter of the negative binomial likelihood, and the two-stage NB-QMLE consisting estimation for both conditional mean and dispersion parameters. Similarly, the two-stage weighted least square estimators (WLSEs) proposed by Aknouche and Francq [57] are general for time series data, and feasible for the INGARCH model. WLSEs can be implemented without fully specifying the conditional distribution or time series structure, and enjoy the same consistency properties as QMLEs when the conditional distribution is mis-specified, even if the conditional variance is mis-specified. Additionally, a data-driven strategy was identified to find asymptotically optimal WLSEs. This actually provides prerequisite support for further relaxation of the conditional distribution assumption for the count time series. For the model in Aknouche and Scotto [46], parameter estimation is conducted by using a two-stage WLSE, and Xu et al. [58] considered a saddlepoint MLE for a special case of this model.

The commonly used MLE method is highly influenced by outliers, so there are several works dedicated to establish robust estimation methods. For Poisson INGARCH models, Li et al. [59] proposed a robust M-estimator by using a new loss function inspired by the Tukey's biweight function. It is the construction of this loss function that contributes to this work. One of the disadvantages of Tukey's function is that it drops to zero so that the effect of very large outliers or leverage points is completely suppressed, which implies the possibility of multiple solutions to the estimated equations. Then, an intuitive idea is to construct a hybrid loss function that does not fully suppress the effects of very large outliers or leverage points. Li et al. [59] proposed a new loss function by twofold improvement. Along similar lines, Xiong and Zhu [60] introduced a robust estimation for the one-parameter exponential family INGARCH(1,1) models.

Moreover, Kim and Lee [61] used the minimum density power dispersion estimator as a robust estimator for INGARCH models whose conditional distribution belongs to the one-parameter exponential family. There are two advantages to this approach: the first is simplicity, i.e., it contains only a single tuning parameter that controls the trade-off between robustness and efficiency; the second is the ability to balance robustness and efficiency, providing considerable robustness while retaining high levels of efficiency as the tuning parameter approaches zero. Further, Xiong and Zhu [62] and Xiong and Zhu [63] used the Mallows' quasi-likelihood estimator and the minimum density power dispersion estimator as a robust estimator for negative binomial INGARCH models, respectively. For negative binomial INGARCH models, Elsaied and Fried [64] also developed several robust estimators including robustifications of method of moments and ML-estimation, one of which was an alternative to the robust estimator proposed by Xiong and Zhu [63].

Another drawback of MLE for INGARCH models is that numerical results are sensitive to the choice of initial values. Hence, Li and Zhu [65] proposed the mean targeting estimation, which is an analogue to variance targeting estimation used in the GARCH model. In addition, there have been some other advances in estimation. Jo and Lee [66] introduced the mean targeting QMLE based on INGARCH models, which provides a new perspective. When shifting to focus on more specific estimation problems, it is a common problem that the estimation performance of the intercept parameter is inferior to that of other parameters, either in the Poisson or negative binomial INGARCH model. Integrating the likelihood function by assuming a conditional distribution is one option to eliminate this obstacle. Hence, Pei and Zhu [67] adopted the marginal likelihood to estimate the intercept parameter in the negative binomial INGARCH model.

There are also some areas that are of interest to scholars. For example, to test the parameter variation of INGARCH, the cumulative sum (CUSUM) statistics has been the most popular method in recent years (Lee and Lee [68], Lee et al. [69], Lee [70], Vanli et al. [71], Weiß and Testik [72]), and Lee and Kim [73] reviewed a recent progress regarding the change point test for integer-valued time series models. Further, as with the techniques employed to present robust estimates, Kim and Lee [74] introduced a robust change point test based on density power divergence. Michel [75] considered the limiting distribution of the INGARCH(1,1) with $\alpha + \beta = 1$.

2.4. Applications

The practical application of INGARCH models has developed considerably in recent years, especially in the period of COVID-19 when there is a strong demand for analysis of count time series data such as daily new infections in various countries or regions. This has also given rise to many valuable research topics. Agosto and Giudici [1] focused on COVID-19 contagion and digital finance, and presented Poisson INGARCH of the daily newly observed cases to understand the contagion dynamics of the COVID-19. In addition, the purpose of Agosto et al. [2] is to monitor COVID-19 contagion growth and came to the interesting conclusion that policy measures aimed at reducing infection are very useful when the it is at its peak and can reduce reproduction rates. Souza et al. [10] considered the tv-DINGARCH model with covariate/exogenous time series and applied to the daily number of deaths due to COVID-19 in Ireland. In contrast, Roy [12] focused specifically on the data for New York City because the epidemic status in this city lasted for a month and with the help of the ongoing blockade, it recovered significantly within about three months. Hence, it is this temporal variability in the data that drew Roy [12] to explore its trends by the time-varying Bayesian INGARCH model. Similarly, Giudici et al. [3] was also concerned about the time-varying features of COVID-19 and proposed Bayesian time-dependent Poisson autoregressive models. Additionally, Gning et al. [76] focused on COVID-19 in Senegal and China. Moreover, the dynamics of COVID-19 infectivity in Saudi Arabia were evaluated in Alzahrani [77] by using two statistical models, namely the log-linear Poisson autoregressive model and the ARIMA model. The results of this study showed that the log-linear Poisson autoregressive model had superior predictive performance. At the same time, many application-oriented works have actually proposed new models to meet the requirements. For example, Xu et al. [78] proposes a comprehensive adaptive log-linear zero-inflated generalized Poisson INGARCH to describe crime counts in Byron and Australia, and the features of this data set include autocorrelation, heteroscedasticity, overdispersion and excessive number of zero observations.

In addition to COVID-19, INGARCH-type models are employed in other areas such as stock trading, co-tracking of commodity marketsand so on (Chen and Khamthong [79], Agosto and Raffinetti [80], Jamaludin et al. [81], Algieri and Leccadito [82], Aknouche et al. [83], Berentsen et al. [84], Cerqueti et al. [85]). It is worth mentioning that the INGARCH-type model has been favored for human influenza research even before the outbreak of COVID-19. Specifically, Chen et al. [86] involves the INGARCH-type model when examining the causal relationship between environmental fine particulate matter and human influenza in Taiwan. The study on traffic forecasting of Kim [87] affirms the value of the INGARCH model for applications. The prediction model of Kim [87] was generated by estimating the parameters of the INGARCH process and predicting the Poisson parameters of the future step ahead process using conditional MLE methods and prediction procedures, respectively. They came to the conclusion: "INGARCH captures the characteristics of network traffic better than other statistical models, it is more tractable than neural networks (NN), overcomes the black-box nature of NN, and some statistical models perform comparable or even better than NN, especially when there is insufficient data to apply deep NN". Another application that tends to be humanistic and social

Anavatan and Kayacan [88], the aim of which was to reveal the relationship between the number of femicide, female unemployment rate, male unemployment rate and the amount of information in Turkey by using INGARCH model.

In summary, the application scenarios of the INGARCH-type models can be as specific as a vehicle prediction at an intersection or as macro as a humanistic exploration of a country, and what is more valuable is that the exploration about their application still remains expected and promising.

3. $\mathcal{Z}$-Valued Time Series

The previous section is concerned with count time series (i.e., $\mathcal{N}$-valued time series). However, time series that allow both non-negative and negative integer values (i.e., $\mathcal{Z}$-valued time series) are also worth investigating, whose research value is reflected in the following aspects: first, there are many typical application areas for $\mathcal{Z}$-valued time series, e.g., describing score gaps in sports, trading changes in finance (Xu and Zhu [20]); secondly, the non-stationary property embodied in such time series is also of interest, such as differenced series that are initially non-stationary (Gonçalves and Mendes-Lopes [89]). Let $\{Z_t\}_{t \in \mathbb{Z}}$ be a $\mathcal{Z}$-valued process, and $z_1, \cdots, z_T$ be a finite set of observations. A recent study on the mixing properties of $\mathcal{Z}$-valued GARCH processes can be found in Doukhan et al. [90]. Below, we review contributions on $\mathcal{Z}$-valued time series in recent years.

Firstly, the Skellam distribution is introduced into the INGARCH-type models as a conditional distribution. Alomani et al. [91] proposed the Skellam GARCH(1, 1) model defined in Definition 10, where the Skellam is the distribution of the difference of two independent Poisson variates and thus allows for both non-negative and negative integer-valued variables. The specific definition of the Skellam distribution is placed in Table 2. The benefits of the Skellam INGARCH model are that they allow time-varying variance and nonstationarity in the mean for time series, and the conditional maximum likelihood and conditional least squares methods have been developed for estimation of the parameters.

Table 2. Summary of basic properties of distributions related to $\mathcal{Z}$-valued INGARCH.

Distribution	Definition
Skellam (Alomani et al. [91])	The PMF of Skellam is $$P_S(Z=z) = exp\{-(\lambda_1 + \lambda_2)\}\left(\frac{\lambda_1}{\lambda_2}\right)^{z/2} I_{|z|}(2\sqrt{\lambda_1\lambda_2}),$$ where $I_r(x)$ is the modified Bessel function of order r and is defined by $I_r(x) = \left(\frac{x}{2}\right)^r \sum_{k=0}^{\infty} \frac{(x^2/4)^k}{k!\Gamma(r+k+1)}$. And $E(Z) = \lambda_1 - \lambda_2$ and $Var(Z) = \lambda_1 + \lambda_2$.
MS (Cui et al. [92])	The PMF of MS is $$P_{MS}(Z=z) = f_{MS}(z|\lambda_1,\lambda_2) = \begin{cases} P_S(Z=z), & z \in \mathcal{Z}/\{0,\pm1\} \\ P_S(Z=z) - \frac{1}{2}\gamma\Delta, & z=-1 \text{ or } 1 \\ P_s(Z=z) + \gamma\Delta, & z=0 \end{cases}$$ where $\Delta = P_s(Z=0) - \min\{P_s(Z=-1), P_s(Z=1)\} > 0$ and $P_s(Z=q) = f_s(q\|\lambda_1,\lambda_2)$ for $q \in \mathcal{Z}$. $E(Z) = \lambda_1 - \lambda_2$ and $Var(Z) = \lambda_1 + \lambda_2 - \gamma\Delta$.
SGe (Xu and Zhu [20])	The PMF of SGe is $$P(X=k) = p(1-p)^{k-1}, k=1,2,3,\cdots$$ $E(Z) = 1/p$ and $Var(Z) = (1-p)/p^2$

Definition 10. $\{Z_t|F_{t-1}\}$ follow symmetric Skellam $(\frac{\sigma^2_{t|t-1}}{2}, \frac{\sigma^2_{t|t-1}}{2})$ with the conditional variance satisfying

$$\sigma^2_{t|t-1} = \omega + \alpha Z^2_{t-1} + \beta \sigma^2_{t-1|t-2}, \ t \geq 2, \tag{14}$$

In (14), the parameters ω, α and β satisfy the following constraints: $\omega > 0, 0 < \alpha < 1, 0 < \beta < 1$ and $\alpha + \beta < 1$, which are necessary and sufficient for stationarity of the process (14). We refer to the above model as Skellam GARCH(1, 1).

The drawback of the symmetric Skellam INGARCH(1,1) in Definition 10 is that only the simplest case is considered, i.e., the conditional expectation value is equal to zero. Cui et al. [92] proposes an asymmetric Skellam INGARCH to eliminate this limitation. Furthermore, the modified Skellam (MS) distribution (see Table 2) is introduced and thus the modified Skellam INGARCH in Definition 11. Specifically, Cui et al. [92] added a new parameter, γ, to the standard Skellam distribution whose mission is to control the distance between the probabilities $P(Z_t = 0)$ and $\min\{P(Z_t = 1), P(Z_t = -1)\}$ by its magnitude and sign. Hence, the modified Skellam INGARCH can flexibly compensate for the over- or under-representation of specific integers $(-1, 0, 1)$.

Definition 11 (Modified Skellam INGARCH model). *Consider the following model:*

$$Z_t|F_{t-1} \sim MS(\gamma, \lambda^2_t, \lambda^{*2}_t),$$
$$\lambda^2_t = \alpha_0 + \alpha_1 Z^2_{t-1} + \beta_1 \lambda^2_{t-1},$$
$$\lambda^{*2}_t = \alpha^*_0 + \alpha_1 Z^2_{t-1} + \beta_1 \lambda^{*2}_{t-1},$$

where $\gamma \in (\max_t\{-\mathbb{P}_{0t}/\Delta_t\}, \min_t\{2\min(\mathbb{P}_{-1t}, \mathbb{P}_{1t})/\Delta_t\})$, $\Delta_t = \mathbb{P}_{0t} - \min(\mathbb{P}_{-1t}, \mathbb{P}_{1t}) > 0$, $\mathbb{P}_{qt} = \mathbb{P}_{MS}(X_t = q), q \in \mathcal{Z}, \alpha_0 > 0, \alpha^*_0 > 0, \alpha_1 \geq 0, \beta_1 \geq 0$. *The above model is denoted by MS-INGARCH(1, 1). Note that for* $\gamma = 0$, *we recover the AS-INGARCH(1, 1) model.*

Another path to modeling the $\mathcal{Z}$-valued time series is to extend the $\mathcal{N}$-valued IN-GARCH models by introducing a sequence of i.i.d. binary random variables independent of $\{Z_t\}$, $\{Q_t\}$, taking values at 1 and -1 with equal probability 0.5, such as a two-sided Poisson distribution. For example, Hu [18] and Hu and Andrews [19] proposed a Poisson $\mathcal{Z}$-valued Glosten–Jagannathan–Runkle GARCH (PZG) model as follows.

Definition 12. *We call* $\{Z_t\}$ *an integer-valued asymmetric GARCH process of orders p and q, if for all* $t \in \mathcal{Z}$,

$$Z_t = Q_t X_t,$$
$$X_t|F_{t-1} \sim Pois(\lambda_t),$$
$$\lambda_t = \left(\sqrt{1 + 4\eta_t} - 1\right)/2,$$
$$\eta_t = \alpha_0 + \sum_{i=1}^{p} \alpha_i (|Z_{t-i}| - \gamma Z_{t-i})^2 + \sum_{j=1}^{q} \beta_j \eta_{t-j}.$$

Hence, $\{X_t, t \in \mathcal{Z}\}$ is a non-negative integer-valued stochastic process; conditioned on past information up to and including time $t - 1$, X_t has Poisson distribution with mean λ_t. So that $\{\eta_t\}$ and $\{\lambda_t\}$ are positive, we assume parameter $\alpha_0 > 0$ and parameters α_i, β_j, for $i = 1, \cdots, p$, $j = 1, \cdots, q$, are all non-negative, with model orders $p \geq 1, q \geq 0$.

It can be seen that a two-sided Poisson distribution is employed and the structure of η_t is inspired by the Glosten–Jagannathan–Runkle GARCH (GJR-GARCH, Glosten et al. [93]). The main role of GJR-GARCH here is to portray asymmetric responses in the volatility of $\mathcal{Z}$-valued time series, such as the presence of leverage effects in financial time series, and is thus the highlight of the PZG model. Furthermore, Xu and Zhu [20] proposed the geometric $\mathcal{Z}$-valued GJR-GARCH model based on the shifted geometric distribution that is more flexible than Poisson distribution. Additionally, Xu and Zhu [20] expanded from a two-side equality probability of 0.5 to a broader form where the ratio of positive, negative and zero values can be controlled by a parameter ρ (Definition 13).

Definition 13. *The geometric $\mathcal{Z}$-valued GJR-GARCH model is defined as*

$$Z_t = Q_t^* X_t,$$
$$X_t | F_{t-1} \sim SGe(p_t),$$
$$p_t = \frac{\sqrt{\rho^2 + 4\rho\lambda_t} - \rho}{\lambda_t},$$
$$\lambda_t = \omega + \sum_{i=1}^{p} \alpha_i (|Z_{t-i}| - \gamma_i Z_{t-i})^2 + \sum_{j=1}^{q} \beta_j \lambda_{t-j},$$

where $SGe(\cdot)$ denotes the shifted geometric distribution, $\omega > 0, |\gamma_i| \leq 1, \alpha_i \geq 0, \beta_j \geq 0$, for $i = 1, \cdots, p, j = 1, \cdots, q, p \geq 1, q \geq 0$. Specially, $\{Q_t^*\}$ is a sequence of i.i.d. random variables taking values at 1, 0 and -1 with probabilities $\rho, 1 - 2\rho$ and ρ, respectively, where $0 < \rho < \frac{1}{2}$.

4. Bounded Count Time Series

In what follows, $\{B_t\}_{t \in \mathcal{Z}}$ consists of bounded counts with range $\{0, \cdots, n\}$ for given $n \in \mathcal{N}$. This type of data also includes categorical time series, with binary time series being a special case. In terms of the INCARCH-type models, the study of $\{B_t\}$ differs from that of unbounded counts $\{X_t\}$ in two ways: one is that the candidates for the conditional distribution are required to be discrete distributions with bounded range of values; the other is that the dynamic structure of the conditional mean needs to be adjusted accordingly to the constraints of parameters of the conditional distribution, which leads to the possibility that the dynamic structure cannot be directly attached to the conditional mean, but some kind of functional transformation of the conditional mean. Accordingly, researchers have been bursting at the seams with recent innovative work in these two areas.

The definition of the bound INGARCH (BINGARCH) model is obtained by the distributional assumption that B_t is generated by a bounded-count distribution and the normalized conditional mean $P_t = \frac{1}{n} E(X_t | X_{t-1}, \cdots)$, where

$$P_t = a_0 + \sum_{i=1}^{p} a_i X_{t-i}/n + \sum_{j=1}^{q} b_j P_{t-j} \tag{15}$$

with the additional constraint $a_0 + \sum_{i=1}^{p} a_i + \sum_{j=1}^{q} b_j < 1$.

First, for convenience, we assume that B_t is generated by the conditional binomial distribution $Bin(n, P_t)$.

$$B_t | F_{t-1} \sim Bin(n, P_t) \tag{16}$$

Then, as with the constraints on the INGARCH-type model with linear dynamic structure mentioned previously, the BINGARCH-type model also requires constraints $a_0 > 0$ and $a_1, \cdots, a_p, b_1, \cdots, b_q \geq 0$ to ensure positive P_t in (15). In line with the idea of Weiß et al. [11], Weiß and Jahn [22] solved this puzzle in the BINGARCH-type model with the help of the soft-clipping functions, which enables the migration and application of this framework proposed by Weiß et al. [11]. Specially, Weiß and Jahn [22] considered the normalized conditional mean as follows:

$$P_t = f\left(\alpha_0 + \sum_{i=1}^{p} \alpha_i X_{t-i}/n + \sum_{j=1}^{q} \beta_j P_{t-j}\right). \tag{17}$$

where $f(\cdot)$ was set to be the soft-clipping function

$$f(x) = sc_{\frac{c}{n}}(x) = \frac{c}{n}\ln\left(\frac{1+\exp(\frac{nx}{c})}{1+\exp\left(\frac{n(x-1)}{c}\right)}\right).\qquad(18)$$

Definition 14. $\{B_t\}$ *is said to follow the soft-clipping binomial INGARCH model if satisfying (16), (17) and (18). The model would be well-defined without any further restrictions, but some reasonable constraints such as $|\alpha_i|, |\beta_j| < 1$ for $i = 1,\cdots,p$ and $j = 1,\cdots,q$, and $\alpha_0 \in (0, 1+p+q)$ are needed.*

The soft-clipping binomial INGARCH was employed in Weiß and Testik [72] once more. The fascinating question considered by Weiß and Testik [72] is how the performance of the control chart is affected if the CUSUM control chart is designed based on the assumption of a completely linear data generation process, while the true one is only approximately linear. Weiß and Testik [72] aptly exploits the fact that the soft-clipping binomial INGARCH is an approximately linear counterpart of BINGARCH, and draw the conclusion that, in general, chart designs are robust to model mis-specification when parameters are specified, whereas the opposite result is obtained when the parameters are estimated.

The binomial distribution is a traditional choice for studying bounded count time series, as in (16), due to its simple form and relatively well-established properties. The motivation for the innovation against the conditional distribution is that there is a fixed relationship between the variance and the mean of the binomial distribution, denoted as the binomial index of dispersion (BID), similar to the equidispersion property of Poisson. The BID for a random variable X taking values in $\mathcal{N}$ is defined as

$$\text{BID} = \frac{n\text{Var}(X)}{E(X)(n-E(X))}.$$

Additionally, the BID of the binomial random variable is calculated to be 1, which indicates that (16) is not competent for modeling data with BID > 1. Hence, Chen et al. [23] proposed a new class of INGARCH models with beta-binomial (BB) variation, which is a generalization of Chen et al. [94], in Definition 15, where the BID of BB distribution takes values in the interval $(1, n)$. The specific form of the beta-binomial distribution is available in Table 3. To analyze the high volatility in time series counts, covariates were further introduced by Chen et al. [24], and thus a covariate-driven beta-binomial INGARCH model was proposed in Definition 16.

Definition 15. *Let $\theta = (\theta_1,\cdots,\theta_d)^\top$ be the vector of parameters. Then, the beta-binomial GARCH(1,1) model is defined as:*

$$B_t | F_{t-1} : BB(n, p_t, \phi), \quad Y_t := np_t := g_\theta(Y_{t-1}, B_{t-1}), t = 1, 2, \cdots,\qquad(19)$$

where $g_\theta(Y_{t-1}, B_{t-1})$ is a non-negative and continuous function in terms of each θ_j for a given Z_{t-1} and $Y_{t-1}, \forall j = 1, 2, \cdots, d$.

Definition 16. *Let $C_t = (C_{1t}, C_{2t},\cdots, C_{dt})$ be a d-dimensional exogenous covariate vector. Then, the logit-BBGARCHX(1,1) model is defined as:*

$$B_t | F_{t-1} : BB(n, P_t, \phi), logit(p_t) = \omega + \alpha logit(p_{t-1}) + \beta B_{t-1} + f(C_{t-1}, \gamma), t \in \mathcal{Z},\qquad(20)$$

where $logit(x) = \log(x/(1-x)), \forall x \in (0,1), f(\cdot,\gamma): \mathcal{R}^d \to \mathcal{R}, (\omega,\alpha,\beta,\phi)$ is the parameter vector with $\omega \in \mathcal{R}, \phi \in (0,1), |\beta| < 4(1-|\alpha|)$ and $|\alpha| < 1$, and γ is the additional parameter vector involving in f.

The beta-binomial distribution employed in Chen et al. [23] and Chen et al. [24] allow to model bounded data with under-dispersion. Then, Chen [95] turned its attention to a rare case, i.e., an under-diversified pseudo-binomial data set. It is the discrete beta (DB) distribution that competently models bounded data with under-dispersion, equiv-dispersion, and over-dispersion. Motivated by this and the soft-clipping function used in Weiß et al. [11], Chen [95] proposed a new soft-clipping discrete beta GARCH model as follows.

Definition 17. *The soft-clipping discrete beta GARCH(1,1) model is defined by*

$$\begin{cases} B_t | F_{t-1} \sim DB(n_{bot}, n_{top}, p_t, \tau), \\ p_t = sc_c(\omega + \alpha_1 p_{t-1} + \beta_1 B_{t-1}/n_{top}), \end{cases} \tag{21}$$

where the definition of DB is placed in Table 3, $sc_c(\cdot)$ is defined in (18), $|\alpha_1| + |\beta_1| < 1$, $n_{bot} = 0$ or 1 and $n_{top} \in \mathcal{N}$ is the predetermined upper limit of the range.

Table 3. Summary of basic properties of distributions related to BINGARCH.

Distribution	Definition
BB (Chen et al. [23])	The PMF of BB is $$P(Z=z) = \binom{n}{z} \frac{B(z+a, n-z+a)}{B(a,b)} \text{ with } B(a,b) = \frac{\Gamma(a)\Gamma(b)}{\Gamma(a+b)}.$$ The beta-binomial distribution approximately reduces to the usual binomial distribution when $a \to \infty$ or $b \to \infty$. $E(Z) = \frac{na}{a+b}$ and $$Var(Z) = \frac{nab}{(a+b)^2}\left(1 + \frac{n-1}{a+b+1}\right)$$
DB (Chen [95])	The PMF of DB is $$P(X = x | \alpha, \beta, n) = \frac{1}{Z(\alpha, \beta)} f\left(\frac{x - n_{bot} + 1}{n_{top} - n_{bot} + 2}\right),$$ where $$f(x) = \frac{1}{B(\alpha,\beta)} x^{\alpha-1}(1-x)^{\beta-1}, Z(\alpha,\beta) = \sum_{x=n_{bot}}^{n_{top}} f\left(\frac{x - n_{bot} + 1}{n_{top} - n_{bot} + 2}\right),$$ $n_{top} \in \mathcal{N}$ is the predetermined upper limit of the range and $n_{bot} = 0$ or 1 is the predetermined lower limit of the range (Z taking values in $\{n_{bot}, n_{bot} + 1, n_{bot} + 2, ..., n_{top}\}$)

Categorical time series are also a type of bounded count-valued time series. In the study of INGARCH-type models, categorical time series are usually presented in vector form, which can be modeled by an autoregressive multinomial logistic time series model with a latent process and is defined by a GARCH-type recursive equation. Suppose that we observe a process with state space $\{0, 1, \cdots, n\}$ and define a $(n-1)$-dimensional vector $Y_t = (Y_{1t}, Y_{2t}, \cdots, Y_{(n-1)t})^\top$, for $1 \leq t \leq n$, such that

$$Y_{kt} = \begin{cases} 1, & \text{if the } k\text{th category is observed at time } t, \\ 0, & \text{otherwise,} \end{cases} \tag{22}$$

for all $k = 1, 2, \cdots, n-1$. Moreover,

$$p_{kt} = \mathbb{P}(Y_{kt} = 1 | F_{t-1}), \ 1 \leq j \leq N-1,$$

is defined as a vector of conditional probabilities and $p_t \equiv (p_{1t}, p_{2t}, \cdots, p_{(n-1)t})'$. For the last category n, set $Y_{nt} = 1 - \sum_{k=1}^{n-1} Y_{kt}$ and $p_{nt} = 1 - \sum_{k=1}^{n-1} p_{kt}$. The dynamic structure is dependent on p_t. For instance, the following linear dynamic structure is a classic:

$$p_t = d + A p_{t-1} + B Y_{t-1},$$

where d is a vector and A, B are matrices of appropriate dimensions. It is easy to see that the disadvantage of this modeling approach in application is the problem of using multidimensional methods to deal with univariate data resulting in increased pressure on parameter estimation and other troubles. However, its theoretical development is being gradually refined. Fokianos and Truquet [21] employed a useful coupling technique to study the ergodicity of infinite-order finite-state stochastic processes, making significant improvements to previous conditions on the stationarity and ergodicity of these models.

In addition to dealing with categorical time series in vector form, Liu et al. [96] proposed a simple and less computationally stressful way of modeling, which was essentially an innovative approach to conditional distributions from an application perspective. Liu et al. [96] first introduced a new zero-one-inflated bounded Poisson (ZOBP) distribution defined as

$$P(B = k) = p_1 I_{\{k=0\}} + p_2 I_{\{k=1\}} + (1 - p_1 - p_2) \frac{\lambda^k / k!}{\sum_{i=0}^{M} \lambda^i / i!}, \quad k = 0, 1, \cdots, M, \tag{23}$$

where $p_1 \geq 0$ and $p_2 \geq 0$ are the inflated parameters for the states 0 and 1, respectively, with the constraint $p_1 + p_2 < 1$, $\lambda > 0$ is the intensity parameter and the integer $M \geq 2$ is a given upper bound. This distribution is suitable for depicting data for air quality classes that are predominantly excellent, where 0 and 1 represent excellent and good air quality, respectively. Liu et al. [96] defined a new INGARCH-type model based on the ZOBP distribution. For ease of presentation, we denote the ZOBP distribution in (23) with $(p_1, p_2) = (0, 0)$ by $P^*(\lambda, M)$. Let $\{D_t\}$ be an i.i.d. sequence with the following probability distribution:

$$P(D_t = 0) = p_1, \quad P(D_t = 1) = p_2, \quad P(D_t = 2) = 1 - p_1 - p_2, \tag{24}$$

where $p_1 \geq 0$, $p_2 \geq 0$ and $p_1 + p_2 \leq 1$. Then, the ZOBP autoregressive (ZOBPAR) model is defined as follows:

Definition 18. $\{B_t\}$ is said to follow the ZOBPAR model, if

$$B_t = (2 - D_t) D_t + (D_t - 1) D_t B_t^* / 2$$

with $B_t^* | F_{t-1} \sim P^*(\lambda_t, M)$ and

$$\lambda_t = d + a \lambda_{t-1} + b B_{t-1} \tag{25}$$

where $d > 0$, $a \geq 0$, $b > 0$, and D_t satisfying (12) is independent of B_t^*.

Compared with the model in vector form, the ZOBPAR model is concise both in terms of estimation and its own form. The follow-up Liu et al. [97] is also based on this and is an application-oriented study completed using Bayesian estimation methods.

5. Multivariate Integer-Valued Time Series

It can be seen that univariate INGARCH models have been well-studied in the literature, but progress in multivariate INGARCH models has lagged somewhat in comparison. It is encouraging to note that there have been some recent developments.

We start with a review of some studies based on further improvements of the bivariate Poisson (BP) INGARCH model. The definition of the BP-INGARCH model is given here.

Definition 19. Let $Y_t = (Y_{t,1}, Y_{t,2})^\top$. $\{Y_t\}$ is said to follow a BP-INGARCH(1,1) model if

$$Y_t | F_{t-1} \sim BP^*(\lambda_t^\top, \phi), \quad \lambda_t = (\lambda_{t,1}, \lambda_{t,2})^\top = \omega + A\lambda_{t-1} + BY_{t-1},$$

where the definition of BP^* is placed in Table 4, $\phi \geq 0, \omega = (\omega_1, \omega_2)^\top \in \mathcal{R}_+^2$ and $A = \{\alpha_{ij}\}_{i,j=1,2}$ and $B = \{\beta_{ij}\}_{i,j=1,2}$ are 2×2 matrices with non-negative entries.

Table 4. Summary of basic properties of distributions related to MINGARCH.

Distribution	Definition
BP* (Lee et al. [25])	Consider random variables X_k, $k = 1, 2, 3$, which follow independent Poisson distributions with parameters $\lambda_1 - \phi, \lambda_2 - \phi, \phi$, respectively, and then the random variables $Y_1 = X_1 + X_3$ and $Y_2 = X_2 + X_3$ jointly follow a bivariate Poisson distribution $BP^*(\lambda_1, \lambda_2, \phi)$ with probability mass function: $P(Y_1 = y_1, Y_2 = y_2) = e^{-(\lambda_1 + \lambda_2 - \phi)} \frac{(\lambda_1 - \phi)^{y_1}(\lambda_2 - \phi)^{y_2}}{y_1! y_2!} \times \sum_{k=0}^{\min(y_1, y_2)} \binom{y_1}{k}\binom{y_2}{k} k! \left(\frac{\phi}{(\lambda_1 - \phi)(\lambda_2 - \phi)} \right)^k$, with $E(Y_1) = Var(Y_1) = \lambda_1$, $E(Y_2) = Var(Y_2) = \lambda_2$ and $Cov(Y_1, Y_2) = \phi$.
BP (Cui and Zhu [27])	A bivariate Poisson distribution is defined as a product of Poisson marginals with a multiplicative factor δ, whose PMF is given by $$P(Y_1 = y_1, Y_2 = y_2) = \frac{\lambda_1^{y_1} \lambda_2^{y_2}}{y_1! y_2!} e^{-(\lambda_1 + \lambda_2)} [1 + \delta(e^{-y_1} - e^{-c\lambda_1})(e^{-y_2} - e^{-c\lambda_2})]$$ where $c = 1 - e^{-1}$, with $E(Y_1) = Var(Y_1) = \lambda_1$, $E(Y_2) = Var(Y_2) = \lambda_2$ and $Cov(Y_1, Y_2) = \delta c^2 \lambda_1 \lambda_2 e^{-c(\lambda_1 + \lambda_2)}$.
BPG (Cui et al. [98])	The PMF of BPG is $$P(Y_1 = y_1, Y_2 = y_2) = \frac{1}{Z(\lambda_1, \lambda_2, \theta)} \frac{\lambda_1^{y_1} \lambda_2^{y_2}}{y_1! y_2!} e^{-(\lambda_1 + \lambda_2)} c_\rho(F_1(y_1), F_2(y_2)),$$ where $Z(\lambda_1, \lambda_2, \theta)$ is the normalizing factor, $$c_\rho(\mu_1, \mu_2) = \frac{1}{\sqrt{1-\rho^2}} \exp\left(-\frac{\rho^2(q_1^2 + q_2^2) - 2\rho q_1 q_2}{2(1-\rho^2)} \right)$$ $q_i = \phi^{-1}(u_i)$, ϕ^{-1} is the inverse of the standard univariate normal distribution, $\mu_i \in [0,1]$ for $i = 1, 2$, $\rho \in (-1,1)$, $\gamma \in (-\infty, \infty)/\{0\}$ and $\sigma \in (-1,1)$ are regarded as dependency parameters for BP. Let $$c_\gamma(\mu_1, \mu_2) = \frac{-\gamma(e^{-\gamma}-1)e^{-(\mu_1+\mu_2)\gamma}}{[(e^{-\gamma}-1) + (e^{-\mu_1\gamma}-1)(e^{-\mu_2\gamma}-1)]^2}$$ and $$c_\sigma(\mu_1, \mu_2) = 1 + \sigma(1 - 2\mu_1)(1 - 2\mu_2),$$ and then replacing $c_\rho(F_1(y_1), F_2(y_2))$ with $c_\gamma(F_1(y_1), F_2(y_2))$ and $c_\sigma(F_1(y_1), F_2(y_2))$ yields the PMF of BPF and that of BPFGM, respectively.

For the BP-INGARCH model, Lee et al. [25] showed the asymptotic normality of the conditional MLE and introduced the CUSUM test for parameter change based on the estimates and residuals. Additionally, Kim et al. [26] focused on a robust estimation method for BP-INGARCH models using the minimum density power divergence estimator.

The limitation of the BP-INGARCH model is that it can only handle positive cross-correlation between two components and is not competent for cross-correlations. A new BP-INGARCH model was proposed by Cui and Zhu [27] allowing for negative cross-correlation. Cui and Zhu [27] enabled an alternative definition of the BP distribution, just denoted as BP with definition given in Table 4, i.e., the product of Poisson marginals and a multiplicative factor δ that can promote positive, zero, or negative cross-correlation.

Further, a class of flexible BP-INGARCH(1,1) model was introduced by Cui et al. [98]. This class of models cover three distributions determined by different special multiplicative factors, making the portrayal of dependence more flexible.

Definition 20. *A new class of BP-INGARCH(1,1) model with flexible multiplicative factor is proposed as follows:*

$$Y_t | F_{t-1} \sim GBP(\lambda_t^\top), \lambda_t = (\lambda_{t,1}, \lambda_{t,2})^\top = \omega + A\lambda_{t-1} + BY_{t-1}, \tag{26}$$

where $GBP(\cdot)$ stands for one of three distributions, denoted as $BPG(\lambda_t^\top, \rho)$, $BPF(\lambda_t^\top, \gamma)$ and $BPFGM(\lambda_t^\top, \sigma)$ in Table 4.

For correlated bivariate count time series data, it is worth mentioning that a new flexible bivariate conditional Poisson INGARCH model was recently proposed by Piancastelli et al. [99] to capture negative and positive cross-correlations as well. Although all of these models are flexible in terms of contemporaneous correlation, the explicit form of the correlation structure of Piancastelli et al. [99] is easier to assess. Piancastelli et al. [99] also provided a detailed comparison of these methods for bivariate count time series for reference.

The next review is no longer limited to bivariate INGARCH models, but multivariate INGARCH models. Multivariate count time series remains an area of research with a vast scope, both in terms of theoretical approaches and application-oriented research. Much of the existing methodological literature does not focus only on INGARCH models, but is more broadly applicable to various types of count time series models; see Fokianos [100] for some recent methodological developments including multivariate INGARCH models. Moreover, Fokianos et al. [101] introduced an overview of statistical analysis for some models for multivariate discrete-valued time series based on higher-order Markov chains, where several extensions are highlighted including non-stationarity, network autoregressions, conditional non-linear autoregressive models, robust estimation, random fields and spatio-temporal models. We only show the contribution made by the latest literature Lee et al. [102] because it was not included in Fokianos [100].

Definition 21. *Let $Y_t = (Y_{t1}, \cdots, Y_{tm})^\top, t \geq 1$, be the time series of counts taking values in $\mathcal{N}^m$, and*

$$p_i(y|\eta) = \exp\{\eta y - A_i(\eta)\} h_i(y), \ y \in \mathcal{N},$$

which stands for the probability mass function of one-parameter exponential family, wherein η is the natural parameter, $A_i(\eta)$ and $h_i(y)$ are known functions, and both A_i and $B_i = A_i'$ stands for the derivative of A_i, are strictly increasing. We then consider the following model.

$$Y_{ti} | F_{t-1} \sim p_i(y|\eta_{ti}), \ i = 1, \cdots, m,$$
$$M_t = E(Y_t | F_{t-1}) = f_\theta(M_{t-1}, Y_{t-1}),$$

where $B_i(\eta_{ti}) = M_{ti}$, f_θ is a non-negative function defined on $[0, \infty)^m \times \mathcal{N}^m$ depending on the parameter $\theta \in \Theta \subset \mathcal{R}^d$ for some $d = 1, 2, \cdots$, and $\eta_t = (\eta_{t1}, \cdots, \eta_{tm})^\top := B^{-1}(M_t) := (B_1^{-1}(M_{t1}), \cdots, B_m^{-1}(M_{tm}))^\top.$

Unlike other authors who have devoted more effort to specifying the joint distribution of multivariate time series and the marginal distributions of their components, Lee et al. [102] argued that the conditional mean equation forms the bulk of the modeling and that the specification of the underlying joint distribution is not a major concern. We think this premise is a reasonable presupposition. It is well-known that the INGARCH-type model is built based on two types of assumptions, i.e., assumptions of conditional distribution and dynamic structure. Then it is intuitively reasonable that the contemporaneous dependence of the multivariate count time series can be reflected in the INGARCH model in two ways: the joint distribution and the multivariate dynamic structure. This presupposition reduces the complexity of modeling. Specially, although each component of Y_t is modeled using a univariate INGARCH model in Definition 21, the dependence structure is imposed by the conditional mean process M_t.

Another emerging area is the development of INGARCH-type models applicable to time-varying network data, which can be considered as a special class of multivariate count time series. Therefore, a part of recent research has been devoted to establish INGARCH models and their methodologies applicable to time-varying discrete network data. Let $\{Y_t = (Y_{1t}, \cdots, Y_{Nt})\}$ be a network with N nodes. The structure of the network is completely described by the adjacency matrix $A = (a_{ij}) \in R^{N \times N}$, where $a_{ii} = 1$ for any $i = 1, \cdots, N$ and $a_{ij} = 1$ that means the presence of a directed edge from i to j, $a_{ij} = 0$ otherwise. In general, any time-varying discrete network data whose relationship can be modeled by an adjacency matrix can be considered as a multivariate time series.

In a similar vein to the development of univariate integer-valued time series, the Poisson distribution and linearity assumptions continue to be used instinctively in order to establish a universal framework. Armillotta and Fokianos [28] considered the Poisson network autoregressive (PNAR) models for count data with a non-random adjacency matrix. We show here only the simplest linear form of order 1:

Definition 22. *The PNAR(1) model is defined as:*

$$Y_{i,t}|F_{t-1} \sim Poisson(\lambda_{i,t}),$$

$$\lambda_{i,t} = \beta_0 + \beta_1 n_i^{-1} \sum_{j=1}^{N} a_{ij} Y_{j,t-1} + \beta_2 Y_{i,t-1}, \qquad (27)$$

where $n_i = \sum_{i \neq j} a_{ij}$ is the out-degree, i.e., the total number of nodes, which i has an edge with.

The PNAR(1) model reduces the inference complexity by incorporating network information into the dependence structure, where the response of each individual can be explained by its lagged values and the average effect of its neighbors in (27). Note that Equation (27) does not include information about the joint dependence structure of the PNAR(1) model. It is then convenient to rewrite (27) in vector form,

$$\mathbf{Y}_t = \mathbf{N}_t(\lambda_t), \quad \lambda_t = \vec{\beta}_0 + \mathbf{G} \mathbf{Y}_{t-1}, \qquad (28)$$

where $\vec{\beta}_0 = \beta_0 \mathbf{1}_N \in \mathcal{R}^N$, with $\mathbf{1} = (1, 1, \cdots, 1)^\top \in \mathcal{R}^N$ and the matrix $\mathbf{G} = \beta_1 \mathbf{W} + \beta_2 \mathbf{I}_N$, where $\mathbf{W} = diag\{n_1^{-1}, \cdots, n_N^{-1}\} \mathbf{A}$ is the row-normalized adjacency matrix, with $\mathbf{A} = (a_{ij})$, so $\mathbf{w}_i = (a_{ij}/n_i, j = 1, \cdots, N)^\top \in \mathcal{R}^N$ is the i-th row vector of the matrix $\mathbf{W}$, satisfying $\|\mathbf{W}\|_\infty = 1$, and $\mathbf{I}_N$ is the $N \times N$ identity matrix. $\{\mathbf{N}_t\}$ is a sequence of independent N-variate copula–Poisson processes. The main methodological contribution of Armillotta and Fokianos [28] was the study of the asymptotic properties of such models by employing L_p-near epoch dependence and α-mixing, rather than based on the assumption that i.i.d. on which the development of all network time series models discussed so far has strongly relied. Further, Armillotta et al. [29] reviewed some of the work by Armillotta and Fokianos [28] and provided a unified framework for the statistical analysis of both continuous and integer-valued data with a known adjacency matrix. Armillotta and Fokianos [28] also specified a log-linear PNAR model for the count processes, and another recent work Armillotta and Fokianos [30] was closely related to this, where a quasi-score linearity test for continuous and count network autoregressive models was developed.

It can be seen in (27) and (28) that the PNAR model assumes that all individuals are homogeneous and they share a common autoregressive coefficient. This is a somewhat detached assumption from reality. Therefore, Tao et al. [31] proposed the grouped PNAR, which divides individuals into different groups and describes heterogeneous node behavior with group-specific parameters. Compared to the original PNAR model, the constraints are relaxed while competently portraying the heterogeneity. Specially, all individuals could be classified into K groups in the setting of Tao et al. [31], and each group was characterized by a specific set of positive parameters $\theta_k = (\omega_k, \alpha_k, \rho_k, \beta_k)' \in \mathcal{R}^4$, for $1 \leq k \leq K$.

Definition 23. *A grouped PNAR model can be constructed as*

$$Y_{i,t}|F_{t-1} \sim Poisson(\lambda_{i,t}), \tag{29}$$

$$\lambda_{i,t} = \sum_{k=1}^{K} z_{ik} \left(\omega_k + \alpha_k Y_{i,t-1} + \rho_k d_i^{-1} \sum_{j \neq i} a_{ij} Y_{j,t-1} + \beta_k \lambda_{i,t-1} \right), \tag{30}$$

for each $i = 1, \cdots, N$, and $t \geq 1$. Following the PNAR model, the parameters $\omega_k, \alpha_k, \rho_k, \beta_k$ represent the group-specific baseline effect, regression coefficient on past observations, network effect, and regression coefficient on past intensity processes, respectively. Note that we assume the adjacency matrix A is asymmetric, which covers the special case of symmetric networks. To distinguish between groups, latent variable $z_{ik} \in \{0,1\}$ was defined for each object i, where $z_{ik} = 1$ if object i is from the k-th group, and $z_{ik} = 0$ otherwise. Assume $\{(z_{i1}, \cdots, z_{iK})', 1 \leq i \leq N\}$ is a sequence of i.i.d. multinomial random vectors with number of events $n = 1$ and probability $\gamma = (\gamma_1, \cdots, \gamma_K)'$. Here, γ_k represents the group proportion satisfying $\gamma_k \geq 0$ and $\sum_{k=1}^{K} \gamma_k = 1$.

Tao et al. [31] explored the accuracy of model estimation and prediction when the group labels were unknown and the number of group K is mis-specified, respectively. There is already a body of mature research on network data, but it is still a relatively emerging topic in the INGARCH field. Therefore, many further attempts to consider time-varying networks from the perspective of count time series are worthwhile, such as optimization of conditional distributions, nonlinear dynamic structures, and related hypothesis testing, or time-varying network data with upper bounds on the number of edges.

In addition to modeling and methodology, INGARCH is popular for application-oriented analysis of time-varying networks. For example, Agosto and Ahelegbey [103] used a financial network model to study the contagion effects between business sectors based on discrete data, and tested the conditional means (and volatilities) of default counts across economic sectors estimated by Poisson INGARCH and their dependence in shocks. Through an empirical analysis of corporate defaults in Italy over the period 1996–2018, a high degree of intersectoral vulnerability was concluded by Agosto and Ahelegbey [103], in particular at the onset of the global financial crisis in 2008 and in subsequent years. Such a wide range of application prospects is accompanied by a desire for theoretical development.

6. Discussion and Conclusions

The purpose of this section is to present some potentially useful research topics based on the methodology and applications reviewed in the previous sections.

(1). First, we focus on the softplus function $sp_c(\cdot)$ and soft-clipping function $sc_{\frac{c}{n}}(\cdot)$, which contribute to the modeling of unbounded counts and bounded counts allowing for negative auto-correlation, respectively. For the sake of clarity, $sc_{\frac{c}{n}}(\cdot)$ is used next as an example. As already mentioned, the advantage of $sc_{\frac{c}{n}}(\cdot)$ is that the support set is $\mathcal{R}$ and is nearly linear on $(0, n]$, which allows the parameters in the dynamic structure of the BINGARCH model not to be restricted to positive numbers. This is certainly an excellent innovation, and one that seems worth exploring further. The images of $sc_{\frac{c}{n}}(x)$ corresponding to different parameters c or n are reported in Figures 3 and 4. It is obvious that the slope of $sc_{\frac{c}{n}}(x)$ is small when $x < 0$ and tends to zero as x decreases. Moreover, the parameter c has a small moderating effect on this tendency, while n even has almost no effect. This insensitivity to negative values may lead to concerns that the corresponding INGARCH models do not fairly model positively and negatively correlated data.

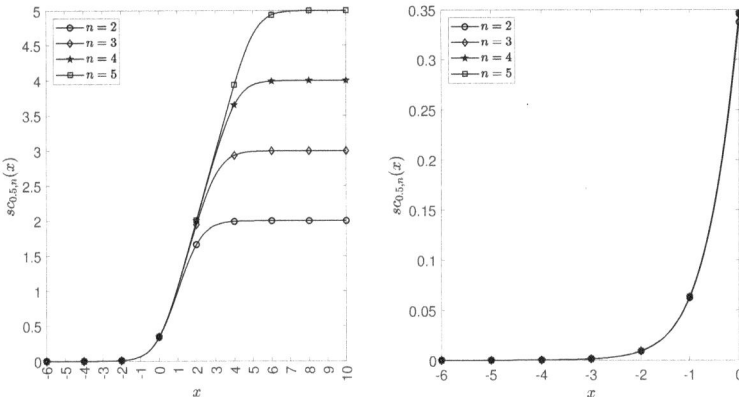

Figure 3. Plots of soft clipping functions $sc_{\frac{c}{n}}(x)$, $x \in [-6, 10]$ (**left**) and $sc_{\frac{c}{n}}(x)$, $x \in [-6, 0]$ (**right**) with $c = 0.5$ and $n = 2, 3, 4, 5$.

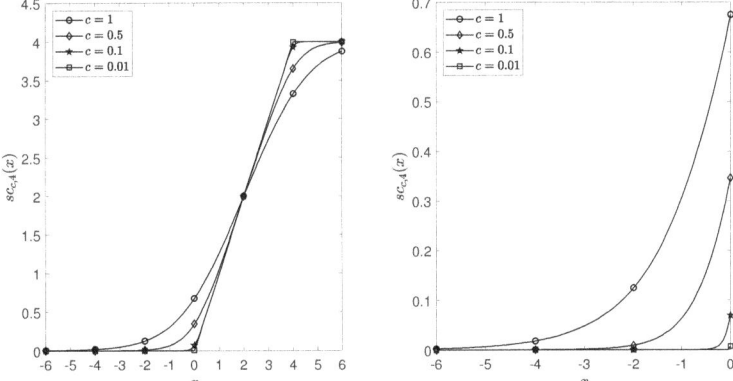

Figure 4. Plots of soft clipping functions $sc_{\frac{c}{n}}(x)$, $x \in [-6, 6]$ (**left**) and $sc_{\frac{c}{n}}(x)$, $x \in [-6, 0]$ (**right**) with $n = 4$ and $c = 1, 0.5, 0.1, 0.01$.

The reasons for this confusion can be summarized as follows. Although negative parameters are allowed to appear in the dynamic structure of the conditional expectation, such leniency seems to be somewhat offset by the fact that P_t (we use $Bin(n, p_t)$ as an example here) that were obtained from $sc_{\frac{c}{n}}(x)$ are concentrated around 0 when $x < 0$, while $sc_{\frac{c}{n}}(x)$ is approximately linear when $x > 0$. The $sc_{\frac{c}{n}}(x)$ does allow for the existence of negative correlation, but the extent of the negative correlation is to be explored further. However, to put it another way, it seems to us that $sc_{\frac{c}{n}}(x)$ provides a new perspective of modeling zero-inflated counts. In contrast to the previous idea of embodying the zero-inflated feature in a conditional distribution, it seems possible to assign this task to a dynamic structure through $sc_{\frac{c}{n}}(x)$.

(2). There is still a large demand for innovations dedicated to bounded count time series. We personally think that there are two feasible directions for exploration: one is the innovation of conditional distribution, and the other is to continue to deepen the research based on vector forms. From an application point of view, it is worth exploring to truncate existing distributions in a reasonable way, or to find distributions that themselves take values in bounded sets of integers. For the study of vector forms, where the theory is relatively well-developed, it is imperative to ease the estimation pressure.

(3). The emergence of new methodologies or research topics in other fields or in the broader field can stimulate the development of the INGARCH-type models. For example, Pedersen and Rahbek [104] presented the theory for testing for reduction of GARCHX type models with an exogenous covariate to standard GARCH type models. Many INGARCH-type models, including some of the recent literature mentioned earlier, are also focusing on covariates. Thus, the tests and methodologies proposed by Pedersen and Rahbek [104] can actually inspire existing INGARCH models to achieve tests of the reasonableness of introducing covariates. Similarly, refer also to Aknouche and Francq [105] and Debaly and Truquet [106].

Funding: Liu's work is supported by National Natural Science Foundation of China (No. 12201233) and Natural Science Foundation of Hubei Province. Zhu's work is supported by National Natural Science Foundation of China (No. 12271206), and Natural Science Foundation of Jilin Province (No. 20210101143JC), and Science and Technology Research Planning Project of Jilin Provincial Department of Education (No. JJKH20231122KJ).

Institutional Review Board Statement: Not applicable.

Data Availability Statement: No new data were created in this review.

Acknowledgments: We thank two reviewers for their insightful and constructive comments that greatly improved the overall presentation.

Conflicts of Interest: The authors declare no conflict of interest.

References

1. Agosto, A.; Giudici, P. COVID-19 contagion and digital finance. *Digit. Financ.* **2020**, *2*, 159–167. [CrossRef]
2. Agosto, A.; Campmas, A.; Giudici, P.; Renda, A. Monitoring COVID-19 contagion growth. *Stat. Med.* **2021**, *40*, 4150–4160. [CrossRef]
3. Giudici, P.; Tarantino, B.; Roy, A. Bayesian time-varying autoregressive models of COVID-19 epidemics. *Biom. J.* **2023**, *8*, 2200054. [CrossRef]
4. Fokianos, K.; Rahbek, A.; Tjøstheim, D. Poisson autoregression. *J. Am. Stat. Assoc.* **2009**, *104*, 1430–1439. [CrossRef]
5. Weiß, C.H. *An Introduction to Discrete-Valued Time Series*; John Wiley & Sons: Chichester, UK, 2018.
6. Davis, R.A.; Fokianos, K.; Holan, S.H.; Joe, H.; Livsey, J.; Lund, R.; Pipiras, V.; Ravishanker, N. Count time series: A methodological review. *J. Am. Stat. Assoc.* **2021**, *535*, 1533–1547. [CrossRef]
7. Gorgi, P. Beta–negative binomial auto-regressions for modelling integer-valued time series with extreme observations. *J. R. Stat. Soc. Ser. B* **2020**, *82*, 1325–1347. [CrossRef]
8. Qian, L.; Zhu, F. A flexible model for time series of counts with overdispersion or underdispersion, zero-inflation and heavy-tailedness. *Commun. Math. Stat.* **2023**. [CrossRef]
9. Silva, R.B.; Souza, W.B. Flexible and robust mixed Poisson INGARCH models. *J. Time Ser. Anal.* **2019**, *40*, 788–814. [CrossRef]
10. Souza, W.B.; Piancastelli, L.S.C.; Fokianos, K.; Ombao, H. Time-varying dispersion integer-valued GARCH models. *arXiv* **2022**, arXiv:2208.02024.
11. Weiß, C.H.; Zhu, F.; Hoshiyar, A. Softplus INGARCH models. *Stat. Sin.* **2022**, *32*, 1099–1120. [CrossRef]
12. Roy, A. Time-varying auto-regressive models for count time-series. *Electron. J. Stat.* **2021**, *15*, 2905–2938. [CrossRef]
13. Liu, M.; Li, Q.; Zhu, F. Threshold negative binomial autoregressive model. *Statistics* **2019**, *53*, 1–25. [CrossRef]
14. Chen, C.W.S.; Khamthong, K. Bayesian modelling of nonlinear negative binomial integer-valued GARCHX models. *Stat. Model.* **2019**, *20*, 537–561. [CrossRef]
15. Liu, M.; Li, Q.; Zhu, F. Self-excited hysteretic negative binomial autoregression. *AStA Adv. Stat. Anal.* **2020**, *104*, 385–415. [CrossRef]
16. Chen, C.W.S.; Lee, S.; Khamthong, K. Bayesian inference of nonlinear hysteretic integer-valued GARCH models for disease counts. *Comput. Stat.* **2021**, *36*, 261–281. [CrossRef]
17. Doukhan, P.; Neumann, M.H. Absolute regularity of semi-contractive GARCH-type processes. *J. Appl. Probab.* **2019**, *56*, 91–115. [CrossRef]
18. Hu, X. Volatility Estimation for Integer-Valued Financial Time Series. Doctoral Dissertation, Northwestern University: Evanston, IL, USA, 2016.
19. Hu, X.; Andrews, B. Integer-valued asymmetric GARCH modeling. *J. Time Ser. Anal.* **2021**, *42*, 737–751. [CrossRef]
20. Xu, Y.; Zhu, F. A new GJR-GARCH model for $\mathbb{Z}$-valued time series. *J. Time Ser. Anal.* **2022**, *43*, 490–500. [CrossRef]
21. Fokianos, K.; Truquet, L. On categorical time series models with covariates. *Stoch. Process. Their Appl.* **2019**, *129*, 3446–3462. [CrossRef]
22. Weiß, C.H.; Jahn, M. Soft-clipping INGARCH models for time series of bounded counts. *Stat. Model.* **2023**. [CrossRef]

23. Chen, H.; Li, Q.; Zhu, F. A new class of integer-valued GARCH models for time series of bounded counts with extra-binomial variation. *AStA Adv. Stat. Anal.* **2022**, *106*, 243–270. [CrossRef]
24. Chen, H.; Li, Q.; Zhu, F. A covariate-driven beta-binomial integer-valued GARCH model for bounded counts with an application. *Metrika* **2023**. [CrossRef]
25. Lee, Y.; Lee, S.; Tjøstheim, D. Asymptotic normality and parameter change test for bivariate Poisson INGARCH models. *Test* **2018**, *27*, 52–69. [CrossRef]
26. Kim, B.; Lee, S.; Kim, D. Robust estimation for bivariate Poisson INGARCH models. *Entropy* **2021**, *23*, 367. [CrossRef] [PubMed]
27. Cui, Y.; Zhu, F. A new bivariate integer-valued GARCH model allowing for negative cross-correlation. *Test* **2018**, *27*, 428–452. [CrossRef]
28. Armillotta, M.; Fokianos, K. Poisson network autoregression. *arXiv* **2022**, arXiv:2104.06296.
29. Armillotta, M.; Fokianos, K.; Krikidis, I. Generalized linear models network autoregression. In *Network Science*; Ribeiro, P., Silva, F., Mendes, J.F., Laureano, R., Eds.; Springer: Berlin/Heidelberg, Germany, 2022; pp. 112–125.
30. Armillotta, M.; Fokianos, K. Testing linearity for network autoregressive models. *arXiv* **2022**, arXiv:2202.03852.
31. Tao, Y.; Li, D.; Niu, X. Grouped network Poisson autoregressive model. *Stat. Sin.* **2023**. [CrossRef]
32. Barreto-Souza, W.; Simas, A.B. General mixed Poisson regression models with varying dispersion. *Stat. Comput.* **2016**, *26*, 1263–1280. [CrossRef]
33. Zhu, F. A negative binomial integer-valued GARCH model. *J. Time Ser. Anal.* **2011**, *32*, 54–67. [CrossRef]
34. Manaa, A.; Bentarzi, M. Periodic negative binomial INGARCH(1,1) model. *Commun. Stat.-Simul. Comput.* **2023**, *52*, 569–595. [CrossRef]
35. Almohaimeed, B.S. A negative binomial autoregression with a linear conditional variance-to-mean function. *Fractals* **2022**, *30*, 2240239. [CrossRef]
36. Almohaimeed, B.S. Asymptotic negative binomial quasi-likelihood inference for periodic integer-valued time series models. *Commun. Stat.-Theory Methods* **2023**. [CrossRef]
37. Cui, Y.; Wang, X. Conditional maximum likelihood estimation in negative binomial INGARCH processes with known number of successes when the true parameter is at the boundary of the parameter space. *Commun. Stat.-Theory Methods* **2019**, *48*, 3388–3401. [CrossRef]
38. Lee, S.; Kim, D. Multiple values-inflated time series of counts: Modeling and inference based on INGARCH scheme. *J. Stat. Comput. Simul.* **2023**, *93*, 1297–1317. [CrossRef]
39. Wechsung, M.; Neumann, M.H. Consistency of a nonparametric least squares estimator in integer-valued GARCH models. *J. Nonparametric Stat.* **2022**, *34*, 491–519. [CrossRef]
40. Aknouche, A.; Demmouche, N. Ergodicity conditions for a double mixed Poisson autoregression. *Stat. Probab. Lett.* **2019**, *147*, 6–11. [CrossRef]
41. Doukhan, P.; Fokianos, K.; Rynkiewicz, J. Mixtures of nonlinear Poisson autoregressions. *J. Time Ser. Anal.* **2021**, *42*, 107–135. [CrossRef]
42. Mao, H.; Zhu, F.; Cui, Y. A generalized mixture integer-valued GARCH model. *Stat. Methods Appl.* **2020**, *29*, 527–552. [CrossRef]
43. Diop, M.L.; Diop, A.; Diongue, A.K. A negative binomial mixture integer-valued GARCH model. *Afr. Stat.* **2018**, *3*, 1645–1666. [CrossRef]
44. Lee, J.; Hwang, E. A generalized regime-switching integer-valued GARCH(1,1) model and its volatility forecasting. *Commun. Stat. Appl. Methods* **2018**, *25*, 29–41.
45. Li, G.; Guan, B.; Li, W.K.; Yu, P.L.H. Hysteretic autoregressive time series models. *Biometrika* **2015**, *102*, 717–723. [CrossRef]
46. Aknouche, A.; Scotto, M. A multiplicative thinning-based integer-valued GARCH model. *J. Time Ser. Anal.* **2023**. [CrossRef]
47. Weiß, C.H.; Zhu, F. Multiplicative error models for count time series. *arXiv* **2022**, arXiv:2212.0583.
48. Sim, T.; Douc, R.; Rouef, F. General-order observation-driven models: Ergodicity and consistency of the maximum likelihood estimator. *Electron. J. Stat.* **2021**, *15*, 3349–3393. [CrossRef]
49. Tsamtsakiri, P.; Karlis, D. On Bayesian model selection for INGARCH models viatrans-dimensional Markov chain Monte Carlo methods. *Stat. Model.* **2023**, *23*, 81–98. [CrossRef]
50. Tian, Y.; Wang, D.; Wang, X. Order shrinkage and selection for the INGARCH(p,q) model. *Int. J. Biomath.* **2021**, *14*, 2150070. [CrossRef]
51. Su, B.; Zhu, F. Temporal aggregation and systematic sampling for INGARCH processes. *J. Stat. Plan. Inference* **2022**, *219*, 120–133. [CrossRef]
52. Neumann, M.H. Bootstrap for integer-valued GARCH(p,q) processes. *Stat. Neerl.* **2021**, *75*, 343–363. [CrossRef]
53. Doukhan, P.; Leucht, A.; Neumann, M.H. Mixing properties of non-stationary INGARCH(1,1) processes. *Bernoulli* **2022**, *28*, 663–688. [CrossRef]
54. Aknouche, A.; Francq, C. Stationarity and ergodicity of markov switching positive conditional mean models. *J. Time Ser. Anal.* **2022**, *43*, 436–459. [CrossRef]
55. Douc, R.; Roueff, F.; Sim, T. Necessary and sufficient conditions for the identifiability of observation-driven models. *J. Time Ser. Anal.* **2021**, *42*, 140–160. [CrossRef]
56. Aknouche, A.; Bendjeddou, S.; Touche, N. Negative binomial quasi-likelihood inference for general integer-valued time series models. *J. Time Ser. Anal.* **2018**, *39*, 192–211. [CrossRef]

57. Aknouche, A.; Francq, C. Two-stage weighted least squares estimator of the conditional mean of observation-driven time series models. *J. Econom.* **2023**. [CrossRef]
58. Xu, Y.; Li, Q.; Zhu, F. A modified multiplicative thinning-based INARCH model: Properties, saddlepoint maximum likelihood estimation and application. *Entropy* **2023**, *25*, 207. [CrossRef] [PubMed]
59. Li, Q.; Chen, H.; Zhu, F. Robust estimation for Poisson integer-valued GARCH models using a new hybrid loss. *J. Syst. Sci. Complex.* **2021**, *34*, 1578–1596. [CrossRef]
60. Xiong, L.; Zhu, F. Robust estimation for the one parameter exponential family integer-valued GARCH(1,1) models based on a modifed Tukey's biweight function. *Comput. Stat.* **2023**. [CrossRef]
61. Kim, B.; Lee, S. Robust estimation for general integer-valued time series models. *Ann. Inst. Stat. Math.* **2020**, *72*, 1371–1396. [CrossRef]
62. Xiong, L.; Zhu, F. Robust quasi-likelihood estimation for the negative binomial integer-valued GARCH(1,1) model with an application to transaction counts. *J. Stat. Plan. Inference* **2019**, *203*, 178–198. [CrossRef]
63. Xiong, L.; Zhu, F. Minimum density power divergence estimator for negative binomial integer-valued GARCH models. *Commun. Math. Stat.* **2022**, *10*, 233–261. [CrossRef]
64. Elsaied, H.; Fried, R. On robust estimation of negative binomial INARCH models. *Metron* **2021**, *79*, 137–158. [CrossRef]
65. Li, Q.; Zhu, F. Mean targeting estimator for the integer-valued GARCH(1, 1) model. *Stat. Pap.* **2020**, *61*, 659–679. [CrossRef]
66. Jo, M.; Lee, S. Mean targeting estimation for integer-valued time series with application to change point test. *Commun. Stat.-Theory Methods* **2022**, *16*, 5549–5565. [CrossRef]
67. Pei, J.; Zhu, F. Marginal likelihood estimation for the negative binomial INGARCH model. *Commun. Stat.-Simul. Comput.* **2023**. [CrossRef]
68. Lee, Y.; Lee, S. CUSUM test for general nonlinear integer-valued GARCH models: Comparison study. *Ann. Inst. Stat. Math.* **2019**, *71*, 1033–1057. [CrossRef]
69. Lee, S.; Kim, D.; Seok, S. Modeling and inference for counts time series based on zero-inflated exponential family INGARCH models. *J. Stat. Comput. Simul.* **2021**, *91*, 2227–2248. [CrossRef]
70. Lee, S. Residual-based CUSUM of squares test for Poisson integer-valued GARCH models. *J. Stat. Comput. Simul.* **2019**, *89*, 3182–3195. [CrossRef]
71. Vanli, O.A.; Giroux, R.; Ozguven, E.E.; Pignatiello, J.J. Monitoring of count data time series: Cumulative sum change detection in Poisson integer valued GARCH models. *Qual. Eng.* **2019**, *31*, 439–452. [CrossRef]
72. Weiß, C.H.; Testik, M.C. Monitoring count time series: Robustness to nonlinearity when linear models are utilized. *Qual. Reliab. Eng. Int.* **2022**, *38*, 4356–4371. [CrossRef]
73. Lee, S.; Kim, S. Recent progress in parameter change test for integer-valued time series models. *J. Korean Stat. Soc.* **2021**, *50*, 730–755. [CrossRef]
74. Kim, B.; Lee, S. Robust change point test for general integer-valued time series models based on density power divergence. *Entropy* **2020**, *22*, 493. [CrossRef]
75. Michel, J. The limiting distribution of a non-stationary integer valued GARCH(1,1) process. *J. Time Ser. Anal.* **2020**, *41*, 351–356. [CrossRef]
76. Gning, L.D.; Diop, A.; Diagne, M.L.; Tchuenche, J. Modelling COVID-19 in Senegal and China with count autoregressive models. *Model. Earth Syst. Environ.* **2022**, *8*, 5713–5721. [CrossRef] [PubMed]
77. Alzahrani, S.M. A log linear Poisson autoregressive model to understand COVID-19 dynamics in Saudi Arabia. *Beni-Suef Univ. J. Basic Appl. Sci.* **2022**, *11*, 118. [CrossRef] [PubMed]
78. Xu, X.; Chen, Y.; Chen, C.W.S.; Lin, X. Adaptive log-linear zero-inflated generalized Poisson autoregressive model with applications to crime counts. *Ann. Appl. Stat.* **2020**, *14*, 1493–1515. [CrossRef]
79. Chen, C.W.S.; Khamthong, K. Markov switching integer-valued generalized auto-regressive conditional heteroscedastic models for dengue counts. *J. R. Stat. Soc. Ser. C* **2019**, *68*, 963–983. [CrossRef]
80. Agosto, A.; Raffinetti, E. Validation of PARX models for default count prediction. *Front. Artif. Intell.* **2019**, *2*, 9. [CrossRef]
81. Jamaludin, A.R.; Yusof, F.; Suhartono, S. Modelling Asthma Cases using count analysis approach: Poisson INGARCH and negative binomial INGARCH. *Matematika* **2020**, *36*, 15–30. [CrossRef]
82. Algieri, B.; Leccadito, A. Extreme price moves: An INGARCH approach to model coexceedances in commodity markets. *Eur. Rev. Agric. Econ.* **2021**, *48*, 878–914. [CrossRef]
83. Aknouche, A.; Almohaimeed, B.S.; Dimitrakopoulos, D. Forecasting transaction counts with integer-valued GARCH models. *Stud. Nonlinear Dyn. Econom.* **2022**, *26*, 529–539. [CrossRef]
84. Berentsen, G.D.; Bulla, J.; Maruotti, A.; Stove, B. Modelling clusters of corporate defaults:Regime-switching models significantly reduce the contagion source. *J. R. Stat. Soc. Ser. C* **2022**, *71*, 698–722. [CrossRef]
85. Cerqueti, R.; D'Urso, P.; Giovanni, L.D.; Mattera, R.; Vitale, V. INGARCH-based fuzzy clustering of count time series with a football application. *Mach. Learn. Appl.* **2022**, *10*, 100417.
86. Chen, C.W.S.; Hsieh, Y.H.; Su, H.C.; Wu, J. Causality test of ambient fine particles and human influenza in Taiwan:Age group-specific disparity and geographic heterogeneity. *Environ. Int.* **2018**, *111*, 354–361. [CrossRef]
87. Kim, M. Network traffic prediction based on INGARCH model. *Wirel. Netw.* **2020**, *26*, 6189–6202. [CrossRef]
88. Anavatan, A.; Kayacan, E.Y. Investigation of femicide in Turkey: Modeling time series of counts. *Qual. Quant.* **2023**. [CrossRef]

89. Gonçalves, E.; Mendes-Lopes, N. Signed compound Poisson integer-valued GARCH processes. *Commun. Stat.-Theory Methods* **2022**, *49*, 5468–5492. [CrossRef]
90. Doukhan, P.; Khan, N.M.; Neumann, M.H. Mixing properties of integer-valued GARCH processes. *ALEA-Lat. Am. J. Probab. Math. Stat.* **2021**, *18*, 401–420. [CrossRef]
91. Alomani, G.A.; Alzaid, A.A.; Omair, M.A. A Skellam GARCH model. *Braz. J. Probab. Stat.* **2018**, *32*, 200–214. [CrossRef]
92. Cui, Y.; Li, Q.; Zhu, F. Modeling $\mathbb{Z}$-valued time series based on new versions of the Skellam INGARCH model. *Braz. J. Probab. Stat.* **2021**, *35*, 293–314. [CrossRef]
93. Glosten, L.; Jagannathan, R.; Runkle, D.E. On the relation between the expected value and the volatility of the nominal excess return on stocks. *J. Financ.* **1993**, *48*, 1779–1801. [CrossRef]
94. Chen, H.; Li, Q.; Zhu, F. Two classes of dynamic binomial integer-valued ARCH models. *Braz. J. Probab. Stat.* **2020**, *34*, 685–711. [CrossRef]
95. Chen, H. A new Soft-Clipping discrete beta GARCH model and its application on measles infection. *Stats* **2023**, *6*, 293–311. [CrossRef]
96. Liu, M.; Zhu, F.; Zhu, K. Modeling normalcy-dominant categorical time series: An application to air quality level. *J. Time Ser. Anal.* **2022**, *43*, 460–478. [CrossRef]
97. Liu, M.; Li, Q.; Zhu, F. A flexible categorical autoregression for modeling air quality level. *Stoch. Environ. Res. Risk Assess.* **2022**, *36*, 2835–2845. [CrossRef]
98. Cui, Y.; Li, Q.; Zhu, F. Flexible bivariate Poisson integer-valued GARCH model. *Ann. Inst. Stat. Math.* **2020**, *72*, 1449–1477. [CrossRef]
99. Piancastelli, L.S.C.; Souza, W.B.; Ombao, H. Flexible bivariate INGARCH process with a broad range of contemporaneous correlation. *J. Time Ser. Anal.* **2022**, *44*, 206–222. [CrossRef]
100. Fokianos, K. Multivariate count time series modelling. *Econom. Stat.* **2023**. [CrossRef]
101. Fokianos, K.; Fried, R.; Kharin, Y.; Voloshko, V. Statistical analysis of multivariate discrete-valued time series. *J. Multivar. Anal.* **2022**, *188*, 104805. [CrossRef]
102. Lee, S.; Kim, D.; Kim, B. Modeling and inference for multivariate time series of counts based on the INGARCH scheme. *Comput. Stat. Data Anal.* **2023**, *177*, 107579. [CrossRef]
103. Agosto, A.; Ahelegbey, D.F. Default count-based network models for credit contagion. *J. Oper. Res. Soc.* **2022**, *73*, 139–152. [CrossRef]
104. Pedersen, R.S.; Rahbek, A. Testing GARCH-X type models. *Econom. Theory* **2019**, *35*, 1012–1047. [CrossRef]
105. Aknouche, A.; Francq, C. Count and duration time series with equal conditional stochastic and mean orders. *Econom. Theory* **2021**, *37*, 248–280. [CrossRef]
106. Debaly, Z.M.; Truquet, L. Iterations of dependent random maps and exogeneity in nonlinear dynamics. *Econom. Theory* **2021**, *37*, 1135–1172. [CrossRef]

Disclaimer/Publisher's Note: The statements, opinions and data contained in all publications are solely those of the individual author(s) and contributor(s) and not of MDPI and/or the editor(s). MDPI and/or the editor(s) disclaim responsibility for any injury to people or property resulting from any ideas, methods, instructions or products referred to in the content.

Article

A Modified Multiplicative Thinning-Based INARCH Model: Properties, Saddlepoint Maximum Likelihood Estimation, and Application

Yue Xu [1], Qi Li [2] and Fukang Zhu [1,*]

[1] School of Mathematics, Jilin University, Changchun 130012, China
[2] College of Mathematics, Changchun Normal University, Changchun 130032, China
* Correspondence: fzhu@jlu.edu.cn

Abstract: In this article, we propose a modified multiplicative thinning-based integer-valued autoregressive conditional heteroscedasticity model and use the saddlepoint maximum likelihood estimation (SPMLE) method to estimate parameters. A simulation study is given to show a better performance of the SPMLE. The application of the real data, which is concerned with the number of tick changes by the minute of the euro to the British pound exchange rate, shows the superiority of our modified model and the SPMLE.

Keywords: INARCH model; saddlepoint approximation; thinning-based model; time series of counts

1. Introduction

In practice, we can often observe a series of integer-valued data that have their own distinguishing characteristics, and many models were proposed for modeling integer-valued time series, such as the integer-valued autoregressive (INAR) process introduced by McKenzie (1985) [1], and Al-Osh and Alzaid (1987) [2]; the integer-valued moving average process proposed by Al-Osh and Alzaid (1988) [3]; the integer-valued autoregressive moving-average model defined by McKenize (1988) [4]; and the integer-valued generalized autoregressive conditional heteroscedasticity (INGARCH) model proposed by Ferland et al. (2006) [5], among others. Here we focus on two kinds of the models above: one is the INAR process, which was introduced as a convenient way to transfer the usual autoregressive structure to a discrete-valued time series, and a p-order model, which is defined as follows:

$$X_t = \sum_{i=1}^{p} \alpha_i \circ X_{t-i} + \varepsilon_t,$$

where $\alpha_i \in [0,1)$ for $i = 1, \ldots, p$, and $\{\varepsilon_t\}$ is a sequence of independent and identically distributed (i.i.d.) non-negative integer-valued random variables with $E(\varepsilon_t) = \mu$ and $\text{Var}(\varepsilon_t) = \sigma_\varepsilon^2$. The binomial thinning operator $\circ$ is defined by Steutel and Van Harn (1979) [6] as:

$$\alpha \circ X = \sum_{i=1}^{X} Y_i, \text{ if } X > 0 \text{ and } 0 \text{ otherwise},$$

where Y_i are i.i.d. Bernoulli random variables, independent of X, with a success probability are defined by α. This model has been generalized by Qian and Zhu (2022) [7], and Huang et al. (2023) [8], among others.

The other is the INGARCH model which was proposed by Ferland et al. (2006) [5] to model the observations of integer-valued time series which exist heteroscedasticity; this INGARCH(p,q) model with a Poisson deviate is defined as:

$$X_t | \mathscr{F}_{t-1} : P(\lambda_t), \quad \lambda_t = \alpha_0 + \sum_{i=1}^{p} \alpha_i X_{t-i} + \sum_{j=1}^{q} \beta_j \lambda_{t-j},$$

where $\alpha_0 > 0, \alpha_i \geq 0, \beta_j \geq 0, i = 1, \ldots, p, j = 1, \ldots, q, p \geq 1, q \geq 0$, and $\mathscr{F}_{t-1}$ is the σ-field generated by $\{X_{t-1}, X_{t-2}, \ldots\}$. This model has been generalized by Hu (2016) [9], Liu et al. (2022) [10], and Weiß et al. (2022) [11], among others. Weiß (2018) [12] and Davis et al. (2021) [13] gave recent reviews. According to definitions of INAR and INGARCH models, we noticed that the INAR model is thinning-based, while the INGARCH model is specified by a conditional distribution with a time-varying mean depending on past observations. Combining the thinning-based stochastic equations and the INGARCH model, Aknouche and Scotto (2022) [14] proposed a multiplicative thinning-based INGARCH (MthINGARCH) model to model the integer-valued time series with high overdispersion and persistence. Furthermore, it fits well with heavy-tailed data regardless of the choice of innovation distribution and does not require recourse to complex random coefficient equations. The MthINGARCH model is denoted by:

$$\begin{cases} X_t = \lambda_t \varepsilon_t, \\ \lambda_t = 1 + \omega \circ m + \sum_{i=1}^{q} \alpha_i \circ X_{t-i} + \sum_{j=1}^{p} \beta_j \circ \lambda_{t-j}, \end{cases} \quad (1)$$

where the symbol $\circ$ stands for the binomial thinning operator, and $0 \leq \omega \leq 1, 0 \leq \alpha_i < 1$ and $0 \leq \beta_j < 1$ ($i = 1, \ldots, q, j = 1, \ldots, p$), m is a fixed positive integer number that was introduced for more flexibility. Since there is no explicit probability mass function for the series $\{X_t\}$, then the traditional maximum likelihood estimation (MLE) cannot be applied to estimate the parameters; therefore, Aknouche and Scotto (2022) [14] used a two-stage weighted least squares estimation instead.

Note that the probability mass function of the random variables cannot be given directly for the likelihood function in some cases; to solve this problem, saddlepoint approximation has been proposed. Daniel (1954) [15] introduced saddlepoint techniques into the statistical field, which have been extended by Field and Ronchetti (1990) [16], Jensen (1995) [17], and Butler (2007) [18]. Saddlepoint techniques have been used successfully in many applications because of the high accuracy with which they can approximate intractable densities and tail probabilities. Pedeli et al. (2015) [19] proposed an alternative approach based on the saddlepoint approximation to log-likelihood, and the saddlepoint maximum likelihood estimation (SPMLE) was used to estimate the parameters of the INAR model, which demonstrates the usefulness of this technique. Thus, through combining the MthINGARCH model of Aknouche and Scotto (2022) [14] and the saddlepoint approximation, we propose a modified multiplicative thinning-based INARCH model for modeling high overdispersion, before applying the saddlepoint method to the estimated parameters. Although the two-stage weighted least squares estimation could be used to estimate the parameters of our modified model, we still adopted the SPMLE as it was still expected to have a better performance than the two-stage weighted least squares estimation in practice. Here, we just consider the INARCH model instead of the INGARCH model because it is difficult and complex to give the conditional cumulant-generating function of random variables for the latter model when applying the saddlepoint approximation.

This article has the following structure. A modified multiplicative thinning-based INARCH model is given, alongside some related properties in Section 2. Moreover, we use the Poisson distribution and geometric distribution for innovations. Section 3 discusses the SPMLE and its asymptotic properties, then simulation studies for both models with SPMLE are also given. A real data example is analyzed with our modified models in Section 4, and comparisons with existing models are made. In-sample and out-of-sample forecasts are used to show the superiority of the SPMLE and our modified model. The conclusion is given in Section 5. Some details of SPMLE and proof of some theorems are presented in the Appendix A.

2. A Multiplicative Thinning-Based INARCH Model

Note that $\mathbb{N} = \{0, 1, 2, \ldots\}$ and $\mathbb{Z} = \{\ldots, -1, 0, 1, \ldots\}$ are the set of non-negative integers and integers, respectively. It can be supposed that $\{\varepsilon_t, t \in \mathbb{Z}\}$ is a sequence of i.i.d. random variables with a mean of one and finite variance of σ^2. The modified multiplicative thinning-based INARCH (denoted by the MthINARCH(q)) model, which we deal with in this paper, is defined by

$$X_t = \lambda_t \varepsilon_t, \quad \lambda_t = w \circ m + \sum_{i=1}^{q} \alpha_i \circ X_{t-i}, \tag{2}$$

where $0 < w \leq 1, 0 \leq \alpha_i < 1, i = 1, \ldots, q$, m is a fixed positive integer number. In real applications, we can set m as the upper integer part of the sample mean. It is assumed that the Bernoulli terms corresponding to the binomial variables $w \circ m$ and $\alpha_i \circ X_{t-i}$ are mutually independent and independent of the sequence $\{\varepsilon_t, t \in \mathbb{Z}\}$. The reason that we defined the new model in this way can be explained as follows. The additive term 1 in λ_t and in (1) is unnatural, and is posed to ensure $\lambda_t > 0$, but we can achieve this by adjusting the range of w; therefore, we adopted a simple version of λ_t in (2).

Now that we discuss the conditional mean and conditional variance of X_t. Note that $\mathscr{F}_{t-1}$ is the σ-field generated by $X_{t-1}, X_{t-2}, \ldots$. For $E(\varepsilon_t) = 1$, let $\mu_t := E(X_t|\mathscr{F}_{t-1}) = E(\lambda_t \varepsilon_t|\mathscr{F}_{t-1}) = E(\varepsilon_t)E(\lambda_t|\mathscr{F}_{t-1}) = E(\lambda_t|\mathscr{F}_{t-1}) = wm + \sum_{i=1}^{q} \alpha_i X_{t-i}$. Then we can obtain the conditional variance; first, let $v_t := \text{Var}(\lambda_t|\mathscr{F}_{t-1})$ and $\sigma_t^2 := \text{Var}(X_t|\mathscr{F}_{t-1})$. For $E(\varepsilon_t) = 1, \text{Var}(\varepsilon_t) = \sigma^2$, so $E(\varepsilon_t^2) = \sigma^2 + 1$. Therefore,

$$v_t := \text{Var}(\lambda_t|\mathscr{F}_{t-1}) = w(1-w)m + \sum_{i=1}^{q} \alpha_i(1-\alpha_i)X_{t-i},$$

$$\sigma_t^2 := \text{Var}(X_t|\mathscr{F}_{t-1}) = E(X_t^2|\mathscr{F}_{t-1}) - [E(X_t|\mathscr{F}_{t-1})]^2 = E(\lambda_t^2|\mathscr{F}_{t-1})E(\varepsilon_t^2) - \mu_t^2$$
$$= [\text{Var}(\lambda_t|\mathscr{F}_{t-1}) + (E(\lambda_t|\mathscr{F}_{t-1}))^2]E(\varepsilon_t^2) - \mu_t^2$$
$$= (\sigma^2 + 1)(v_t + \mu_t^2) - \mu_t^2 = (\sigma^2 + 1)v_t + \sigma^2 \mu_t^2.$$

Proposition 1. *The necessary and sufficient condition for the first-order stationarity of X_t defined in (2) is that all roots of $1 - \sum_{i=1}^{q} \alpha_i z^i = 0$ should lie outside the unit circle.*

Proposition 2. *The necessary and sufficient condition for the second-order stationarity of X_t defined in (2) is that $(\sigma^2 + 1) \sum_{i=1}^{q} \alpha_i^2 < 1$.*

Proofs of Propositions 1 and 2 are similar to the proofs of Theorems 2.1 and 2.2 in Aknouche and Scotto (2022) [14], so we omit the details.

For convenience, we need to specify the distribution of $\{\varepsilon_t\}$ in (2). First, we let $\varepsilon_t \sim P(1)$, then $E(\varepsilon_t) = \text{Var}(\varepsilon_t) = 1$, and this model is denoted by PMthINARCH(q). It is easy to obtain

$$\mu_t = wm + \sum_{i=1}^{q} \alpha_i X_{t-i}, \quad \sigma_t^2 = 2v_t + \mu_t^2.$$

Second, let $\varepsilon_t \sim Ge(p^*)$. The mean of ε_t is $(1 - p^*)/p^* = 1$, so we have $p^* = 0.5$ and the variance is $\text{Var}(\varepsilon_t) = 2$. This model is denoted by GMthINARCH(q), then we have

$$\mu_t = wm + \sum_{i=1}^{q} \alpha_i X_{t-i}, \quad \sigma_t^2 = 3v_t + 2\mu_t^2.$$

3. Parameter Estimation

In this section, we will consider the SPMLE and its asymptotic properties, and a simulation study will be conducted to assess the performance of this estimator.

3.1. Saddlepoint Maximum Likelihood Estimation

Let $\theta = (\omega, \alpha_1, \ldots, \alpha_q)^T$ be the unknown parameter vector. Note that according to the condition on ε_t, σ^2 is no longer an unknown parameter. The maximum likelihood estimator of θ was obtained by maximizing the conditional log-likelihood function

$$l(\theta) = \sum_{t=1}^{n} \log P(X_t = x_t | X_{t-1} = x_{t-1}, \ldots, X_{t-q} = x_{t-q}), \qquad (3)$$

giving $\hat{\theta} = \arg\max_\theta l(\theta)$. But the above procedure is challenging to implement because it is difficult to give the likelihood function due to the thinning operations.

Now we discuss the SPMLE. The conditional moment generating function of X_t is

$$E(e^{uX_t}|X_{t-1} = x_{t-1}, \ldots, X_{t-q} = x_{t-q}) = E(e^{u\lambda_t \varepsilon_t}|X_{t-1} = x_{t-1}, \ldots, X_{t-q} = x_{t-q})$$

$$= E(e^{u(\omega \circ m + \sum_{i=1}^{q} \alpha_i \circ X_{t-i})\varepsilon_t}|X_{t-1} = x_{t-1}, \ldots, X_{t-q} = x_{t-q})$$

$$= E(e^{u(\omega \circ m)\varepsilon_t}) \prod_{i=1}^{q} E(e^{u(\alpha_i \circ X_{t-i})\varepsilon_t}).$$

Remark 1. *Here we just consider the INARCH model instead of the INGARCH model because for the INGARCH model, the conditional cumulant-generating function of X_t should be given by $E(e^{uX_t}|X_{t-1} = x_{t-1}, \ldots, X_{t-q} = x_{t-q}) = E(e^{u(\omega \circ m + \sum_{i=1}^{q} \alpha_i \circ X_{t-i} + \sum_{j=1}^{p} \beta_j \circ \lambda_{t-i})\varepsilon_t}|X_{t-1} = x_{t-1}, \ldots, X_{t-q} = x_{t-q})$. Notice that X_t and λ_t are correlated, it is difficult and complex to show the conditional cumulant-generating function.*

Using the binomial theorem $(a+b)^n = \sum_{k=0}^{n} C_n^k a^{n-k} b^k$, we have

$$E(e^{u(\omega \circ m)\varepsilon_t}) = E\left[E(e^{u(\omega \circ m)\varepsilon_t}|\varepsilon_t)\right] = E(\omega e^{u\varepsilon_t} + (1-\omega))^m$$

$$= E\left[\sum_{r=0}^{m} C_m^r (1-\omega)^r \omega^{m-r} e^{u(m-r)\varepsilon_t}\right] = \sum_{r=0}^{m} C_m^r (1-\omega)^r \omega^{m-r} E(e^{u(m-r)\varepsilon_t}).$$

Similarly, we also have

$$E(e^{u(\alpha_i \circ X_{t-i})\varepsilon_t}) = \sum_{r=0}^{x_{t-i}} C_{x_{t-i}}^r (1-\alpha_i)^r \alpha_i^{x_{t-i}-r} E(e^{u(x_{t-i}-r)\varepsilon_t}).$$

Therefore, for the PMthINARCH(q) model, we have

$$E(e^{u(\omega \circ m)\varepsilon_t}) = \sum_{r=0}^{m} C_m^r (1-\omega)^r \omega^{m-r} e^{(e^{u(m-r)}-1)},$$

$$E(e^{u(\alpha_i \circ X_{t-i})\varepsilon_t}) = \sum_{r=0}^{x_{t-i}} C_{x_{t-i}}^r (1-\alpha_i)^r \alpha_i^{x_{t-i}-r} e^{(e^{u(x_{t-i}-r)}-1)},$$

while for the GMthINARCH(q) model, we have

$$E(e^{u(\omega \circ m)\varepsilon_t}) = \sum_{r=0}^{m} C_m^r (1-\omega)^r \omega^{m-r} \frac{1}{2 - e^{u(m-r)}},$$

$$E(e^{u(\alpha_i \circ X_{t-i})\varepsilon_t}) = \sum_{r=0}^{x_{t-i}} C_{x_{t-i}}^r (1-\alpha_i)^r \alpha_i^{x_{t-i}-r} \frac{1}{2 - e^{u(x_{t-i}-r)}}.$$

Thus the conditional cumulant-generating function of X_t is:

$$K_t(u) = \log[E(e^{uX_t}|X_{t-1} = x_{t-1}, \ldots, X_{t-q} = x_{t-q})] = \log E(e^{u(\omega \circ m)\varepsilon_t}) + \sum_{i=1}^{q} \log E(e^{u(\alpha_i \circ x_{t-i})\varepsilon_t}).$$

A highly accurate approximation to the conditional mass function of X_t at x_t is provided by the saddlepoint approximation:

$$\tilde{f}_{X_t|X_{t-1}=x_{t-1},\ldots,X_{t-q}=x_{t-q}}(x_t) = \left(2\pi K_t''(\tilde{u}_t)\right)^{-\frac{1}{2}} \exp\{K_t(\tilde{u}_t) - \tilde{u}_t x_t\}, \qquad (4)$$

where $\tilde{u}_t$ is the unique value of u which satisfies the saddlepoint equation $K_t'(u) = x_t$, with K_t' and K_t'' represent the first and second order derivatives of K_t with respect to u. Notice that it is difficult to solve the saddlepoint equation $K_t'(u) = x_t$ analytically; similar to that mentioned in Pedeli et al. (2015) [19], we can use the Newton–Raphson method to solve this equation.

The log-likelihood function (3) can be approximated by summing the logarithms of the corresponding density approximations (4), yielding:

$$\tilde{L}_n(\theta) = \sum_{t=1}^{n} \tilde{l}_t(\theta) := \sum_{t=1}^{n} \log \tilde{f}_{X_t|X_{t-1}=x_{t-1},\ldots,X_{t-q}=x_{t-q}}(x_t). \qquad (5)$$

The value θ maximizing this expression is called the saddlepoint maximum likelihood estimator (SPMLE).

3.2. Asymptotic Properties of the SPMLE

Now we discuss the asymptotic properties of the SPMLE. First we give the first-order Taylor expansion of $K_t'(u)$ at $u = 0$ yields,

$$K_t'(u) = K_t'(0) + uK_t''(0) + o(u) = \mu_t(\theta) + u\sigma_t^2(\theta) + o(u), \qquad (6)$$

where $\mu_t(\theta)$ and $\sigma_t^2(\theta)$ are the conditional mean and conditional variance of X_t. Notice that $\tilde{u}_t$ can be given by $K_t'(\tilde{u}_t) = x_t$, so with the Taylor series expansion of $K_t'(u)$ in (6), we have:

$$\tilde{u}_t = \frac{x_t - \mu_t(\theta)}{\sigma_t^2(\theta)} + o(1), \quad t = q+1, \ldots, n. \qquad (7)$$

Then, we can obtain the second-order Taylor expansion of $K_t(u)$ at $u = 0$, which is:

$$K_t(u) \approx uK_t'(0) + \frac{u^2}{2}K_t''(0) = u\mu_t(\theta) + \frac{u^2}{2}\sigma_t^2(\theta). \qquad (8)$$

Focusing on the exponent of the saddlepoint approximation (4), Equation (8) gives

$$K_t(u) - ux_t \approx u(\mu_t(\theta) - x_t) + \frac{u^2}{2}\sigma_t^2(\theta).$$

Then using Equation (7), we have

$$K_t(\tilde{u}_t) - \tilde{u}_t x_t \approx -\frac{[x_t - \mu_t(\theta)]^2}{2\sigma_t^2(\theta)}. \qquad (9)$$

Hence, we can derive from (8) and (9) that the first-order saddlepoint approximation to the conditional probability mass function is approximately:

$$\tilde{f}_{X_t|X_{t-1}=x_{t-1},\ldots,X_{t-q}=x_{t-q}}(x_t) = (2\pi K_t''(\tilde{u}_t))^{-\frac{1}{2}}$$

$$\times \exp\left[-\frac{(x_t - \omega m - \sum_{i=1}^q \alpha_i x_{t-i})^2}{2\left[(\sigma^2+1)(\omega(1-\omega)m + \sum_{i=1}^q \alpha_i(1-\alpha_i)x_{t-i}) + \sigma^2(\omega m + \sum_{i=1}^q \alpha_i x_{t-i})^2\right]}\right].$$

Therefore, $\tilde{L}_n(\theta) = \sum_{t=1}^n \tilde{l}_t(\theta) = \sum_{t=1}^n \log \tilde{f}_{X_t|X_{t-1}=x_{t-1},\ldots,X_{t-q}=x_{t-q}}(x_t)$ is the quasi-likelihood function for the estimation of θ. To establish the large-sample properties, we have

$$L_n(\theta) = \sum_{t=1}^n l_t(\theta) = \sum_{t=1}^n \log f_{X_t|X_{t-1}=x_{t-1},\ldots,X_{t-q}=x_{t-q}}(x_t),$$

which is the ergodic approximation of $\tilde{L}_n(\theta)$. The first and second derivatives of the quasi-likelihood function are given in the Appendix A. The strong convergence and asymptotic normality for the SPMLE $\hat{\theta}_n$ are established in the following theorems.

First of all, the assumptions for Theorems 1 and 2 are listed as follows.

Assumption 1. *The solution of the MthINARCH process is strictly stationary and ergodic.*

Assumption 2. *Θ is compact and $\theta_0 \in \mathring{\Theta}$, where $\mathring{\Theta}$ denotes the interior of Θ. For technical reasons, we assumed the lower and upper values of each component of parameters as $0 < \omega_L \leq \omega \leq \omega_U \leq 1$ and $0 \leq \alpha_L \leq \alpha_i \leq \alpha_U < 1, i = 1, \ldots, q$.*

Theorem 1. *Let $\hat{\theta}_n$ be a sequence of SPMLEs satisfying $\hat{\theta}_n = \arg\max_{\theta \in \Theta} \tilde{L}_n(\theta)$, then under Assumptions 1 and 2, $\hat{\theta}_n$ converges to θ_0 almost as surely, as $n \to \infty$.*

Theorem 2. *Under Assumptions 1 and 2, there exists a sequence of maximizers $\hat{\theta}_n$ of $\tilde{L}_n(\theta)$ such as that of $n \to \infty$,*

$$\sqrt{n}(\hat{\theta}_n - \theta_0) \xrightarrow{d} N(0, \Sigma^{-1}),$$

where

$$\Sigma = -E_{\theta_0}\left(\frac{\partial^2 l_t(\theta_0)}{\partial \theta \partial \theta^T}\right),$$

and Σ is positively definite.

3.3. Simulation Study

In this section, simulation studies of PMthINARCH(q) and GMthINARCH(q) models for finite sample size are given, where $q = 2$. Here, we used several combinations to show the performance of SPMLE, and the mean absolute deviation error (MADE) $\frac{1}{s}\sum_{j=1}^s |\hat{\theta}_j - \theta_j|$ was used as the evaluation criterion; here, s is the number of replications. The sample size is $n = 100, 200, 500$, and the number of replications is $s = 200$. We used the following combinations of $(\omega, \alpha_1, \alpha_2)^T$ as the true values to generate the random sample: A1 = $(0.65, 0.4, 0.4)^T$, A2 = $(0.9, 0.5, 0.3)^T$ for the PMthINARCH(2) model, and B1 = $(0.8, 0.4, 0.4)^T$, B2 = $(0.65, 0.3, 0.5)^T$ for the GMthINARCH(2) model. Tables 1 and 2 show the results of these simulations. Notice that as the sample sizes become larger, the MADEs become smaller, and the estimates seem to be close to the true values. Therefore, the SPMLE performs well.

Table 1. Mean and MADE of estimates for PMthINARCH(2) model with SPMLE.

Model				ω	α_1	α_2
A1	$m = 3$	$n = 100$	Mean	0.6069	0.5356	0.3569
			MADE	0.3681	0.2866	0.2510
		$n = 200$	Mean	0.5722	0.5026	0.3952
			MADE	0.3557	0.2434	0.2243
		$n = 500$	Mean	0.6436	0.4888	0.4140
			MADE	0.2724	0.1287	0.1005
A2	$m = 8$	$n = 100$	Mean	0.7782	0.5076	0.4750
			MADE	0.2533	0.2752	0.3007
		$n = 200$	Mean	0.7935	0.5161	0.4701
			MADE	0.2318	0.2527	0.2778
		$n = 500$	Mean	0.8703	0.5170	0.4677
			MADE	0.1752	0.2155	0.2390

Table 2. Mean and MADE of estimates for GMthINARCH(2) model with SPMLE.

Model				ω	α_1	α_2
B1	$m = 4$	$n = 100$	Mean	0.7821	0.2930	0.2870
			MADE	0.1195	0.1499	0.1766
		$n = 200$	Mean	0.8190	0.3611	0.3185
			MADE	0.1121	0.1425	0.1640
		$n = 500$	Mean	0.8456	0.3610	0.3298
			MADE	0.0601	0.1331	0.1414
B2	$m = 6$	$n = 100$	Mean	0.4718	0.2086	0.3811
			MADE	0.1965	0.1466	0.1463
		$n = 200$	Mean	0.5186	0.2632	0.5080
			MADE	0.1607	0.1198	0.1412
		$n = 500$	Mean	0.5468	0.2874	0.4896
			MADE	0.1415	0.1050	0.0770

4. A Real Example

Here, we considered the number of tick changes by the minute of the euro to the British pound exchange rate (ExRate for short) on December 12th from 9.00 a.m. to 9.00 p.m. The dataset is available at the website http://www.histdata.com/ (accessed on 17 January 2023). The series comprises of 720 observations with a sample mean of 13.2153 and a sample variance of 224.2498. Obviously, the sample variance is much larger than the sample mean, which shows high overdispersion, and this high overdispersion can also be seen in Figure 1a. Figure 1b,c are the plots of the autocorrelation function (ACF), and the partial autocorrelation function (PACF) means that we know the tick changes are correlated.

We analyzed the data using the PMthINARCH(3) model, GMthINARCH(3) model, Poisson INAR(3) (here denoted by PINAR(3) for short) model, and the INARCH(3) model. The Poisson INAR model is mentioned in Pedeli et al. (2015) [19], and the SPMLE was used to estimate the parameters. Here, the innovations in the PINAR model were assumed to be Poisson with a mean of one. The INARCH model with a Poisson deviate was proposed by Ferland et al. (2006) [5], and the MLE was used to estimate the parameters. According to Aknouche and Scotto (2022) [14], in real applications, we can set m as the upper integer part of the sample mean. Here the sample mean is 13.2153, so m is set to the value of 14. Table 3 gives the estimates of SPMLE and the values of the Akaike information criterion (AIC) and Bayesian information criterion (BIC). According to Table 3, it is clear to see that the values of AIC and BIC of PMthINARCH(3) and GMthINARCH(3) are smaller than those of the PINAR(3) and INARCH(3) models, the values of AIC and BIC of INARCH(3) are smaller than those of the PINAR(3) model. Moreover, the values of AIC and BIC of PMthINARCH(3) are smaller than those of GMthINARCH(3). In summary, the INARCH model performed better than the PINAR model; meanwhile, the PMthINARCH model and GMthINARCH model performed better than the PINAR model and INARCH model.

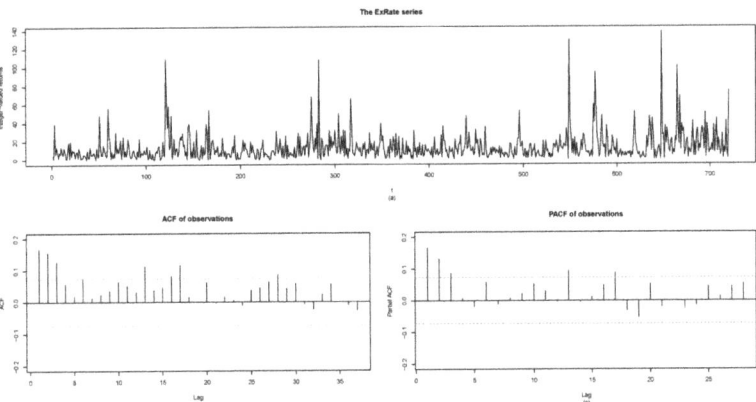

Figure 1. (a) The plot of integer-valued series of ExRate. (b) The plot of ACF of observations. (c) The plot of PACF of observations.

Table 3. Estimation results: AIC and BIC values for PMthINARCH(3), GMthINARCH(3), PINAR(3) and INARCH(3) models.

PMthINARCH(3)	ω 0.3242	α_1 0.5214	α_2 0.1945	α_3 0.0842	AIC 1395.296	BIC 1413.613
GMthINARCH(3)	ω 0.4904	α_1 0.2532	α_2 0.2155	α_3 0.2392	AIC 1402.472	BIC 1420.789
PINAR(3)		α_1 0.1335	α_2 0.4116	α_3 0.3901	AIC 1572.806	BIC 1586.544
INARCH(3)	ω 8.5670	α_1 0.1140	α_2 0.1379	α_3 0.1009	AIC 1524.638	BIC 1542.955

According to Aknouche and Scotto (2022) [14], the two-stage weighted least squares estimation (2SWLSE) was used to estimate the parameters of the MthINGARCH model. Therefore, to compare the performance of 2SWLSE and SPMLE, and the performance of PMthINARCH, GMthINARCH, and PINAR models, to consider the in-sample and out-of-sample forecasts of these two estimation methods and the three models above, respectively. First, we considered the in-sample forecast. We used all of the observations to estimate the model, and then we could forecast the last 10 observations 711–720, the last 15 observations 706–720, and the last 20 observations 701–720; these three-time horizons of in-sample forecast are denoted by C1, C2, and C3, respectively. Similar to the in-sample forecast process, we also considered the out-of-sample forecast and divided all the observations into three-time horizons: the first one was 1–710 and 711–720, the second one was 1–705 and 706–720, and the third one was 1–700 and 701–720, which are denoted by D1, D2, and D3, respectively.

Here we illustrate the performance of the considered models by comparing the MADEs of each forecast. The MADEs of in-sample forecasts and out-of-sample forecasts for three models with SPMLE are shown in Table 4. The MADEs of the in-sample forecasts and out-of-sample forecasts for the PMthINARCH model with 2SWLSE and SPMLE are shown in Table 5, and the in-sample forecasts and out-of-sample forecasts for the GMthI-NARCH model with 2SWLSE and SPMLE are shown in Table 6. According to Table 4, the MADEs of PMthINARCH(3) and GMthINARCH(3) are smaller than those of PINAR(3), Tables 5 and 6 show that the MADEs of PMthINARCH(3) and GMthINARCH(3) of SPMLE are smaller than those of 2SWLSE; meanwhile, in these three Tables, the MADEs of in-sample forecasts were smaller than those of out-of-sample forecasts. In summary, the PMthINARCH model and GMthINARCH model were superior to the PINAR model in modeling this real data set, and the PMthINARCH model performed better than the GMthI-NARCH model. Meanwhile, the performance of SPMLE was better than 2SWLSE for MthINARCH models.

Table 4. MADEs of in-sample forecasts and out-of-sample forecasts for PMthINARCH(3), GMthINARCH(3), and PINAR(3) models with SPMLE.

Methods of Forecast		PMthINARCH	GMthINARCH	PINAR
In-sample	C1	15.30	16.80	17.40
	C2	15.87	17.67	18.40
	C3	16.65	20.70	21.90
Out-of-sample	D1	17.50	17.70	22.50
	D2	19.47	19.80	23.80
	D3	20.50	25.25	27.50

Table 5. MADEs of in-sample forecasts and out-of-sample forecasts for PMthINARCH(3) model with SPMLE and 2SWLSE.

Methods of Forecast		SPMLE	2SWLSE
In-sample	C1	15.30	16.20
	C2	15.87	17.20
	C3	16.65	18.55
Out-of-sample	D1	17.50	18.60
	D2	19.47	21.67
	D3	20.50	22.70

Table 6. MADEs of in-sample forecasts and out-of-sample forecasts for GMthINARCH(3) model with SPMLE and 2SWLSE.

Methods of Forecast		SPMLE	2SWLSE
In-sample	C1	16.80	17.20
	C2	17.67	18.07
	C3	20.70	21.05
Out-of-sample	D1	17.70	19.90
	D2	19.80	22.87
	D3	25.25	26.50

5. Conclusions

In this paper, we modified a multiplicative thinning-based INARCH model. The probability mass function of random variables is provided by saddlepoint approximation. We used the SPMLE to estimate the parameters and obtain the asymptotic distribution of the SPMLE. Moreover, to show the superiority of the MthINARCH models and the

SPMLE, we used the PMthINARCH(q) process and GMthINARCH(q) process for discussion and comparison. The SPMLE performs well in the simulation studies. A real dataset indicates that the PMthINARCH model and the GMthINARCH model are able to describe the overdispersed integer-valued data, and the real data example leads to a superior performance of the MthINARCH models compared with the PINAR and INARCH models. In addition, the results also show a superior performance of SPMLE compared with 2SWLSE.

For further discussion, more research is needed for some aspects. Here we used the Poisson distribution and geometric distribution for ε_t; however, we could use the negative binomial distribution or some zero-inflated distributions as well. Moreover, we just considered the INARCH model, so the corresponding INGARCH model should be considered as well.

Author Contributions: Conceptualization, F.Z.; methodology, Y.X.; software, Y.X. and Q.L.; validation, Y.X. and Q.L.; formal analysis, Y.X. and Q.L.; investigation, Y.X. and F.Z.; resources, Q.L.; data curation, Y.X. and Q.L.; writing—original draft preparation, Y.X., Q.L. and F.Z.; writing—review and editing, Y.X., Q.L. and F.Z.; visualization, Y.X.; supervision, F.Z.; project administration, F.Z.; funding acquisition, Q.L. and F.Z. All authors have read and agreed to the published version of the manuscript.

Funding: Li's work is supported by the National Natural Science Foundation of China (No. 12201069), the Natural Science Foundation of Jilin Province (No. 20210101160JC), the Science and Technology Research Project of Education Bureau of Jilin Province (No. JJKH20220820KJ), and Natural Science Foundation Projects of CCNU (CSJJ2022006ZK). Zhu's work is supported by the National Natural Science Foundation of China (No. 12271206) and the Natural Science Foundation of Jilin Province (No. 20210101143JC).

Data Availability Statement: The dataset is available at the website http://www.histdata.com/ (accessed on 17 January 2023).

Acknowledgments: The authors are very grateful to three reviewers for their constructive suggestions and comments, leading to a substantial improvement in the presentation and contents.

Conflicts of Interest: The authors declare no conflict of interest.

Appendix A

Appendix A.1. Details of SPMLE

Here, we give the derivatives of $K_t(u)$ mentioned in Section 3.1 of PMthINARCH(q) and GMthINARCH(q). Now we give $K'_t(u)$ and $K''_t(u)$ of PMthINARCH(q). In Section 3.1, we have

$$K_t(u) = \log E(e^{u(\omega \circ m)\varepsilon_t}) + \sum_{i=1}^{q} \log E(e^{u(\alpha_i \circ x_{t-i})\varepsilon_t}) = \log a_1 + \sum_{i=1}^{q} \log b_1,$$

so the derivatives of $K_t(u)$ are given by

$$K'_t(u) = \frac{c_1}{a_1} + \sum_{i=1}^{q} \frac{d_1}{b_1}, \quad K''_t(u) = \frac{e_1 a_1 - c_1^2}{a_1^2} + \sum_{i=1}^{q} \frac{f_1 b_1 - d_1^2}{b_1^2},$$

where

$$a_1 = \sum_{r=0}^{m} C_m^r (1-\omega)^r \omega^{m-r} e^{e^{u(m-r)} - 1},$$

$$b_1 = \sum_{r=0}^{x_{t-i}} C_{x_{t-i}}^r (1-\alpha_i)^r \alpha_i^{x_{t-i}-r} e^{e^{u(x_{t-i}-r)} - 1},$$

$$c_1 = \sum_{r=0}^{m} C_m^r (1-\omega)^r \omega^{m-r} e^{u(m-r)} e^{e^{u(m-r)} - 1},$$

$$d_1 = \sum_{r=0}^{x_{t-i}} C_{x_{t-i}}^r (1-\alpha_i)^r \alpha_i^{x_{t-i}-r} e^{u(x_{t-i}-r)} e^{e^{u(x_{t-i}-r)}-1},$$

$$e_1 = \sum_{r=0}^{m} C_m^r (1-\omega)^r \omega^{m-r} e^{u(m-r)} (m-r)^2 e^{e^{u(m-r)}-1} [1+e^{u(m-r)}],$$

$$f_1 = \sum_{r=0}^{x_{t-i}} C_{x_{t-i}}^r (1-\alpha_i)^r \alpha_i^{x_{t-i}-r} (x_{t-i}-r)^2 e^{u(x_{t-i}-r)} e^{e^{u(x_{t-i}-r)}-1} [1+e^{u(x_{t-i}-r)}].$$

Then we give $K_t'(u)$ and $K_t''(u)$ of GMthINARCH(q). In Section 3.1, we have

$$K_t(u) = \log E(e^{u(\omega \circ m)\varepsilon_t}) + \sum_{i=1}^{q} \log E(e^{u(\alpha_i \circ x_{t-i})\varepsilon_t}) = \log a_2 + \sum_{i=1}^{q} \log b_2,$$

so the derivatives of $K_t(u)$ are given by

$$K_t'(u) = \frac{c_2}{a_2} + \sum_{t=1}^{q} \frac{d_2}{b_2}, \quad K_t''(u) = \frac{e_2 a_2 - c_2^2}{a_2^2} + \sum_{t=1}^{q} \frac{f_2 b_2 - d_2^2}{b_2^2},$$

where

$$a_2 = \sum_{r=0}^{m} C_m^r (1-\omega)^r \omega^{m-r} \frac{1}{2-(2-e^{u(m-r)})},$$

$$b_2 = \sum_{r=0}^{x_{t-i}} C_{x_{t-i}}^r (1-\alpha_i)^r \alpha_i^{x_{t-i}-r} \frac{1}{2-(2-e^{u(x_{t-i}-r)})},$$

$$c_2 = \frac{1}{4} \sum_{r=0}^{m} C_m^r (1-\omega)^r \omega^{m-r} (m-r) \frac{e^{u(m-r)}}{[1-(1-\frac{1}{2}e^{u(m-r)})]^2},$$

$$d_2 = \frac{1}{4} \sum_{r=0}^{x_{t-i}} C_{x_{t-i}}^r (1-\alpha_i)^r \alpha_i^{x_{t-i}-r} (x_{t-i}-r) \frac{e^{u(x_{t-i}-r)}}{[1-(1-\frac{1}{2}e^{u(x_{t-i}-r)})]^2},$$

$$e_2 = \frac{1}{4} \sum_{r=0}^{m} C_m^r (1-\omega)^r \omega^{m-r} (m-r)^2 e^{u(m-r)} \frac{1+\frac{1}{2}e^{u(m-r)}}{[1-(1-\frac{1}{2}e^{u(m-r)})]^3},$$

$$f_2 = \frac{1}{4} \sum_{r=0}^{x_{t-i}} C_{x_{t-i}}^r (1-\alpha_i)^r \alpha_i^{x_{t-i}-r} (x_{t-i}-r)^2 e^{u(x_{t-i}-r)} \frac{1+\frac{1}{2}e^{u(x_{t-i}-r)}}{[1-(1-\frac{1}{2}e^{u(x_{t-i}-r)})]^3}.$$

Appendix A.2. Derivatives of the Quasi-Likelihood Function

The conditional log-quasi-likelihood function $l_t(\theta)$ is continuous on Θ: for $1 \leq t \leq n$,

$$\frac{\partial l_t(\theta)}{\partial \theta} = m_1 \frac{\partial \mu_t(\theta)}{\partial \theta} + m_2 \frac{\partial \sigma_t^2(\theta)}{\partial \theta},$$

$$\frac{\partial^2 l_t(\theta)}{\partial \theta \partial \theta^T} = (m_1 - m_3) \frac{\partial^2 \mu_t(\theta)}{\partial \theta \partial \theta^T} - 2m_1 m_3 \frac{\partial \mu_t(\theta)}{\partial \theta} \frac{\partial \sigma_t^2(\theta)}{\partial \theta^T} + \left(m_2 + \frac{m_3^2}{2} - m_1^2 m_3\right) \frac{\partial^2 \sigma_t^2(\theta)}{\partial \theta \partial \theta^T},$$

where

$$m_1 = \frac{X_t - \mu_t(\theta)}{\sigma_t^2(\theta)}, \quad m_2 = \frac{(X_t - \mu_t(\theta))^2 - \sigma_t^2(\theta)}{2\sigma_t^4(\theta)}, \quad m_3 = \frac{1}{\sigma_t^2(\theta)}.$$

Then the first and second derivatives of $\mu_t(\theta)$ and $\sigma_t^2(\theta)$ can be easily expressed by

$$\frac{\partial \mu_t(\theta)}{\partial \omega} = m, \quad \frac{\partial \mu_t(\theta)}{\partial \alpha_i} = X_{t-i},$$

$$\frac{\partial \sigma_t^2(\theta)}{\partial \omega} = (\sigma^2 + 1)(m - 2\omega m) + 2\sigma^2 \left(m^2 \omega + m \sum_{i=1}^{q} \alpha_i X_{t-i}\right),$$

$$\frac{\partial \sigma_t^2(\theta)}{\partial \alpha_i} = (\sigma^2 + 1)(X_{t-i} - 2\alpha_i X_{t-i}) + 2\sigma^2(m\omega X_{t-i} + \alpha_i X_{t-i}^2),$$

$$\frac{\partial^2 \mu_t(\theta)}{\partial \omega^2} = 0, \quad \frac{\partial^2 \mu_t(\theta)}{\partial \alpha_i^2} = 1, \quad \frac{\partial^2 \mu_t(\theta)}{\partial \omega \alpha_i} = 0,$$

$$\frac{\partial^2 \sigma_t^2(\theta)}{\partial \omega^2} = -2m(\sigma^2 + 1) + 2m^2\sigma^2, \quad \frac{\partial^2 \sigma_t^2(\theta)}{\partial \alpha_i^2} = -2X_{t-i}(\sigma^2 + 1) + 2X_{t-i}^2 \sigma^2,$$

$$\frac{\partial^2 \sigma_t^2(\theta)}{\partial \omega \alpha_i} = 2m\sigma^2 X_{t-i}.$$

Appendix A.3. Proof of Theorem 1

The techniques used here are mainly based on Francq and Zakoïan (2004) [20]. We will establish the following intermediate results:

(i) $\lim_{n\to\infty} \sup_{\theta\in\Theta} \left|\frac{1}{n}(L_n(\theta) - \tilde{L}_n(\theta))\right| = 0$ a.s.
(ii) $E(l_t(\theta))$ is continuous in θ.
(iii) It exists $t \in \mathbb{Z}$ such that $\sigma_t^2(\theta) = \sigma_t^2(\theta_0)$ a.s., then $\Rightarrow \theta = \theta_0$.
(iv) Any $\theta \neq \theta_0$ has a neighbourhood $V(\theta)$ such that

$$\limsup_{n\to\infty} \sup_{\theta^* \in V_k(\theta) \cap \Theta} \frac{1}{n}\tilde{L}_n(\theta^*) > E_{\theta_0} l_1(\theta_0) \text{ a.s.}$$

First we prove (i). Let $a_t := \sup_{\theta\in\Theta} |\tilde{\mu}_t(\theta) - \mu_t(\theta)|$, $b_t := \sup_{\theta\in\Theta} |\tilde{\sigma}_t^2(\theta) - \sigma_t^2(\theta)|$. Standard arguments from Corollary 2.2 in Aknouche and Francq (2023) [21] show that $a_t(1 + X_t + \sup_{\theta\in\Theta} \mu_t(\theta)) \to 0, a.s.$ and $b_t(1 + X_t^2 + \sup_{\theta\in\Theta} \mu_t^2(\theta)) \to 0, a.s., t \to \infty$, so we obtain the inequality

$$\sup_{\theta\in\Theta} \left|\frac{1}{n}(L_n(\theta) - \tilde{L}_n(\theta))\right| = \sup_{\theta\in\Theta} \left|\frac{1}{2n}\sum_{t=1}^{n} \log \frac{\tilde{\sigma}_t^2(\theta)}{\sigma_t^2(\theta)} + \left(\frac{(x_t - \tilde{\mu}_t)^2}{\tilde{\sigma}_t^2} - \frac{(x_t - \mu_t(\theta))^2}{\sigma_t^2}\right)\right|$$

$$\leq \sup_{\theta\in\Theta} \left|\frac{1}{2n}\sum_{t=1}^{n} \frac{\tilde{\sigma}_t^2(\theta) - \sigma_t^2(\theta)}{\sigma_t^2(\theta)} + \left(\frac{(x_t - \tilde{\mu}_t(\theta))^2}{\tilde{\sigma}_t^2(\theta)} - \frac{(x_t - \mu_t(\theta))^2}{\sigma_t^2}\right)\right|$$

$$\leq \sup_{\theta\in\Theta} \frac{1}{2n}\sum_{t=1}^{n} \frac{|\tilde{\sigma}_t^2(\theta) - \sigma_t^2(\theta)|}{\sigma_t^2(\theta)} + \frac{|\tilde{\mu}_t(\theta) - \mu_t(\theta)||\mu_t(\theta) + \tilde{\mu}_t(\theta) - 2X_t|}{\tilde{\sigma}_t^2(\theta)}$$

$$+ \frac{|\tilde{\sigma}_t^2(\theta) - \sigma_t^2(\theta)||X_t - \mu_t(\theta)|^2}{\sigma_t^2(\theta)\tilde{\sigma}_t^2(\theta)}$$

$$\leq \frac{1}{2n}\sum_{t=1}^{n} \frac{2}{\sigma_t^2(\theta)} a_t(1 + X_t + \sup_{\theta\in\Theta}\mu_t(\theta)) + \frac{1 + \tilde{\sigma}_t^2(\theta)}{\sigma_t^2(\theta)\tilde{\sigma}_t^2(\theta)} c_t(1 + X_t^2 + \sup_{\theta\in\Theta}\mu_t^2(\theta)).$$

The a.s. limit holds because of the Cesàro lemma.

We prove (ii) now. For any $\theta \in \Theta$, let $V_\eta(\theta) = B(\theta, \eta)$ be an open ball centered at θ with radius η,

$$|l_t(\tilde{\theta}) - l_t(\theta)| \leq |\sigma_t^2(\tilde{\theta}) - \sigma_t^2(\theta)| \left|\frac{X_t^2 + \mu_t^2(\theta) + \sigma_t^2(\tilde{\theta})}{\sigma_t^2(\theta)\sigma_t^2(\tilde{\theta})}\right| + \frac{|\mu_t(\tilde{\theta}) - \mu_t(\theta)||\mu_t(\theta) + \mu_t(\tilde{\theta}) - 2X_t|}{\sigma_t^2(\tilde{\theta})}.$$

Then

$$E\left(\sup_{\theta\in\tilde{V}_\eta(\theta)} |l_t(\tilde{\theta}) - l_t(\theta)|\right) \leq \|\sigma_t^2(\tilde{\theta}) - \sigma_t^2(\theta)\|_2 \left\|\frac{X_t^2 + \mu_t^2(\theta) + \sigma_t^2(\tilde{\theta})}{\sigma_t^2(\theta)\sigma_t^2(\tilde{\theta})}\right\|_2$$

$$+ \frac{\|\mu_t(\tilde{\theta}) - \mu_t(\theta)\|_2 \|\mu_t(\theta) + \mu_t(\tilde{\theta}) - 2X_t\|_2}{\sigma_t^2(\tilde{\theta})} \to 0, \text{ as } \eta \to 0.$$

Next, we check (iii). By Jensen's inequality, we have

$$E[l_t(\theta) - l_t(\theta_0)] = E\left[E\left(\frac{1}{2}\log\frac{\sigma_t^2(\theta_0)}{\sigma_t^2(\theta)} + \frac{(x_t - \mu_t(\theta_0))^2}{2\sigma_t^2(\theta_0)} - \frac{(x_t - \mu_t(\theta))^2}{2\sigma_t^2(\theta)}\bigg|\mathscr{F}_{t-1}\right)\right]$$
$$\leq E\left[\log E\left(\frac{\sigma_t^2(\theta_0)}{\sigma_t^2(\theta)}\bigg|\mathscr{F}_{t-1}\right)\right]$$
$$= E(\log(1)) = 0.$$

The equality holds if $\frac{\sigma_t^2(\theta_0)}{\sigma_t^2(\theta)} = 1$ a.s. $\mathscr{F}_{t-1}$, i.e. $\theta = \theta_0$.

Then the proof of (iv) is similar to that in the Supplementary Material A.4 in Xu and Zhu (2022) [22]. Here we omit the details.

Appendix A.4. Proof of the Positive Definiteness of Σ

Here, we prove the positive definiteness of Σ. By definition of positive definiteness, we need to prove for any $\xi = (\xi_0, \xi_1, \ldots, \xi_q)^T \in \mathbb{R}^{q+1}$, if $\xi^T \Sigma \xi = 0$, then $\xi = 0$.

$$\xi^T \Sigma \xi = \xi^T E\left[\frac{1}{2\sigma_t^4(\theta_0)}\frac{\partial \sigma_t^2(\theta_0)}{\partial \theta}\frac{\partial \sigma_t^2(\theta_0)}{\partial \theta^T} + \frac{1}{\sigma_t^2(\theta_0)}\frac{\partial \mu_t(\theta_0)}{\partial \theta}\frac{\partial \mu_t(\theta_0)}{\partial \theta^T}\right]\xi$$
$$= E\left[\frac{1}{2\sigma_t^4(\theta_0)}(\xi^T\frac{\partial \sigma_t^2(\theta_0)}{\partial \theta})^2 + \frac{1}{\sigma_t^2(\theta_0)}(\xi^T\frac{\partial \mu_t(\theta_0)}{\partial \theta})^2\right].$$

Suppose the left-hand side is 0, then under Assumption 1, the expectation in the right-hand side is 0 for any $t \in \mathbb{Z}$. Because $\sigma_t^2(\theta_0) > 0$, this expectation is always greater than or equal to 0. It equals 0 only when $\xi^T\frac{\partial \sigma_t^2(\theta_0)}{\partial \theta} = 0$ and $\xi^T\frac{\partial \mu_t(\theta_0)}{\partial \theta} = 0$ almost surely.

Thus, $\xi^T \Sigma \xi = 0$ yields $\xi^T\frac{\partial \sigma_t^2(\theta_0)}{\partial \theta} = 0$ and $\xi^T\frac{\partial \mu_t(\theta_0)}{\partial \theta} = 0$ a.s. for $t \in \mathbb{Z}$, and vice versa.

Using vector form of $\frac{\partial \sigma_t^2(\theta_0)}{\partial \theta}$, we have

$$\xi_a^T\frac{\partial \sigma_t^2(\theta_0)}{\partial \theta} = \xi^T \begin{pmatrix} (\sigma^2+1)(m-2\omega m) + 2\sigma^2(\omega m^2 + m\sum_{i=1}^q \alpha_i X_{t-i}) \\ (\sigma^2+1)(X_{t-1} - 2\alpha_1 X_{t-1}) + 2\sigma^2(\omega m X_{t-1} + \alpha_1 X_{t-1}^2) \\ \vdots \\ (\sigma^2+1)(X_{t-q} - 2\alpha_q X_{t-q}) + 2\sigma^2(\omega m X_{t-q} + \alpha_q X_{t-q}^2) \end{pmatrix}.$$

Suppose the left-hand side is 0 almost surely, then the right-hand side is also 0 almost surely, which can be written as

$$\xi_0(\sigma^2+1)(m-2\omega m) + 2\sigma^2 \xi_0(\omega m^2 + m\sum_{i=1}^q \alpha_i X_{t-i})$$
$$+ \xi_1(\sigma^2+1)(X_{t-1} - 2\alpha_1 X_{t-1}) + 2\sigma^2 \xi_1(\omega m X_{t-1} + \alpha_1 X_{t-1}^2) + M_{t-2} = 0 \text{ a.s.,}$$

where

$$M_{t-2} = \sum_{k=2}^p \xi_k\left[(\sigma^2+1)(X_{t-k} - 2\alpha_k X_{t-k}) + 2\sigma^2(\omega m X_{t-k} + \alpha_k X_{t-k}^2)\right].$$

So the coefficients of the above equation must satisfy

$$\xi_i(\sigma^2+1) = 0, \quad 2\sigma^2 \xi_i = 0, \quad i = 0, \ldots, q.$$

For $\sigma^2 > 0$, we must have $\xi_i = 0, i = 0, \ldots, q$. Thus, $\xi = (\xi_0, \xi_1, \ldots, \xi_q)^T = 0$, which completes the proof of the positive definiteness of Σ.

Appendix A.5. Lemmas for the Proof of Theorem 2

Similar to the proof of Theorem 1.2 in Hu (2016) [9], we give some related lemmas for the proof of Theorem 2. According to the derivatives of the quasi-likelihood function, we have

$$\frac{\partial \mu_t(\theta)}{\partial \omega} = m,$$

$$\frac{\partial \sigma_t^2(\theta)}{\partial \omega} = (\sigma^2 + 1)(m - 2\omega m) + 2\sigma^2 \left(m^2 \omega + m \sum_{i=1}^{q} \alpha_i X_{t-i} \right),$$

$$\leq (\sigma^2 + 1)m(1 - 2\omega_L) + 2\sigma^2 \left(m^2 \omega_U + m \sum_{i=1}^{q} \alpha_U X_{t-i} \right),$$

thus, $E(\frac{\partial \mu_t(\theta)}{\partial \omega})^2 < \infty$ and $E(\frac{\partial \sigma_t^2(\theta)}{\partial \omega})^2 < \infty$. Likewise for the other terms of parameters.

Lemma A1. *Under Assumptions 1 and 2, when $n \to \infty$,*

$$\frac{1}{\sqrt{n}} \sum_{t=1}^{n} \frac{\partial \tilde{l}_t(\theta_0)}{\partial \theta_i} \xrightarrow{d} N(0, \Sigma), \quad \frac{1}{n} \sum_{t=1}^{n} \frac{\partial^2 \tilde{l}_t(\theta_0)}{\partial \theta_i \partial \theta_j} \xrightarrow{P} -\Sigma.$$

Proof of Lemma A1. First, we show that

$$n^{-1/2} \sum_{t=1}^{n} \left| \frac{\partial l_t(\theta_0)}{\partial \theta_i} - \frac{\partial \tilde{l}_t(\theta_0)}{\partial \theta_i} \right| \xrightarrow{P} 0, \quad n^{-1} \sum_{t=1}^{n} \left| \frac{\partial^2 l_t(\theta_0)}{\partial \theta_i \partial \theta_j} - \frac{\partial^2 \tilde{l}_t(\theta_0)}{\partial \theta_i \partial \theta_j} \right| \xrightarrow{P} 0.$$

Notice that $\tilde{\mu}_t(\theta)$ and $\tilde{\sigma}_t^2(\theta)$ are stationary approximations of $\mu_t(\theta)$ and $\sigma_t^2(\theta)$, since X_t is stationary and ergodic, using arguments similar to Proposition 2.1.1 in Straumann (2005) [23], for fixed $\theta \in \Theta$, $\tilde{\mu}_t(\theta)$ and $\tilde{\sigma}_t^2(\theta)$, $\mu_t(\theta)$ and $\sigma_t^2(\theta)$ are also stationary and ergodic. Hence, similar to the proof of Lemma A2 in Hu and Andrews (2021) [24], it is easy to have

$$n^{-1/2} \sum_{t=1}^{n} \left| \frac{\partial l_t(\theta_0)}{\partial \theta_i} - \frac{\partial \tilde{l}_t(\theta_0)}{\partial \theta_i} \right| \xrightarrow{P} 0, \quad n^{-1} \sum_{t=1}^{n} \left| \frac{\partial^2 l_t(\theta_0)}{\partial \theta_i \partial \theta_j} - \frac{\partial^2 \tilde{l}_t(\theta_0)}{\partial \theta_i \partial \theta_j} \right| \xrightarrow{P} 0.$$

Therefore, it suffices to show that

$$\frac{1}{\sqrt{n}} \sum_{t=1}^{n} \frac{\partial l_t(\theta_0)}{\partial \theta} \xrightarrow{d} N(0, \Sigma), \quad \frac{1}{n} \sum_{t=1}^{n} \frac{\partial^2 l_t(\theta_0)}{\partial \theta \partial \theta^T} \xrightarrow{P} -\Sigma.$$

First, we should guarantee that

$$E_{\theta_0} \left\| \frac{\partial l_t(\theta_0)}{\partial \theta} \frac{\partial l_t(\theta_0)}{\partial \theta^T} \right\| < \infty, \quad E_{\theta_0} \left\| \frac{\partial^2 l_t(\theta_0)}{\partial \theta \partial \theta^T} \right\| < \infty. \tag{A1}$$

Now we prove the first part of (A1).

$$E_{\theta_0} \left(\frac{\partial l_t(\theta_0)}{\partial \omega} \right)^2 = E_{\theta_0} \left[\frac{1}{2\sigma_t^4(\theta_0)} \left(\frac{\partial \sigma_t^2(\theta_0)}{\partial \omega} \right)^2 + \frac{1}{\sigma_t^2(\theta_0)} \left(\frac{\partial \mu_t(\theta_0)}{\partial \omega} \right)^2 \right] < \infty.$$

Similarly, we can prove other terms, thus, the first part of (A1) holds. The proof of the second part of (A1) is similar, here we omit the details.

Under (A1), $\left\{\frac{\partial l_t(\theta_0)}{\partial \theta}\right\}$ is a martingale difference sequence with respect to $\{\mathcal{F}_t\}$, it follows that at $\theta = \theta_0$, $E_{\theta_0}\left(\frac{\partial l_t(\theta_0)}{\partial \theta}\Big|\mathcal{F}_{t-1}\right) = 0$, so $E_{\theta_0}\left(\frac{\partial l_t(\theta_0)}{\partial \theta}\right) = 0$. Moreover, we have shown that $\Sigma = E_{\theta_0}\left(\frac{\partial l_t(\theta_0)}{\partial \theta}\frac{\partial l_t(\theta_0)}{\partial \theta^{\mathrm{T}}}\right)$ in Section 3.2. Hence $\frac{1}{\sqrt{n}}\sum_{t=1}^{n}\frac{\partial \tilde{l}_t(\theta_0)}{\partial \theta} \xrightarrow{d} N(0,\Sigma)$ holds by the central limit theorem for martingale difference sequence in Billingsley (1961). Similarly, we have $E_{\theta_0}\left(\frac{\partial l_t^2(\theta_0)}{\partial \theta \partial \theta^{\mathrm{T}}}\right) = -\Sigma$.

Under Assumption 1, $\frac{1}{n}\sum_{t=1}^{n}\frac{\partial^2 \tilde{l}_t(\theta_0)}{\partial \theta_i \partial \theta_j} \xrightarrow{P} -\Sigma$ follows from the ergodic theorem. Thus, Lemma A1 is proved. □

Before showing Lemma A2, we have

$$\tilde{T}_n(u) \equiv \tilde{l}_n\left(\theta_0 + \frac{u}{\sqrt{n}}\right) - \tilde{l}_n(\theta_0), \quad u \in \mathbb{R}^{q+1},$$

we use $\tilde{T}_n$ to derive the asymptotic distribution of $\hat{\theta}_n$.

For any $u \in \mathbb{R}^{q+1}$, the Taylor series expansion of $\tilde{T}_n(u)$ at θ_0 is

$$\tilde{T}_n(u) = \frac{1}{\sqrt{n}}\sum_{t=1}^{n}u^{\mathrm{T}}\frac{\partial \tilde{l}_t(\theta_0)}{\partial \theta} + \frac{1}{2n}\sum_{t=1}^{n}u^{\mathrm{T}}\frac{\partial^2 \tilde{l}_t(\theta_0)}{\partial \theta \partial \theta^{\mathrm{T}}}u + \frac{1}{2n}\sum_{t=1}^{n}u^{\mathrm{T}}\left[\frac{\partial^2 \tilde{l}_t(\theta^*)}{\partial \theta \partial \theta^{\mathrm{T}}} - \frac{\partial^2 \tilde{l}_t(\theta_0)}{\partial \theta \partial \theta^{\mathrm{T}}}\right]u, \quad \text{(A2)}$$

where $\theta^* = \theta_n^*(u)$ is on the line segment connecting θ_0 and $\theta_0 + \frac{u}{\sqrt{n}}$. For Euclidean distance $\|\cdot\|$ and any compact set $K \subset \mathbb{R}^{q+1}$, $\sup_{u \in K}\|\theta^* - \theta_0\| \to 0$, as $n \to \infty$.

Lemma A2. *Under Assumptions 1 and 2, when $n \to \infty$,*

$$\frac{1}{n}\sum_{t=1}^{n}\left[\frac{\partial^2 \tilde{l}_t(\theta^*)}{\partial \theta \partial \theta^{\mathrm{T}}} - \frac{\partial^2 \tilde{l}_t(\theta_0)}{\partial \theta \partial \theta^{\mathrm{T}}}\right] \xrightarrow{P} 0.$$

Proof. Similar to Lemma A1, for any $1 \leq i,j \leq q+1$,

$$\frac{1}{n}\sum_{t=1}^{n}\left\|\frac{\partial^2 l_t(\theta_0)}{\partial \theta_i \partial \theta_j} - \frac{\partial^2 \tilde{l}_t(\theta_0)}{\partial \theta_i \partial \theta_j}\right\| \xrightarrow{P} 0. \quad \text{(A3)}$$

Using arguments similar to the proof of Theorem 2.2 of Francq and Zakoïan (2004) [20], it suffices to show

$$\frac{1}{n}\sum_{t=1}^{n}\left[\frac{\partial^2 l_t(\theta^*)}{\partial \theta_i \partial \theta_j} - \frac{\partial^2 l_t(\theta_0)}{\partial \theta_i \partial \theta_j}\right] \xrightarrow{P} 0. \quad \text{(A4)}$$

By the Taylor series expansion, we have

$$\frac{1}{n}\sum_{t=1}^{n}\frac{\partial^2 l_t(\theta^*)}{\partial \theta_i \partial \theta_j} = \frac{1}{n}\sum_{t=1}^{n}\frac{\partial^2 l_t(\theta_0)}{\partial \theta_i \partial \theta_j} + \frac{1}{n}\sum_{t=1}^{n}\frac{\partial}{\partial \theta_k}\left(\frac{\partial^2 l_t(\theta^{**})}{\partial \theta_i \partial \theta_j}\right)(\theta^* - \theta_0),$$

here $\theta^{**} = \theta_n^{**}(u)$ is on the line segment connecting θ_0 and θ^*, such that for any u, we have $\|\theta^{**} - \theta_0\| \to 0$ a.s., $n \to \infty$.

From (A2), $\|\theta^* - \theta_0\| \to 0$ a.s, so

$$\frac{1}{n}\sum_{t=1}^{n}\frac{\partial}{\partial \theta_k}\left(\frac{\partial^2 l_t(\theta^{**})}{\partial \theta_i \partial \theta_j}\right)(\theta^* - \theta_0) \to 0, \text{ a.s.}$$

if
$$\limsup_{n\to\infty}\left\|\frac{1}{n}\sum_{t=1}^{n}\frac{\partial}{\partial\theta_k}\left(\frac{\partial^2 l_t(\theta^{**})}{\partial\theta_i\partial\theta_j}\right)\right\|<\infty, \text{ a.s.} \tag{A5}$$

Then we have
$$\frac{1}{n}\sum_{t=1}^{n}\frac{\partial^2 l_t(\theta^*)}{\partial\theta_i\partial\theta_j} \to \frac{1}{n}\sum_{t=1}^{n}\frac{\partial^2 l_t(\theta_0)}{\partial\theta_i\partial\theta_j} \text{ a.s.,}$$

so (A4) is proved.

Using arguments similar to the proof of Theorem 2.2 of Francq and Zakoïan (2004) [20], there exists a neighborhood $\nu(\theta_0)$, that

$$E_{\theta_0}\sup_{\theta\in\nu(\theta_0)\cap\Theta}\left\|\frac{\partial}{\partial\theta_k}\left(\frac{\partial^2 l_t(\theta)}{\partial\theta_i\partial\theta_j}\right)\right\|<\infty, \quad \sup_{\theta\in\nu(\theta_0)}\left\|\frac{1}{n}\sum_{t=1}^{n}\left[\frac{\partial^2 l_t(\theta)}{\partial\theta_i\partial\theta_j} - \frac{\partial^2 \tilde{l}_t(\theta)}{\partial\theta_i\partial\theta_j}\right]\right\| \xrightarrow{P} 0. \tag{A6}$$

Therefore, by the ergodic theorem, we have

$$\limsup_{n\to\infty}\left\|\frac{1}{n}\sum_{t=1}^{n}\frac{\partial}{\partial\theta_k}\left(\frac{\partial^2 l_t(\theta^{**})}{\partial\theta_i\partial\theta_j}\right)\right\| \leq \limsup_{n\to\infty}\frac{1}{n}\sum_{t=1}^{n}\sup_{\theta\in\nu(\theta_0)\cap\Theta}\left\|\frac{\partial}{\partial\theta_k}\left(\frac{\partial^2 l_t(\theta)}{\partial\theta_i\partial\theta_j}\right)\right\|$$
$$= E_{\theta_0}\sup_{\theta\in\nu(\theta_0)\cap\Theta}\left\|\frac{\partial}{\partial\theta_k}\left(\frac{\partial^2 l_t(\theta)}{\partial\theta_i\partial\theta_j}\right)\right\| < \infty,$$

so (A5) is proved.

In view of (A3), (A4) and (A6), we obtain Lemma A2. □

Lemma A3. *For any compact set $K \in \mathbb{R}^{q+1}$ and any $\varepsilon > 0$,*

$$\lim_{\sigma\to 0}\limsup_{n\to\infty}\mathbb{P}\left(\sup_{u,v\in K, \|u-v\|<\sigma}\left|\tilde{T}_n(u) - \tilde{T}_n(v)\right| \geq \varepsilon\right) = 0.$$

Proof. For any $\epsilon > 0$, by (A2) we have

$$\lim_{\delta\to 0}\limsup_{n\to\infty}\mathbb{P}\left(\sup_{u,v\in K, \|u-v\|<\delta}\left|\tilde{T}_n(u) - \tilde{T}_n(v)\right| \geq \varepsilon\right)$$
$$\leq \lim_{\delta\to 0}\limsup_{n\to\infty}\mathbb{P}\left(\sup_{u,v\in K, \|u-v\|<\delta}\left|\frac{1}{\sqrt{n}}\sum_{t=1}^{n}(u-v)^{\mathrm{T}}\frac{\partial \tilde{l}_t(\theta_0)}{\partial\theta}\right| \geq \frac{\epsilon}{3}\right)$$
$$+ \lim_{\delta\to 0}\limsup_{n\to\infty}\mathbb{P}\left(\sup_{u,v\in K, \|u-v\|<\delta}\left|\frac{1}{n}\left(\sum_{t=1}^{n}u^{\mathrm{T}}\frac{\partial^2 \tilde{l}_t(\theta_0)}{\partial\theta\partial\theta^{\mathrm{T}}}u - \sum_{t=1}^{n}v^{\mathrm{T}}\frac{\partial^2 \tilde{l}_t(\theta_0)}{\partial\theta\partial\theta^{\mathrm{T}}}v\right)\right| \geq \frac{2\epsilon}{3}\right)$$
$$+ \lim_{\delta\to 0}\limsup_{n\to\infty}\mathbb{P}\left\{\sup_{u,v\in K, \|u-v\|<\delta}\left|\frac{1}{n}\left[\sum_{t=1}^{n}u^{\mathrm{T}}\left(\frac{\partial^2 \tilde{l}_t(\theta^*)}{\partial\theta\partial\theta^{\mathrm{T}}} - \frac{\partial^2 \tilde{l}_t(\theta_0)}{\partial\theta\partial\theta^{\mathrm{T}}}\right)u \right.\right.\right.$$
$$\left.\left.\left. - \sum_{t=1}^{n}v^{\mathrm{T}}\left(\frac{\partial^2 \tilde{l}_t(\theta^*)}{\partial\theta\partial\theta^{\mathrm{T}}} - \frac{\partial^2 \tilde{l}_t(\theta_0)}{\partial\theta\partial\theta^{\mathrm{T}}}\right)v\right]\right| \geq \frac{2\epsilon}{3}\right\}.$$

Because of Lemmas A1 and A2, we have

$$\frac{1}{\sqrt{n}}\sum_{t=1}^{n}\frac{\partial \tilde{l}_t(\theta_0)}{\partial\theta} = O_p(1), \quad \frac{1}{n}\sum_{t=1}^{n}\frac{\partial^2 \tilde{l}_t(\theta_0)}{\partial\theta\partial\theta^{\mathrm{T}}} = O_p(1),$$

$$\frac{1}{n}\sum_{t=1}^{n}\left[\frac{\partial^2 \tilde{l}_t(\theta^*)}{\partial\theta\partial\theta^{\mathrm{T}}} - \frac{\partial^2 \tilde{l}_t(\theta_0)}{\partial\theta\partial\theta^{\mathrm{T}}}\right] = o_p(1),$$

where $O_p(1)$ and $o_p(1)$ for vector and matrix means $O_p(1)$ and $o_p(1)$ for every elements. By the compactness of K, we have

$$\lim_{\delta \to 0} \limsup_{n \to \infty} \mathbb{P}\left(\sup_{u,v \in K, \|u-v\| < \delta} \left| \frac{1}{\sqrt{n}} \sum_{t=1}^{n} (u-v)^{\mathrm{T}} \frac{\partial \tilde{l}_t(\theta_0)}{\partial \theta} \right| \geq \frac{\epsilon}{3} \right) = 0,$$

$$\lim_{\delta \to 0} \limsup_{n \to \infty} \mathbb{P}\left(\sup_{u,v \in K, \|u-v\| < \delta} \left| \frac{1}{n} \left(\sum_{t=1}^{n} u^{\mathrm{T}} \frac{\partial^2 \tilde{l}_t(\theta_0)}{\partial \theta \partial \theta^{\mathrm{T}}} u - \sum_{t=1}^{n} v^{\mathrm{T}} \frac{\partial^2 \tilde{l}_t(\theta_0)}{\partial \theta \partial \theta^{\mathrm{T}}} v \right) \right| \geq \frac{2\epsilon}{3} \right) = 0,$$

$$\lim_{\delta \to 0} \limsup_{n \to \infty} \mathbb{P}\left\{ \sup_{u,v \in K, \|u-v\| < \delta} \left\| \frac{1}{n} \left[\sum_{t=1}^{n} u^{\mathrm{T}} \left(\frac{\partial^2 \tilde{l}_t(\theta^*)}{\partial \theta \partial \theta^{\mathrm{T}}} - \frac{\partial^2 \tilde{l}_t(\theta_0)}{\partial \theta \partial \theta^{\mathrm{T}}} \right) u \right. \right. \right.$$
$$\left. \left. \left. - \sum_{t=1}^{n} v^{\mathrm{T}} \left(\frac{\partial^2 \tilde{l}_t(\theta^*)}{\partial \theta \partial \theta^{\mathrm{T}}} - \frac{\partial^2 \tilde{l}_t(\theta_0)}{\partial \theta \partial \theta^{\mathrm{T}}} \right) v \right] \right\| \geq \frac{2\epsilon}{3} \right\} = 0,$$

which completes our proof. □

Appendix A.6. Proof of Theorem 2

Proof. Let $T(u) = u^{\mathrm{T}} N(0, \Sigma) - \frac{1}{2} u^{\mathrm{T}} \Sigma u$, where N is a multivariate Gaussian random vector with mean 0 and covariance matrix Σ. By Lemmas A1 and A2, for any $u \in \mathbb{R}^{q+1}$ and $n \to \infty$, the finite dimensional distributions of $\tilde{T}_n$ converge to those of T: $\tilde{T}_n(u) \to T(u)$.

By Lemma A3, similar to Hu (2016) [9], $\tilde{T}_n(u)$ is tight on the continuous function space $C(K)$ for any compact set $K \in \mathbb{R}^{q+1}$. So by Theorem 7.1 in Billingsley (1999) [25], $\tilde{T}_n(\cdot) \to T(\cdot)$ on $C(K)$. From Appendix A.4 and Lemma A1, Σ is positive finite and invertible, meanwhile, $T(\cdot)$ is concave with the unique maximum $\Sigma^{-1} N(0, \Sigma) = N(0, \Sigma^{-1})$. $\tilde{T}_n(\cdot)$ is maximized at $u_{\max} = \sqrt{n}(\hat{\theta}_n - \theta_0)$. Thus, the result of Theorem 2 can be proved by the proof of Lemma 2.2 and Remark 1 in Davis et al. (1992) [26]. □

References

1. McKenzie, E. Some simple models for discrete variate time series. *Water Resour. Bull.* **1985**, *21*, 645–650. [CrossRef]
2. Al-Osh, M.A.; Alzaid, A.A. First-order integer-valued autoregressive (INAR(1)) process. *J. Time Ser. Anal.* **1987**, *8*, 261–275. [CrossRef]
3. Al-Osh, M.A.; Alzaid, A.A. Integer-valued moving average (INMA) process. *Stat. Pap.* **1988**, *29*, 281–300. [CrossRef]
4. McKenzie, E. Some ARMA models for dependent sequences of Poisson counts. *Adv. Appl. Probab.* **1988**, *20*, 822–835. [CrossRef]
5. Ferland, R.; Latour, A.; Oraichi, D. Integer-valued GARCH process. *J. Time Ser. Anal.* **2006**, *27*, 923–942. [CrossRef]
6. Steutel, F.W.; van Harn, K. Discrete analogues of self-decomposability and stability. *Ann. Probab.* **1979**, *7*, 893–899. [CrossRef]
7. Qian, L.; Zhu, F. A new minification integer-valued autoregressive process driven by explanatory variables. *Aust. N. Z. J. Stat.* **2022**, *64*, 478–494. [CrossRef]
8. Huang, J.; Zhu, F.; Deng, D. A mixed generalized Poisson INAR model with applications. *J. Stat. Comput. Simul.* **2023**, forthcoming. [CrossRef]
9. Hu, X. Volatility Estimation for Integer-Valued Financial Time Series. Ph.D. Thesis, Northwestern University, Evanston, IL, USA, 2016. [CrossRef]
10. Liu, M.; Zhu, F.; Zhu, K. Modeling normalcy-dominant ordinal time series: An application to air quality level. *J. Time Ser. Anal.* **2022**, *43*, 460–478. [CrossRef]
11. Weiß, C.H.; Zhu, F.; Hoshiyar, A. Softplus INGARCH models. *Stat. Sin.* **2022**, *32*, 1099–1120.
12. Weiß, C.H. *An Introduction to Discrete-Valued Time Series*; John Wiley & Sons: Chichester, UK, 2018. [CrossRef]
13. Davis, R.A.; Fokianos, K.; Holan, S.H.; Joe, H.; Livsey, J.; Lund, R.; Pipiras, V.; Ravishanker, N. Count time series: A methodological review. *J. Am. Stat. Assoc.* **2021**, *116*, 1533–1547. [CrossRef]
14. Aknouche, A.; Scotto, M. A multiplicative Thinning-Based Integer-Valued GARCH Model. Working Paper. 2022. Available online: https://mpra.ub.uni-muenchen.de/112475 (accessed on 17 January 2023).
15. Daniels, H.E. Saddlepoint approximations in statistics. *Ann. Math. Stat.* **1954**, *25*, 631–650. [CrossRef]
16. Field, C.; Ronchetti, E. Small sample asymptotics. In *Institute of Mathematical Statistics Lecture Notes—Monograph Series*; Institute of Mathematical Statistics: Hayward, CA, USA, 1990.
17. Jensen, J.L. *Saddlepoint Approximations*; Oxford University Press: Oxford, UK, 1995. [CrossRef]
18. Butler, R.W. *Saddlepoint Approximations with Applications*; Cambridge University Press: Cambridge, UK, 2007.

19. Pedeli, X.; Davison, A.C.; Fokianos, K. Likelihood estimation for the INAR(p) model by saddlepoint approximation. *J. Am. Stat. Assoc.* **2015**, *110*, 1229–1238.
20. Francq, C.; Zakoïan, J.M. Maximum likelihood estimation of pure GARCH and ARMA-GARCH processes. *Bernoulli* **2004**, *10*, 605–637.
21. Aknouche, A.; Francq, C. Two-stage weighted least squares estimator of the conditional mean of observation-driven time series models. *J. Econom.* **2023**, *forthcoming*. [CrossRef]
22. Xu, Y.; Zhu, F. A new GJR-GARCH model for $\mathbb{Z}$-valued time series. *J. Time Ser. Anal.* **2022**, *43*, 490–500. [CrossRef]
23. Straumann, D. *Estimation in Conditionally Heteroscedastic Time Series Models*; Springer: Berlin, Germany, 2005. [CrossRef]
24. Hu, X.; Andrews, B. Integer-valued asymmetric GARCH modeling. *J. Time Ser. Anal.* **2021**, *42*, 737–751. [CrossRef]
25. Billingsley, P. *Convergence of Probability Measures*, 2nd ed.; Wiley: New York, NY, USA, 1999.
26. Davis, R.A.; Knight, K.; Liu, J. M-estimation for autoregressions with infinite variance. *Stoch. Process. Their Appl.* **1992**, *40*, 145–180. [CrossRef]

Disclaimer/Publisher's Note: The statements, opinions and data contained in all publications are solely those of the individual author(s) and contributor(s) and not of MDPI and/or the editor(s). MDPI and/or the editor(s) disclaim responsibility for any injury to people or property resulting from any ideas, methods, instructions or products referred to in the content.

Article

State Space Modeling of Event Count Time Series

Sidratul Moontaha [1,*], Bert Arnrich [1] and Andreas Galka [2]

[1] Digital Health—Connected Healthcare, Hasso Plattner Institute, University of Potsdam, 14482 Potsdam, Germany
[2] Bundeswehr Technical Centre for Ships and Naval Weapons, Maritime Technology and Research (WTD 71), 24340 Eckernförde, Germany
* Correspondence: sidratul.moontaha@hpi.de

Abstract: This paper proposes a class of algorithms for analyzing event count time series, based on state space modeling and Kalman filtering. While the dynamics of the state space model is kept Gaussian and linear, a nonlinear observation function is chosen. In order to estimate the states, an iterated extended Kalman filter is employed. Positive definiteness of covariance matrices is preserved by a square-root filtering approach, based on singular value decomposition. Non-negativity of the count data is ensured, either by an exponential observation function, or by a newly introduced "affinely distorted hyperbolic" observation function. The resulting algorithm is applied to time series of the daily number of seizures of drug-resistant epilepsy patients. This number may depend on dosages of simultaneously administered anti-epileptic drugs, their superposition effects, delay effects, and unknown factors, making the objective analysis of seizure counts time series arduous. For the purpose of validation, a simulation study is performed. The results of the time series analysis by state space modeling, using the dosages of the anti-epileptic drugs as external control inputs, provide a decision on the effect of the drugs in a particular patient, with respect to reducing or increasing the number of seizures.

Keywords: n onlinear state space model; iterated extended Kalman filter; Bayesian filtering; count time series; singular value decomposition

---

Citation: Moontaha, S.; Arnrich, B.; Galka, A. State Space Modeling of Event Count Time Series. *Entropy* **2023**, *25*, 1372. https://doi.org/10.3390/e25101372

Academic Editors: Christian H. Weiss and Leonardo Ricci

Received: 24 July 2023
Revised: 2 September 2023
Accepted: 19 September 2023
Published: 23 September 2023

Copyright: © 2023 by the authors. Licensee MDPI, Basel, Switzerland. This article is an open access article distributed under the terms and conditions of the Creative Commons Attribution (CC BY) license (https://creativecommons.org/licenses/by/4.0/).

1. Introduction

Temporal data sequences resulting from counting discrete events over a given time interval represent a particular variant of time series called discrete-valued or event count time series. These count data arise in various fields, such as physics, epidemiology, finance, econometrics, or medicine. Much of the existing framework on time series analysis relies on the assumptions of Gaussianity and linearity. The non-negativity and integrity constraints inherent in count time series have led to the development of alternative modeling approaches that instead employ probability distributions of Poisson type. As prominent examples we mention generalized linear models (GLMs) [1], dynamic generalized linear models [2,3], integer-valued autoregressive (INAR) models [4], and integer-valued generalized autoregressive conditional heteroscedasticity (INGARCH) models [5–7].

However, in this paper we aim at modeling count time series within the linear Gaussian regime as long as possible, while introducing non-negativity and integrity only at the last stage of modeling, namely, at the stage of modeling the observation process. The suitable framework for this agenda is given by classical state space modeling [8]. In state space modeling, the dynamical process underlying the data is explicitly separated from the observation process, such that the former can be kept Gaussian and linear, which considerably simplifies state estimation and model identification. Further advantages of the state space approach are given by the possibility to discriminate between dynamical noise and observation noise, and by the option of straightforward generalization to the multivariate case.

It is also possible to include explanatory factors and external control inputs into state space models, and to incorporate conditional heteroscedasticity [9], with little additional effort.

Since the dynamics is kept Gaussian and linear, temporal correlations in the given data can be modeled by standard models from linear time series analysis, such as linear autoregressive moving average (ARMA) models [10]; ARMA models can be rewritten as (components of) linear state space models. Finally, non-negativity is modeled by employing a nonlinear observation function, while integrity is interpreted as the effect of a suitable additive observation noise term, which performs a kind of quantization of the non-integer output of the nonlinear observation function.

In linear state space modeling, tasks such as prediction, filtering, and smoothing may be performed by algorithms based on the linear Kalman filter (KF) [8,11]. However, in the case of a nonlinear observation function, generalizations of the linear KF need to be employed. Nowadays, a variety of algorithms is available, such as the extended KF (EKF), the iterated extended KF (IEKF), the unscented KF (UKF), and particle filtering [11,12].

The EKF is based on applying a linear KF by forming the local derivative of the nonlinear observation function (and the dynamical process, if it is also nonlinear) at the current estimate of the predicted state. The IEKF extends the EKF by an additional iteration that aims at finding consistent estimates for predicted and filtered states. It has recently been shown that the IEKF iteration can be interpreted as an application of Gauss–Newton optimization [12]. The UKF generalizes the EKF by propagating an ensemble of deterministically chosen points, thus improving Gaussian approximation and eliminating the explicit calculation of the Jacobian matrices. In principle, in Kalman filtering the error covariance matrices of the state estimates should be bounded and should converge to a steady solution, irrespective of the error in the initial state estimate and the corresponding initial error covariance matrix. However, the opposite phenomenon called *theoretical divergence* of classical KF is also well known. In order to keep the covariance matrix positive definite and improve the numerical behavior of the KF, square-root variants of the linear KF and its nonlinear generalizations have been introduced [13].

The mentioned algorithms have played a prominent role in applications in biomedical research and adjacent fields, such as public health [14]. Examples of corresponding count time series include epidemiological data (such as the famous U.S. poliomyelitis incidence time series, consisting of monthly counts starting in 1970) [15], sleep stage sequences, erythrocyte counts, infectious disease data [16], and epileptic seizure counts [17,18]. While most epilepsies respond well to anti-epileptic treatment, modeling the effects of anti-epileptic drugs (AEDs) on seizure frequency is essential for patients with difficult-to-treat or treatment-resistant epilepsies [19,20].

Several quantitative approaches to the analysis of seizure count time series have been proposed, which suffer from significant deficiencies. The *mixed-effects models* employed by Tharayil et al. cannot assess the effects of the changes in AEDs [21]. The epilepsy seizure management tool (*EpiSAT*) proposed by Chiang et al. does not account for observation errors caused by missed seizures or misinterpreted non-seizure events [22]. The Bayesian negative binomial dynamic linear model, recently proposed by Wang et al., cannot model the interaction effect between AEDs [23].

Application of state space modeling and Kalman filtering to seizure count time series has the potential to solve these deficiencies by quantifying the effect of an AED in the presence of other AEDs, describing delays in the effect of each AED, and modeling interaction effects between several AEDs. Moreover, the state space approach allows for the presence of temporal correlations in the seizure count time series that are unrelated to the current AED medication, but may result from other unknown influences on the probability of seizures. Furthermore, the state space approach is robust with respect to observation errors, such as missed seizures, events misclassified as seizures, outliers, missing data, or other observer-related errors. Therefore, the primary objective of this contribution is to explore and develop state space modeling algorithms tailored explicitly for event count time series, with a particular focus on the modeling of seizure count time series, as an illustrative example. To the best of our knowledge, apart from our previous work [24–26], this approach is novel on its own.

This paper is organized as follows. In Section 2, "Materials and Methods", we discuss the main algorithms for Kalman filtering and for parameter estimation; we also describe the simulated data and the real data from a patient. In Section 3, "Results", we show the performance of the proposed algorithm, provide some comparison with respect to the numerical problems arising in previous algorithms, and show results from application to both simulated and real data. Section 4, "Discussion", concludes the paper. Additional information on the proposed state space models and Kalman filtering algorithms is provided in Appendices A and B.

2. Materials and Methods

2.1. Independent Components Linear State Space (IC-LSS) Models

The independent components linear state space (IC-LSS) model is a distinct category of the linear state space (LSS) models proposed by Galka et al. [27]. Let the data vector observed at time t be denoted by $\mathbf{y}_t$, where $t = 1, \ldots, T$ denotes discrete time, and let the dimension of $\mathbf{y}_t$ be denoted by n. The examples for analysis of actual data that are shown below are for scalar data only, i.e., $n = 1$, but we choose to keep the presentation of the methodology more general.

In linear state space (LSS) modeling, the observed data are linked to an unobserved m-dimensional state vector, $\mathbf{x}_t$, as described by an observation equation

$$\mathbf{y}_t = C\mathbf{x}_t + \epsilon_t, \quad \epsilon_t \sim \mathcal{N}(\mathbf{0}, R) \qquad (1)$$

where C and ϵ_t represent the *observation matrix* and the *observation noise*, respectively. The observation noise is a white Gaussian noise with zero mean and *observation noise covariance matrix* R. By including this noise, the model acknowledges the limitations and uncertainties of real-world observations, and the information loss about the true state of the system. Within LSS models, the temporal evolution of the state vector, $\mathbf{x}_t$, is described by a discrete-time dynamical equation

$$\mathbf{x}_t = A\mathbf{x}_{t-1} + \eta_t, \quad \eta_t \sim \mathcal{N}(\mathbf{0}, Q) \qquad (2)$$

where A and η_t represent the *state transition matrix* and the *dynamical noise*, respectively. Also, the dynamical noise is a white Gaussian noise with zero mean and *dynamical noise covariance matrix* Q. An additional control input term, depending on a known external control input, $\mathbf{u}_t$, may be added to the dynamical equation:

$$\mathbf{x}_t = A\mathbf{x}_{t-1} + B_u \mathbf{u}_t + \eta_t \qquad (3)$$

where B_u represents the *control gain matrix*. The respective dimensions of the matrices and vectors are given in Table 1. The IC-LSS model is a specific subset of the LSS model family that characterizes data as a combination of independent source processes through a weighted sum and chooses specific structures of parameter matrices of the general state space model. The model depends on a set of parameters matrices, collected in the set $\Theta = \{C, A, R, Q, B_u\}$. Both the state transition matrix, A, and the dynamical noise covariance matrix, Q, are constructed as block-diagonal matrices with identical sets of block dimensions, as described in Appendix A.

Our modeling approach assumes that the impact of each control input, i.e., each component of the vector $\mathbf{u}_t$, is independent of the other components, and that the corresponding processes can be modeled as deterministic first-order autoregressive models, to be denoted by AR(1). These AR(1) models are made deterministic by setting the corresponding elements of Q to zero. To account for temporally correlated fluctuations in the data caused by factors other than the control input, a stochastic process is also included. This stochastic process is modeled by an autoregressive moving average model with orders p and $p - 1$, to be denoted as ARMA($p, p-1$). In the current implementation, we usually choose $p = 2$, which represents a trade-off between the stochasticity and stability of the model. AR(1) and ARMA(2,1) models can easily be incorporated into IC-LSS models (for technical details, see Appendix A).

Table 1. Quantities arising in state space models.

Notation	Meaning and/or Name	Dimension
$\mathbf{y}_t$	data vector	n
$\boldsymbol{\epsilon}_t$	observation noise vector	n
$\mathbf{x}_t$	state vector	m
$\boldsymbol{\eta}_t$	dynamical noise vector	m
$\mathbf{u}_t$	external control vector	u
C	observation matrix	$n \times m$
A	state transition matrix	$m \times m$
R	observation noise covariance matrix	$n \times n$
Q	dynamical noise covariance matrix	$m \times m$
$\mathbf{B}_u$	control gain matrix	$m \times u$

2.2. Nonlinear State Space (NLSS) Models

Count time series do not follow Gaussian probability distributions, therefore the class of linear Gaussian state space models is not well suited for modeling such time series; rather it would be beneficial to employ appropriate nonlinear state space (NLSS) models. In this paper, we keep the dynamical equation linear, as in Equation (3), while defining a nonlinear observation equation by

$$\mathbf{y}_t = \mathbf{f}(\mathbf{x}_t) + \boldsymbol{\epsilon}_t, \quad \boldsymbol{\epsilon}_t \sim \mathcal{N}(\mathbf{0}, \mathbf{R}) \quad (4)$$

where we have assumed that the observation noise, $\boldsymbol{\epsilon}_t$, can be kept Gaussian; $\mathbf{f}(.)$ represents a nonlinear observation function. We employ two different nonlinear functions, namely

$$\mathbf{f}_1(\mathbf{x}) = \exp(\mathbf{Cx}) \quad (5)$$

and

$$\mathbf{f}_2(\mathbf{x}) = \frac{\mathbf{Cx}}{2} + \sqrt{\frac{(\mathbf{Cx})^2}{4} + k} \quad (6)$$

In the case of multivariate data, $n > 1$, these functions are to be applied component-wise.

The first of these observation functions, simply an exponential function, has been chosen with the intention of achieving non-negativity of the observed data, as it may also be achieved in Poisson regression. The disadvantage of this choice for the observation function is the fact that it diverges exponentially for large positive arguments. In previous work, we have occasionally encountered numerical breakdown of Kalman filtering algorithms due to the resulting extremely large values [25,26]; details will be provided below in Section 3.1. For this reason, we propose a different nonlinear observation function, to be called the "affinely distorted hyperbolic" function, as given in Equation (6); while for negative arguments it behaves like the exponential function, for positive arguments it converges to the linear function, rising with a slope of C, see Figure 1.

Recently, Weiß and coworkers [7,28] have introduced a nonlinear function called the "softplus" function, given by

$$\mathbf{f}(\mathbf{x}) = k \log\left(1 + \exp(\mathbf{x}/k)\right)$$

where k is a real positive parameter. This function could be employed instead of our "affinely distorted hyperbolic" function for the purpose of constraining observations to be non-negative, since it has similar behavior.

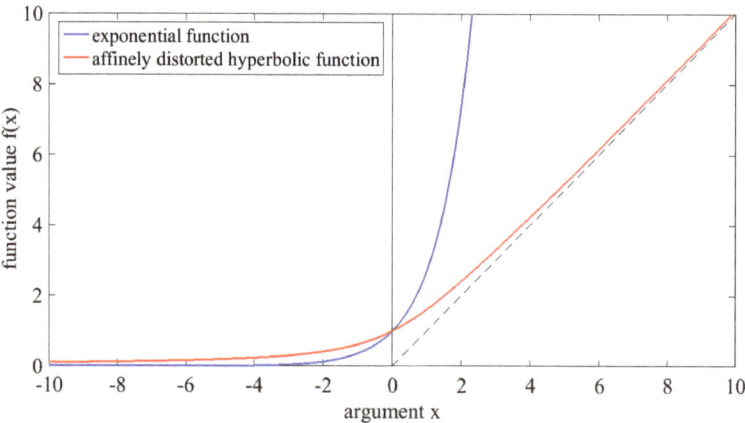

Figure 1. Nonlinear observation functions $\mathbf{f}_1(x)$ (blue) and $\mathbf{f}_2(x)$ (red), with $C = 1$ and $k = 1$; the dashed vertical line on the right side represents the linear function $\mathbf{f}(x) = x$.

2.3. Iterated Extended Kalman Filter (IEKF)

In this paper, we are dealing with a two-fold estimation problem: estimation of states and estimation of parameters. The extended Kalman filter (EKF) has been used as a popular tool for estimating the states in nonlinear state space models. It was first proposed by Kalman and Bucy in 1961 [29]. The EKF results from extending the original Kalman filter developed for linear systems to nonlinear systems by linearizing the dynamical equation and the observation equation around the current state estimate. As an improved variant of EKF, the iterated extended Kalman filter (IEKF) has been proposed, in order to improve the accuracy and stability of the EKF [30]. The IEKF has been applied to various fields, including robotics, aerospace, and control systems.

For a given time series of length T and given initial state estimate $\mathbf{x}_{0|0}$ and corresponding covariance matrix $\mathbf{P}_{0|0}$, the forward temporal recursion begins at time $t = 1$ with the predicted state estimate [25], which is computed by

$$\mathbf{x}_{t|t-1} = \mathbf{A}\mathbf{x}_{t-1|t-1} + \mathbf{B}_u \mathbf{u}_t \tag{7}$$

The notation $\mathbf{x}_{t_1|t_0}$ is used throughout the paper to indicate that an estimate at time t_1 is obtained by using all data available at time t_0. The corresponding predicted state covariance is computed by

$$\mathbf{P}_{t|t-1} = \mathbf{A}\mathbf{P}_{t-1|t-1}\mathbf{A}^T + \mathbf{Q} \tag{8}$$

At each time point, the IEKF iteration starts, after Equations (7) and (8) have been evaluated, with iteration index $i = 1, \ldots, i_m$. First, the derivative of the chosen nonlinear function is computed as

$$\mathbf{H}_t^{(i)} = \left.\frac{\partial \mathbf{f}}{\partial \mathbf{x}}\right|_{\mathbf{x}_{t|t}^{(i-1)}} \tag{9}$$

In the multivariate case, this derivative will be a matrix (Jacobian matrix). At each iteration, the prediction errors (also known as *innovations*), the innovation variance, the Kalman gain, and the filtered state estimates are computed, according to

$$v_t^{(i)} = y_t - f(x_{t|t}^{(i-1)}) \tag{10}$$

$$V_t^{(i)} = H_t^{(i)} P_{t|t-1} (H_t^{(i)})^T + R \tag{11}$$

$$K_t^{(i)} = P_{t|t-1} (H_t^{(i)})^T (V_t^{(i)})^{-1} \tag{12}$$

$$x_{t|t}^{(i)} = x_{t|t-1} + K_t^{(i)} \left(v_t^{(i)} - H_t^{(i)} \left(x_{t|t-1} - x_{t|t}^{(i-1)} \right) \right) \tag{13}$$

The iteration ends when a stopping criterion is fulfilled. We use the stopping criterion of either the norm of the relative change of $x_{t|t}^{(i)}$ falling below 10^{-10}, or the iteration index reaching a maximal value of $i_m = 100$. After the iteration, the filtered state covariance is computed by

$$P_{t|t} = (I_m - K_t^{(i)} H_t^{(i)}) P_{t|t-1} \tag{14}$$

where I_m denotes the m-dimensional identity matrix.

The IEKF algorithm, as presented above, is summarized in Appendix B. In an earlier paper [25], we have presented results from actual application of the IEKF, for the case of an exponential observation function.

2.4. Singular Value Decomposition Iterated Extended Kalman Filter (SVD-IEKF)

At each time step, the covariance matrices $P_{t|t-1}$, V_t, and $P_{t|t}$ arising in the Kalman filter recursion have to be positive definite, or at least, positive semi-definite. However, it is well known that during the recursion due to numerical effects these covariance matrices may lose the property of positive definiteness. As a consequence, computation of the likelihood becomes unreliable; furthermore, the iteration of the IEKF may converge only with delays, or it may entirely fail to converge [25].

As a remedy, one may work with matrix square roots of the covariance matrices, instead of the covariance matrices themselves. There are, at least, two ways to define the square root of a matrix: by Cholesky decomposition (CD) and by singular value decomposition (SVD).

For a given real square matrix M with dimension $m \times m$, the Cholesky decomposition is given by

$$M = S_M^T S_M \tag{15}$$

where S_M denotes a $(m \times m)$-dimensional upper triangular matrix with non-negative diagonal elements. The Cholesky Decomposition is only possible for matrices that are positive (semi-)definite; for semi-definiteness, the decomposition may not be unique.

For a given real matrix, M, with dimension $m \times n$, the SVD decomposition is given by

$$M = U_M \Sigma_M W_M^T \tag{16}$$

where U_M and W_M denote two orthogonal matrices with dimensions $m \times m$ and $n \times n$, respectively, and Σ_M is a diagonal matrix with dimension $m \times n$, which has non-negative real numbers, σ_i, on its diagonal, called the singular values of M [31]. The singular values are the positive square roots of the eigenvalues of $M^T M$ [13]. SVD can be applied to any matrix, without the condition of positive (semi-)definiteness.

The SVD of a positive (semi-)definite square matrix, M, e.g., a covariance matrix, can be formulated as

$$M = W_M \Sigma_M^2 W_M^T \tag{17}$$

such that we have $U_M = W_M$, and the definition of the matrix of singular values has been changed to Σ_M^2 instead of Σ_M; this is a reasonable change of definition, since the singular values are non-negative.

Square-root variants of the Kalman filter that employ CD were proposed already in the 1960s. However, SVD represents a matrix decomposition that offers superior numerical properties, compared with CD. Square-root variants of the Kalman filter that employ SVD were proposed in 1992 by Wang et al. [13], and in 2017 by Kulikova and Tsyganova [32]. In an earlier paper, we proposed a square-root variant of the IEKF that employs SVD [26], based on the algorithm of Kulikova and Tsyganova.

In the proposed nonlinear Kalman filter algorithm, the initial covariance matrix, $P_{0|0}$, and the noise covariance matrices, Q and R, are factorized by SVD as follows:

$$P_{0|0} = W_{0|0} \Sigma_{0|0}^2 W_{0|0}^T$$
$$Q = W_Q \Sigma_Q^2 W_Q^T$$
$$R = W_R \Sigma_R^2 W_R^T$$

These factorizations are performed once outside of the forward recursion of the Kalman filter. The recursion then begins with the same update equation for the predicted state as before, Equation (7). However, now the corresponding predicted state covariance matrix is updated by performing a factorization step by applying SVD to a *pre-array*, as follows [26]:

$$U_1 \underbrace{\begin{bmatrix} \Sigma_{P_{t|t-1}} \\ 0 \end{bmatrix} W_{P_{t|t-1}}^T}_{\text{singular values array}} = \underbrace{\begin{bmatrix} \Sigma_{P_{t-1|t-1}} W_{P_{t-1|t-1}}^T A^T \\ \Sigma_Q W_Q^T \end{bmatrix}}_{\text{pre-array}} \quad (18)$$

$$P_{t|t-1} = W_{P_{t|t-1}} \Sigma_{P_{t|t-1}}^2 W_{P_{t|t-1}}^T \quad (19)$$

where U_1 is an orthogonal matrix that can be discarded. Also, in the SVD-IEKF algorithm, at each time point an iteration is performed, with the equations for the Jacobi matrix and the innovations given by Equations (9) and (10), respectively. However, Equation (11) is replaced by an array factorization step:

$$U_2 \underbrace{\begin{bmatrix} \Sigma_{V_t}^{(i)} \\ 0 \end{bmatrix} (W_{V_t}^{(i)})^T}_{\text{singular values array}} = \underbrace{\begin{bmatrix} \Sigma_R W_R^T \\ \Sigma_{P_{t|t-1}}^{(i)} (W_{P_{t|t-1}}^{(i)})^T (H_t^{(i)})^T \end{bmatrix}}_{\text{pre-array}} \quad (20)$$

$$V_t^{(i)} = W_{V_t}^{(i)} (\Sigma_{V_t}^{(i)})^2 (W_{V_t}^{(i)})^T \quad (21)$$

where U_2 is an orthogonal matrix which can be discarded. As the iteration proceeds, the normalized innovation, the normalized gain, the optimal Kalman gain, and the filtered state estimate are computed as follows:

$$\hat{v}_t^{(i)} = (W_{V_t}^{(i)})^T v_t^{(i)} \quad (22)$$
$$k_t^{(i)} = P_{t|t-1} (H_t^{(i)})^T W_{V_t}^{(i)} \quad (23)$$
$$K_t^{(i)} = k_t^{(i)} (\Sigma_{V_t}^{(i)})^{-2} (W_{V_t}^{(i)})^T \quad (24)$$
$$x_{t|t}^{(i+1)} = x_{t|t-1} + k_t^{(i)} (\Sigma_{V_t}^{(i)})^{-2} \left(\hat{v}_t^{(i)} - (W_{V_t}^{(i)})^T H_t^{(i)} (x_{t|t-1} - x_{t|t}^{(i)}) \right). \quad (25)$$

After reaching the stopping criterion of the iteration, the filtered state covariance matrix, $P_{t|t}$, is computed by a third array factorization step:

$$U_3 \underbrace{\begin{bmatrix} \Sigma_{P_{t|t}} \\ 0 \end{bmatrix} W_{P_{t|t}}^T}_{\text{singular values array}} = \underbrace{\begin{bmatrix} \Sigma_{P_{t|t-1}} W_{P_{t|t-1}}^T (I_m - K_t^{(i)} H_t^{(i)})^T \\ \Sigma_R W_R^T (K_t^{(i)})^T \end{bmatrix}}_{\text{pre-array}} \qquad (26)$$

$$P_{t|t} = W_{P_{t|t}} \Sigma_{P_{t|t}}^2 W_{P_{t|t}}^T \qquad (27)$$

where U_3 denotes another orthogonal matrix that can be discarded, and i denotes the index at which the iteration had stopped. The SVD-IEKF algorithm, as presented above, is summarized in Appendix B.

2.5. Non-Dynamic Regression Model: Gaussian Case

For the purpose of comparison, we also employ two types of non-dynamic regression models. The first of these models represents the classical linear Gaussian case; it is defined as follows:

$$\mathbf{y}_t = \tilde{B}_u \mathbf{u}_t + \mathbf{n}_t \qquad (28)$$

where $\tilde{B}_u$ denotes a matrix of regression coefficients, and $\mathbf{n}_t$ denotes a time series of regression residuals, assumed to have a Gaussian distribution with zero mean and covariance matrix Σ_n. $\tilde{B}_u$ can be estimated by ordinary least squares; the covariance matrix, Σ_n, can then be computed as

$$\Sigma_n = \frac{1}{T} \sum_{t=1}^{T} (\mathbf{y}_t - \tilde{B}_u \mathbf{u}_t)(\mathbf{y}_t - \tilde{B}_u \mathbf{u}_t)^T \qquad (29)$$

Note that the regression model of Equation (28) implicitly assumes that the effects of the different components of the control vector, $\mathbf{u}_t$, are uncorrelated.

From the regression model of Equation (28), a logarithmic likelihood can be computed by

$$\log L(\tilde{B}_u) = -\frac{T}{2}(\log |\Sigma_n| + n \log 2\pi + n) \qquad (30)$$

2.6. Non-Dynamic Regression Model: Poisson Case

The second non-dynamic regression model is an example of a generalized linear model, representing the Poisson case; it is defined as follows:

$$\log y_t = \tilde{B}_u \mathbf{u}_t + n_t \qquad (31)$$

where $\tilde{B}_u$ denotes another matrix of regression coefficients, and n_t denotes another time series of regression residuals; in this case, y_t is assumed to have a Poisson distribution. Note that, for simplicity, we formulate this model only for the case of scalar data.

From the regression model of Equation (31), a logarithmic likelihood can be computed by

$$\log L(\tilde{B}_u) = \sum_{t=1}^{T} \left(y_t \tilde{B}_u \mathbf{u}_t - \exp(\tilde{B}_u \mathbf{u}_t) - \log(y_t!) \right). \qquad (32)$$

2.7. Parameter Estimation and Ensembles of Models

As mentioned earlier, when fitting a state space model to given data we have to solve a two-fold estimation problem: estimation of states and estimation of parameters. In this paper, estimation of the model parameters, denoted by Θ in Section 2.1, is performed by numerical maximization of the logarithmic innovation likelihood, denoted by $\log L(\Theta)$, employing the Broyden–Fletcher–Goldfarb–Shanno (BFGS) quasi-Newton and Nelder–Mead simplex algorithms [33]. Apart from filtered state estimates, the forward recursion of the Kalman filter also provides the corresponding contributions to the logarithmic innovation likelihood, which may then be summed up:

$$\log L(\Theta) = -\frac{1}{2}\sum_{t=1}^{T}\left(\log|V_t| + v_t^T V_t^{-1} v_t\right) - \frac{nT}{2}\log 2\pi \qquad (33)$$

In this expression, the effect of the initial state has been ignored. If the innovations, v_t, have been obtained by the IEKF or SVD-IEKF algorithms, their values correspond to the final values resulting from the iteration performed at time t, until the stopping criterion is fulfilled.

In order to avoid parameter redundancy with respect to the control gain matrix, B_u, the observation matrix, C, has not been included in the set of parameters to be estimated by optimization; instead constant values of 1.0 are employed, except for the ARMA(2,1) component, for which the control gain parameters are fixed at zero.

For given data, comparison of the performance of state space models, as discussed above in Sections 2.2–2.4, with the performance of non-dynamic regression models, as discussed above in Sections 2.5 and 2.6, can be performed by comparison of the corresponding values of an information criterion, such as the (corrected) Akaike information criterion (AICc) [27]. The AICc can be computed from the logarithmic likelihood according to

$$\text{AICc} = -2\log L(\Theta) + \frac{2 N_{\text{par}} T}{T - N_{\text{par}} - 1} \qquad (34)$$

where N_{par} denotes the number of data-adaptive parameters of the corresponding model; in the case of state space models, this would be the total number of parameters in Θ; if the nonlinear observation function, $f(.)$, according to Equation (6), is chosen, also the parameter k becomes part of Θ.

The second term in Equation (34) represents a penalty term for the complexity of the model; through this term, the values of AICc for different models can be directly compared, while the corresponding values of the logarithmic likelihood cannot, since they would be biased in favor of the more complex model. The task of parameter estimation is then given by finding parameters that minimize the AICc. Due to the complicated dependence of the AICc on Θ, including the possible existence of numerous local minima, this task can only be approached by numerical optimization. The chosen algorithm for numerical optimization may converge to one of these local minima, instead of the global minimum, or to other minima with smaller AICc values.

This issue can be resolved by employing an ensemble of models. Each model in the ensemble is initialized with randomly selected initial parameter values and optimized using the same minimum AICc approach until convergence. Finally, the model with the smallest AICc value is retained. However, although the probability of actually finding the global minimum will be improved by this ensemble approach, no assurance of finding the global minimum is provided.

2.8. Simulated Data

To validate the performance and effectiveness of our proposed NLSS modeling approach, we conduct simulations, before applying it to the patient data. These simulations serve to demonstrate the algorithm's capability in accurately capturing the dynamics of event count time series. We assume that three anti-epileptic drugs are given, named AED1, AED2 and AED3, and that, for the particular simulated "patient", AED1 and AED3 reduce the daily number of seizures, while AED2 increases it. The (arbitrarily chosen) time-dependent dosages of the drugs during a time interval of 500 days are shown in the upper panel of Figure 2a, while the resulting time series of the daily numbers of seizures is shown in the lower panel of the figure.

The simulated time series is generated by employing Equations (1) and (3), utilizing the affinely distorted hyperbolic function $f_2(x_t)$, as discussed in Equation (6). In addition to the contributions of the three anti-epileptic drugs, a stochastic ARMA(2,1) process is added. The resulting time series is rounded to integer values, representing the simulated daily seizure counts. The daily seizure counts assume values from 0 to 6, which is a realistic interval. The MATLAB code for recreating the simulated data is provided in Appendix C.

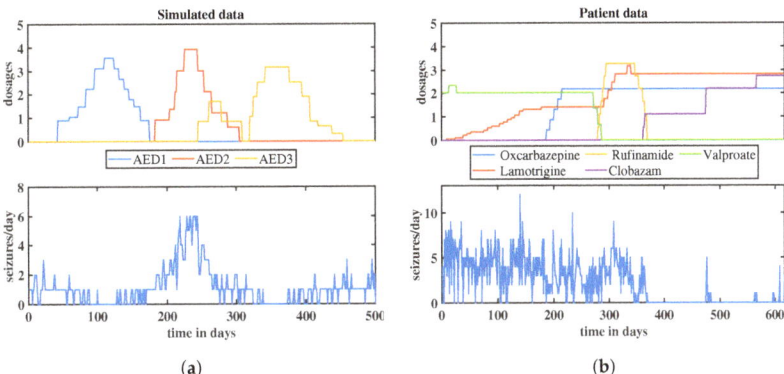

Figure 2. Time-dependent dosages of anti-epileptic drugs (upper panels) and corresponding count time series of daily epileptic seizures (lower panels) for a simulation (**a**) and for a real patient (**b**); in the upper panels, the anti-epileptic drugs are discriminated by colors (see legends below the panels, which also give the names of the drugs).

2.9. Patient Data

We demonstrate the practical application of NLSS modeling through the analysis of a real-world data set obtained from a patient suffering from symptomatic epilepsy. The data set utilized in this study was collected from an electronic seizure diary called EPI-Vista (http://www.epivista.de, accessed on 20 September 2023), which has been in routine use at the North German Epilepsy Center for Children and Adolescents in Schwentinental-Raisdorf, Germany, since 2007. EPI-Vista is a freely available therapy management tool that records information about dosages of administered anti-epileptic drugs and about seizure events.

The time-dependent dosages of the drugs during the chosen time interval of 618 days are shown in the upper panel of Figure 2b, while the recorded time series of the daily numbers of seizures is shown in the lower panel of the figure. During the chosen time interval, five different anti-epileptic drugs were administered: oxcarbazepine, lamotrigine, rufinamide, clobazam, and valproate.

3. Results

3.1. Convergence Behavior of Iteration

We will briefly comment on the convergence behavior of the iteration of the SVD-IEKF. Within the recursion of the SVD-IEKF through the data, the iteration takes place at each time point. We may plot the norm of the relative change of the state estimate as a function of the iteration index, obtaining a set of curves; examples are shown in Figures 3 and 4. These examples refer to the analysis of the simulated data, as described in Section 2.8. In Figure 3a, it can be seen that for all 500 time points the iteration converges according to a power law within, at most, 35 iterations, thereby confirming that the SVD-IEKF works properly.

However, within an ensemble of models, cases may also occur that show less favorable behavior. An example is shown in Figure 3b. In this case, we see convergence only for the first time point; the iteration loop stops after 500 iterations. State estimates obtained from this iteration were extremely large (in the order of e^{100}), and therefore the Kalman filter recursion was not continued, and no further iterations were performed. This problem can be resolved by replacing the exponential function with the affinely distorted hyperbolic function in the observation equation, Equation (1).

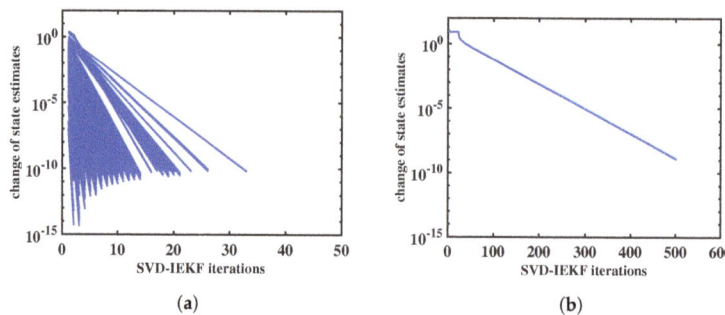

Figure 3. Norm of relative change of state estimate vs. iteration index for the application of the SVD-IEKF to a simulated count time series, using a model initialized with two different sets of random model parameters; in (**a**) iterations for all 500 time points are shown, while in (**b**) only the first iteration is shown, since the recursion of the Kalman filter was aborted afterwards due to numerical failure.

In Figure 4, we illustrate the effects of loss of positive definiteness of covariance matrices on the convergence behavior. Also, this example refers to the analysis of the simulated data. The standard IEKF was used, as described above in Section 2.3, and the affinely distorted hyperbolic function was employed.

In Figure 4a, it can be seen that for most time points the iteration completely fails to converge, instead norms of relative state changes stay approximately constant or oscillate; note the extremely large, or small, values on the vertical axis. In Figure 4b, the effect of switching from the IEKF to the SVD-IEKF is illustrated, for the same simulated data and the same set of model parameters: now, for almost all time points, good convergence within 60 iterations is obtained. For a single time point, convergence is slower and seems not to follow a power law.

Within an ensemble of 1000 random initial models, we find that for 18 models the IEKF encounters numerical problems resulting from covariance matrices losing positive definiteness or becoming singular, and another 11 models exhibit poor convergence behavior. For the SVD-IEKF, all models of the ensemble display good or satisfactory convergence behavior, without any numerical instability of the Kalman filter.

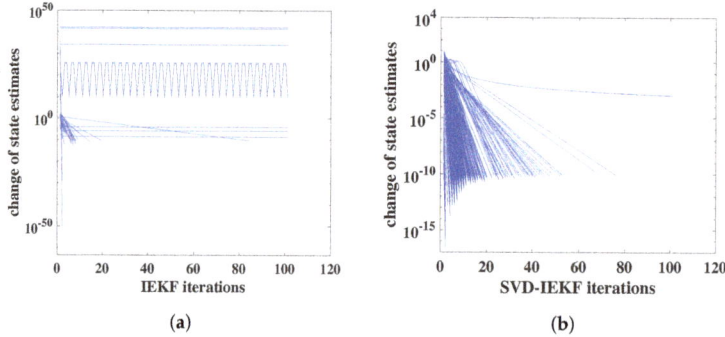

Figure 4. Norm of relative change of state estimate vs. iteration index for the application of the IEKF (**a**) or SVD-IEKF (**b**) to a simulated count time series, using a model initialized with a set of random model parameters; iterations for all 500 time points are shown.

3.2. Results of Ensemble Approach: Simulated Data

Following earlier work [24], we plot the control gain parameters, i.e., the diagonal elements of the control gain matrix, B_u, as defined in Equation (3), against the corresponding values of the AICc for all models of the ensemble; for each diagonal element, i.e., for each anti-epileptic drug, a separate plot is created. For the simulated data, the resulting plots are shown in Figure 5; blue dots represent the results from the 1000 models of the ensemble. The model with the lowest AICc value is highlighted in green. In addition, the results obtained by the two non-dynamic regression models are also shown, and are represented by red (Gaussian) and deep purple (Poisson); error bars are also shown, although they are mostly very small.

From Figure 5 it can be seen that most of the models of the ensemble achieve a lower AICc, compared with the non-dynamic regression models. For AED1 and AED2, the clouds of blue dots scatter over both positive and negative values of the control gain parameter; however, the model with the smallest control gain parameter is negative for AED1, and positive for AED2. For AED3, literally all models yield negative control gain parameters.

Since the simulation was designed such that AED1 and AED3 reduce the daily number of seizures, while AED2 increases it, the ensemble approach has succeeded in retrieving the correct result. As can be seen from Figure 5, in this case also the non-dynamic regression models reproduce the correct result.

In Table 2, the correct values of the observation parameters, which were used for creating the simulated data, and the estimated values of these parameters, obtained for the model with the lowest AICc value, are shown. The table also lists estimated errors for the estimated parameters; these errors can be estimated by computing the Hessian of the local likelihood at the optimal point. However, it is obvious that, at least for AED1 and AED3, these estimated errors are much too small to describe the actual deviation of the estimated values from the correct values; the probable reason for this is that in nonlinear filtering algorithms with an iteration at each time point, such as the IEKF and SVD-IEKF, the local likelihood often has a complicated shape with discontinuous behavior, such that numerically computed Hessians do not provide reliable estimates of the estimation errors.

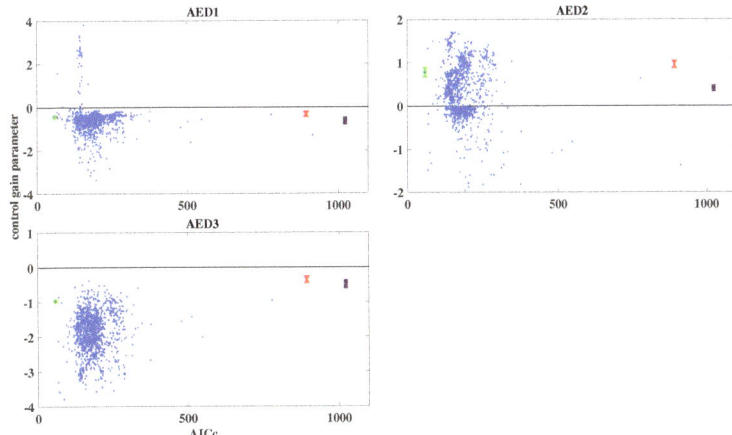

Figure 5. Estimated control gain parameters vs. AICc for simulated data, using an ensemble of 1000 randomly initialized models. The blue dots represent the results for the ensemble, with the model with the lowest AICc highlighted in green. The red (Gaussian case) and deep purple (Poisson case) dots represent the results obtained by the non-dynamic regression models. For the green, red, and deep purple dots, error bars are added, but they are mostly very small. The three panels refer to the three anti-epileptic drugs that were used in the simulation.

Table 2. Correct and estimated values of the observation parameters for the simulation study; the third row gives estimated errors for the estimated values, obtained from the Hessian of the local likelihood.

Anti-Epileptic Drug	AED1	AED2	AED3
correct values	−0.40	0.95	−0.70
estimated values (best model)	−0.43	0.79	−0.96
estimated errors (best model)	$\pm 1.86 \times 10^{-7}$	$\pm 4.57 \times 10^{-2}$	$\pm 4.33 \times 10^{-8}$

3.3. Result of Ensemble Approach: Patient Data

For the patient data, an ensemble of 700 randomly initialized models was employed. When using the SVD-IEKF and the affinely distorted hyperbolic function, optimization of all models proceeded without cases of numerical problems. The resulting plots of estimated control gain parameters against the corresponding values of the AICc for all models of the ensemble are shown in Figure 6; again, blue dots represent the results from the models of the ensemble, the model with the lowest AICc value is highlighted in green, and the results obtained by the two non-dynamic regression models are represented by red (Gaussian) and deep purple (Poisson).

From Figure 6 it can be seen that, again, most of the models of the ensemble achieve a lower AICc, compared with the non-dynamic regression models. The error bars of the red dots are somewhat larger now than in the case of the simulated data; for clobazam, the error interval includes the value of zero for the control gain parameter, such that it would become impossible to decide whether the effect of this anti-epileptic drug on the seizure count would be increasing, decreasing, or zero. Furthermore, based on the non-dynamic regression models, we would conclude that oxcarbazepine and lamotrigine would decrease the seizure count, while rufinamide and valproate would increase it.

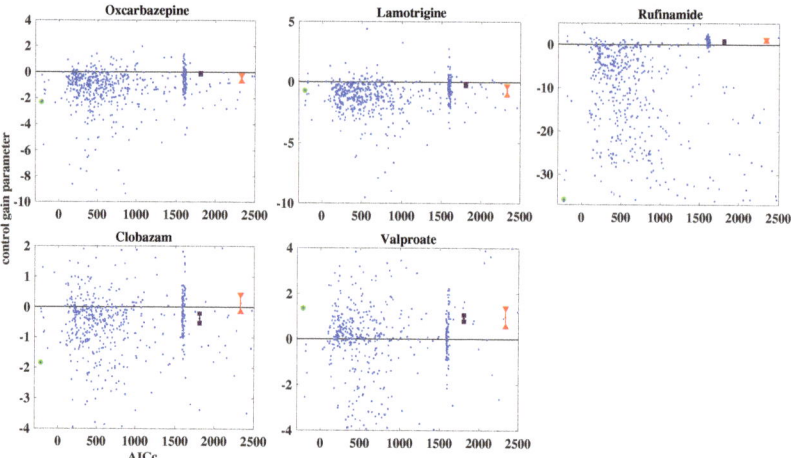

Figure 6. Estimated control gain parameters vs. AICc for patient data, using an ensemble of 700 randomly initialized models. The blue dots represent the results for the ensemble, with the model with the lowest AICc highlighted in green. The red (Gaussian case) and deep purple (Poisson case) dots represent the results obtained by the non-dynamic regression models. For the green, red, and deep purple dots, error bars are added. The five panels refer to the five anti-epileptic drugs that were administered during the chosen time interval of 618 days.

On the other hand, according to the results of the analysis by the SVD-IEKF, we would conclude that all anti-epileptic drugs, except for valproate, would decrease the seizure count; consequently, for two anti-epileptic drugs, the conclusions would differ from the conclusions based on the non-dynamic regression models.

We add a comment on the results of the ensemble approach, both for simulated and patient data, as shown in Figures 5 and 6. In this paper, we are dealing with time series of fairly short length, at most a few hundred values, since this is the typical situation for time series of a daily number of certain events, such as epileptic seizures. As a consequence of this scarcity of data, relative to the dimension of the space of model parameters, the AICc (i.e., the negative likelihood), as a function of these parameters, will display many local minima, and the task of finding the global minimum may not be well defined. For this reason, we consider it necessary to choose an ensemble approach. In the figures showing the results of the ensemble approach, the clouds of results that have a higher value of the AICc than the best model also contain some useful information.

As an example, consider again Figure 6. For each of the five anti-epileptic drugs displayed, it can be seen that the majority of the blue dots correspond to the same sign of the control gain parameter as the best model (i.e., the dot highlighted in green), thereby lending additional credibility to the resulting conclusions on the effects of these drugs. If, in such a case, the blue dots were distributed about equally over positive and negative values of the control gain parameter, we would be less inclined to regard the sign of this parameter for the best model as significant.

The underlying problem with respect to estimating model parameters from small data sets is unrelated to the particular choice of a state space model with linear dynamics and nonlinear observation; it has been observed quite similarly in an earlier study based on purely linear modeling [24]. The task to be addressed here is to draw the optimal conclusion, based on the scarce available data.

3.4. Innovation Whiteness Test

The aim of modeling a given time series by a parametric model, such as a state space model, is given by extracting temporal correlations as much as possible, such that the remaining prediction errors, i.e., the innovations, do not contain any residual correlations; this is equivalent to the innovations being white noise. In order to demonstrate that our modeling of the given data has been successful, with respect to this whiteness criterion, we will now show the autocorrelation function of the innovations of the best state space models from the ensembles.

In Figure 7, the autocorrelation functions for the innovations of the simulated data (left panel) and the patient data (right panel) are shown (green curves); for comparison, the autocorrelation functions of the raw data are also shown (blue curves). The red dashed lines correspond to one standard deviation. Note that, at lag zero, autocorrelation functions always assume a value of 1.0. By comparison of the blue and the green curves it can be seen that, both for simulated data and for patient data, very good whitening has been achieved, with only a few outliers exceeding one standard deviation. This result confirms the validity, in a statistical sense, of the chosen approach of modeling the data by a state space model with a nonlinear observation function.

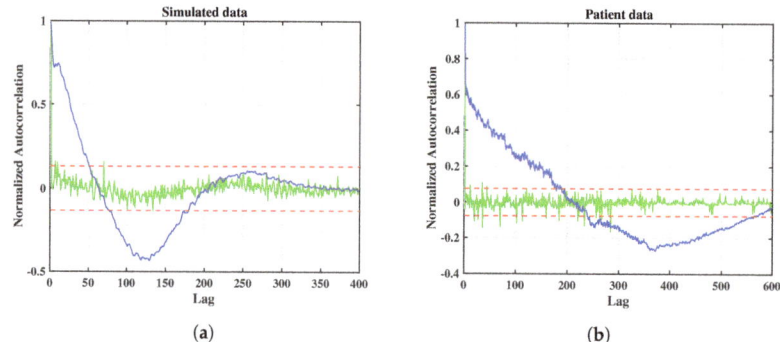

Figure 7. Autocorrelation functions of raw data (blue curves) and innovations after state space modeling (green curves) for simulated data (**a**) and patient data (**b**), for the best models of the respective ensembles. The red dashed lines correspond to one standard deviation of the innovations.

4. Discussion

In this paper, we propose an algorithm for analyzing event count time series by state space modeling and Kalman filtering. Within the larger field of time series analysis, the analysis of event count time series represents a special case, which is characterized by the need to model data that are both non-negative and integer, i.e., data given by natural numbers. The classical linear Gaussian state space model seems less suited for such data, since count data are described by Poisson distributions, rather than Gaussian distributions.

However, as we demonstrate, it is possible to keep the dynamics of the state space model Gaussian and linear, while introducing nonlinearity only at the last stage of modeling, namely, at the stage of modeling the observation process, thereby enforcing non-negativity and integrity of the observed data. The step from the non-integer output of the observation function to the integer data can then be roughly interpreted as addition of "quantization noise". By this device, Poisson distributions need not be employed explicitly. Nevertheless, due to the choice of a nonlinear observation function, it is necessary to employ nonlinear Kalman filters.

The present paper summarizes and extends earlier work, according to a sequence of steps that can be described as follows:

- In an initial study, we had proposed to analyze event count time series by purely linear Gaussian state space modeling, using the standard linear Kalman filter [24].
- Then, as a generalization, we employed nonlinear state space modeling, such that a linear dynamical equation was combined with a nonlinear observation function; the exponential function was chosen. State space modeling was performed by the iterated extended Kalman filter (IEKF) [25].
- As a further step of improvement, the standard IEKF was replaced by the numerically superior singular value decomposition variant of the IEKF [26].
- In the present paper, we replaced the exponential function with the "affinely distorted hyperbolic" function; alternatively, the "softplus function" of Weiß and coworkers could also have been employed. We have not yet systematically compared both functions, but expect that they would display similar performance.

The classical linear Kalman filter consists of a recursion in the forward direction through time, which represents the optimal state estimator for linear Gaussian state space models [8]. As soon as nonlinearities are introduced into the dynamical equation or the observation equation, no closed-form optimal recursion exists, such that approximations and additional iterations have to be employed. The iterated extended Kalman filter (IEKF) represents a well-established example of such approximative nonlinear state estimators.

It is well known that, in the practical application of both the classical linear Kalman filter and its nonlinear generalizations, numerical problems may arise, which result from covariance matrices losing the property of positive definiteness. The usual remedy for such problems is given by expressing the recursion, not directly for the covariance matrices, but for square roots of these matrices, which is known as *square root filtering*. Matrix square roots may be defined either by Cholesky decomposition, or by singular value decomposition (SVD). Since the latter decomposition represents the more general and numerically more robust decomposition [34], we have chosen to employ it for our state estimation algorithm. The resulting algorithm is, to the best our knowledge, the first SVD variant of the IEKF that has been proposed [26].

When defining a state space model with a linear dynamical equation, the use of an exponential observation function represents a natural choice, in order to keep the data non-negative. However, the disadvantage of the exponential function is its exponential divergence for positive arguments, which may lead to numerical failure of the SVD-IEKF algorithm. For this reason, we propose a new nonlinear observation function that converges to zero for negative arguments, just like the exponential function, while converging to a linear function for positive arguments. This function can be derived by affinely distorting a (negative) hyperbolic function, such that the vertical axis becomes the desired linear function for positive arguments. As we have demonstrated in the present paper by analyzing both simulated and real data, employing the affinely distorted hyperbolic function within the SVD-IEKF algorithm finally removes the risk of numerical failure.

We emphasize that the proposed algorithm can be applied to any event count time series that one may wish to analyze under a non-negativity constraint; here, as an example, we have applied the algorithm to the analysis of time series of the daily number of seizures of drug-resistant epilepsy patients undergoing treatment with several, simultaneously administered anti-epileptic drugs. The time-dependent dosages of these drugs are inserted into the state space model as an external control input. The simultaneous presence of several drugs, their potential superposition effects, delay effects, and further unknown factors influencing the daily number of seizures make the objective analysis of seizure count time series arduous. As we have demonstrated, both in a simulation study and in the analysis of data from a real patient, state space modeling provides a powerful and flexible framework for analyzing the effects of anti-epileptic drugs in epilepsy patients. By comparison of an information criterion, such as the AICc, it can be proven that state space modeling provides a modeling of the data that is superior to modeling by conventional non-dynamic regression models.

There exist other model classes that take temporal correlations into account, such as the dynamic GLM, INAR, and INGARCH model classes. Our approach to modeling event count time series, as proposed in the present paper, represents an alternative to these model classes, but we do not intend to replace these classes, but rather provide an additional tool for the analysis of event count time series. According to a well-known proverb, "all models are wrong, but some models are useful". Then, the justification of introducing a new class of models can come only from their usefulness in practical work, which needs to be explored by application to real data. Within the likelihood framework, the comparison of different model classes should be performed by quantifying their performance in predicting the data, preferably by using an information criterion, such as the AICc. It would be very interesting to perform a systematic study of the performance of our approach to modeling event count time series, in comparison with the already available model classes; this, however, is beyond the scope of the present paper and remains a task for future work.

With respect to the chosen field of application, namely, the analysis of time series of the daily number of seizures of drug-resistant epilepsy patients, in the future we intend to apply the presented algorithm to data from larger cohorts of patients. By discriminating between different types of seizures in the same patient, it is also intended to generalize the analysis to the case of vector data. Furthermore, the analysis should be generalized such that not only the effects of individual anti-epileptic drugs are modeled, but also the interaction effects between pairs, or groups, of drugs. Finally, in order to reduce the computational time consumption, work should be devoted to developing more efficient algorithms for fitting ensembles of nonlinear state space models by numerical maximum-likelihood procedures.

Author Contributions: Conceptualization, S.M. and A.G.; methodology, S.M. and A.G.; software, S.M. and A.G.; validation, S.M., B.A. and A.G.; formal analysis, S.M.; investigation, S.M.; resources, S.M. and A.G.; writing—original draft preparation, S.M.; writing—review and editing, S.M. and A.G.; visualization, S.M. and A.G.; supervision, B.A. and A.G.; project administration, S.M., B.A. and A.G.; funding acquisition, B.A. All authors have read and agreed to the published version of the manuscript.

Funding: This research was (partially) funded by the Hasso Plattner Institute, Research School on Data Science and Engineering.

Institutional Review Board Statement: The study was conducted in accordance with the Ethics Committee of Faculty of Medicine, Kiel University, Germany (D 420/17).

Data Availability Statement: The anonymized patient data set used in this paper is available for research purposes from the authors upon specific request.

Acknowledgments: We appreciate the contribution of Sarah von Spiczak, DRK North German Epilepsy Center for Children and Adolescents, for leading the data gathering process and enabling us to work with the patient data. We also acknowledge her guidance on the medical expertise in our previous publications.

Conflicts of Interest: The authors declare no conflict of interest.

Abbreviations

The following abbreviations are used in this paper:

AED	Anti-epileptic drug
IC-LSS	Independent components linear state space
NLSS	Nonlinear state space
AR	Autoregressive
ARMA	Autoregressive moving average
IEKF	Iterated extended Kalman filter
SVD	Singular value decomposition
CD	Cholesky decomposition
BFGS	Broyden–Fletcher–Goldfarb–Shanno
AICc	corrected Akaike information criterion

Appendix A. Block Diagonal Structure of A and Q

The IC-LSS model discussed in Section 2.1 is based on a block-diagonal structure [35] of the state transition matrix, A, and the dynamical noise covariance matrix, Q, as follows [27]:

$$A = \begin{pmatrix} A_1 & 0 & 0 & \cdots & 0 \\ 0 & A_2 & 0 & \cdots & 0 \\ \vdots & \vdots & \vdots & \ddots & 0 \\ 0 & 0 & 0 & \cdots & A_J \end{pmatrix}, \quad Q = \begin{pmatrix} Q_1 & 0 & 0 & \cdots & 0 \\ 0 & Q_2 & 0 & \cdots & 0 \\ \vdots & \vdots & \vdots & \ddots & 0 \\ 0 & 0 & 0 & \cdots & Q_J \end{pmatrix} \quad \text{(A1)}$$

where J denotes the number of independent components and also represents the number of blocks. If we assume that the jth block represents an autoregressive moving average (ARMA) model with model orders $(p, p-1)$, consisting of an autoregressive part with parameters $a_\tau^{(j)}, \tau = 1, \ldots, p$, and a moving average part with parameters $b_\tau^{(j)}, \tau = 0, \ldots, p-1$, the block matrix, A_j, has left companion form, given by [27]

$$A_j = \begin{pmatrix} a_1^{(j)} & 1 & 0 & \cdots & 0 \\ a_2^{(j)} & 0 & 1 & \cdots & 0 \\ \vdots & \vdots & \vdots & \ddots & 0 \\ a_{p-1}^{(j)} & 0 & 0 & \cdots & 1 \\ a_p^{(j)} & 0 & 0 & \cdots & 0 \end{pmatrix} \qquad (A2)$$

and the block matrix, Q_j, has outer product form, given by [27]

$$Q_j = \begin{pmatrix} (b_0^{(j)})^2 & b_0^{(j)} b_1^{(j)} & \cdots & b_0^{(j)} b_q^{(j)} \\ b_1^{(j)} b_0^{(j)} & (b_1^{(j)})^2 & \cdots & b_1^{(j)} b_q^{(j)} \\ \vdots & \vdots & \vdots & \ddots \\ b_{p-2}^{(j)} b_0^{(j)} & b_{p-2}^{(j)} b_1^{(j)} & \cdots & b_{p-2}^{(j)} b_q^{(j)} \\ b_{p-1}^{(j)} b_0^{(j)} & b_{p-1}^{(j)} b_1^{(j)} & \cdots & (b_{p-1}^{(j)})^2 \end{pmatrix} \qquad (A3)$$

We usually employ the scaling convention $b_0^{(j)} = 1$ for all blocks.

Appendix B. Iterated Extended Kalman Filter (IEKF) Algorithms

Algorithm A1 Iterated Extended Kalman Filter (IEKF)

Require: Initial state estimate: $x_{0|0}$
Require: Initial covariance matrix: $P_{0|0}$
1: **for** $t = 1$ to T **do**
2:     **Prediction Step:**
3:     Compute the predicted state estimate: $x_{t|t-1} = Ax_{t-1|t-1} + B_u u_t$
4:     Compute the predicted state covariance matrix: $P_{t|t-1} = AP_{t-1|t-1}A^T + Q$
5:     **Update Step:**
6:     Initialize iteration: $i = 1$, $x_{t|t}^{(0)} = x_{t|t-1}$
7:     **repeat**
8:         Compute the Jacobian matrix of the observation function: $H_t^{(i)} = \left.\frac{\partial f}{\partial x}\right|_{x_{t|t}^{(i-1)}}$
9:         Compute the innovation: $v_t^{(i)} = y_t - f(x_{t|t}^{(i-1)})$
10:        Compute the innovation covariance matrix: $V_t^{(i)} = H_t^{(i)} P_{t|t-1} (H_t^{(i)})^T + R$
11:        Compute the Kalman gain: $K_t^{(i)} = P_{t|t-1}(H_t^{(i)})^T (V_t^{(i)})^{-1}$
12:        Update the state estimate: $x_{t|t}^{(i)} = x_{t|t-1} + K_t^{(i)}\left(v_t^{(i)} - H_t^{(i)}\left(x_{t|t-1} - x_{t|t}^{(i-1)}\right)\right)$
13:        Increment iteration: $i = i + 1$
14:     **until** Stopping criterion met or maximum number of iterations (i_m) reached
15:     Compute the filtered state covariance matrix: $P_{t|t} = (I_m - K_t^{(i)} H_t^{(i)}) P_{t|t-1}$
16: **end for**

Algorithm A2 Singular Value Decomposition Iterated Extended Kalman Filter (SVD-IEKF)

Require: Initial state estimate: $x_{0|0}$
Require: Initial covariance matrix: $P_{0|0} = W_{0|0} \Sigma^2_{0|0} W^T_{0|0}$
Require: Factorized dynamic noise covariance matrix: $Q = W_Q \Sigma^2_Q W^T_Q$
Require: Factorized observation noise covariance matrix: $R = W_R \Sigma^2_R W^T_R$

1: **for** $t = 1$ to T **do**
2:    **Prediction Step:**
3:    Compute the predicted state estimate: $x_{t|t-1} = Ax_{t-1|t-1} + B_u u_t$
4:    Compute the predicted state covariance matrix:
$$U_1 \begin{bmatrix} \Sigma_{P_{t|t-1}} \\ 0 \end{bmatrix} W^T_{P_{t|t-1}} = \begin{bmatrix} \Sigma_{P_{t-1|t-1}} W^T_{P_{t-1|t-1}} A^T \\ \Sigma_Q W^T_Q \end{bmatrix}$$
$$P_{t|t-1} = W_{P_{t|t-1}} \Sigma^2_{P_{t|t-1}} W^T_{P_{t|t-1}}$$
5:    **Update Step:**
6:    Initialize iteration: $i = 1$, $x^{(0)}_{t|t} = x_{t|t-1}$
7:    **repeat**
8:       Compute the Jacobian matrix of the observation function: $H^{(i)}_t = \left. \frac{\partial f}{\partial x} \right|_{x^{(i-1)}_{t|t}}$
9:       Compute the innovation: $v^{(i)}_t = y_t - f(x^{(i-1)}_{t|t})$
10:      Compute the innovation covariance matrix:
$$U_2 \begin{bmatrix} \Sigma^{(i)}_{V_t} \\ 0 \end{bmatrix} (W^{(i)}_{V_t})^T = \begin{bmatrix} \Sigma_R W^T_R \\ \Sigma^{(i)}_{P_{t|t-1}} (W^{(i)}_{P_{t|t-1}})^T (H^{(i)}_t)^T \end{bmatrix}$$
$$V^{(i)}_t = W^{(i)}_{V_t} (\Sigma^{(i)}_{V_t})^2 (W^{(i)}_{V_t})^T$$
11:      Compute normalized innovation: $\hat{v}^{(i)}_t = (W^{(i)}_{V_t})^T v^{(i)}_t$
12:      Compute normalized gain: $k^{(i)}_t = P_{t|t-1} (H^{(i)}_t)^T W^{(i)}_{V_t}$
13:      Compute the optimal Kalman gain: $K^{(i)}_t = k^{(i)}_t (\Sigma^{(i)}_{V_t})^{-2} (W^{(i)}_{V_t})^T$
14:      Update the state estimate:
$$x^{(i+1)}_{t|t} = x_{t|t-1} + k^{(i)}_t (\Sigma^{(i)}_{V_t})^{-2} \left(\hat{v}^{(i)}_t - (W^{(i)}_{V_t})^T H^{(i)}_t (x_{t|t-1} - x^{(i)}_{t|t}) \right)$$
15:      Increment iteration: $i = i + 1$
16:    **until** Stopping criterion met or maximum number of iterations (i_m) reached
17:    Compute the filtered state covariance matrix:
$$U_3 \begin{bmatrix} \Sigma_{P_{t|t}} \\ 0 \end{bmatrix} W^T_{P_{t|t}} = \begin{bmatrix} \Sigma_{P_{t|t-1}} W^T_{P_{t|t-1}} (I_m - K^{(i)}_t H^{(i)}_t)^T \\ \Sigma_R W^T_R (K^{(i)}_t)^T \end{bmatrix}$$
$$P_{t|t} = W_{P_{t|t}} \Sigma^2_{P_{t|t}} W^T_{P_{t|t}}$$
18: **end for**

Appendix C. Matlab Code for Generating Simulated Data

Listing A1. MATLAB code for generating simulated data

```matlab
x = zeros(m, n); % state vector according to Equation (3).
x_initial = zeros(m, 1); % initializing sate vector with zero.
k = 1; % parameter defined in Equation (6).
C = [-0.40, 0.95, -0.70, 0.75, 0];
A = [0.5, 0, 0, 0, 0;
     0, 0.25, 0, 0, 0;
     0, 0, 0.25, 0, 0;
     0, 0, 0, 0.9, 1;
     0, 0, 0, -0.5, 0];
ϵ = [zeros(1, u), randn(1), zeros(1, 1)]; % observation noise from
    Equation (6).


for(time = 1: T)   % Begin Kalman forward loop

    external_input = [u_t(:, time:-1:T, 1)' , zeros(1, (m - u))]; %
        external input defined in Equation (4)
    if t == 1
        x(:, t) = A * x_initial + external_input' + ϵ';
    else
        x(:, t) = A * x(:, t - 1) + external_input' + ϵ';
    end

    for seizure_type = 1:n % for multi-dimensional data
        srexpr = sqrt(((C(seizure_type, :) * x(:, t))^2) / 4 + k); %
            intermediate variable
        y(seizure_type, t) = round((C(seizure_type, :) * x(:, t)) /
            2 + srexpr);
    end
end % End Kalman forward loop
```

References

1. Nelder, J.; Wedderburn, R. Generalized linear models. *J. R. Stat. Soc. Ser. A* **1972**, *135*, 370–384. [CrossRef]
2. West, M.; Harrison, P.J.; Migon, H.S. Dynamic generalized linear models and Bayesian forecasting. *J. Am. Stat. Assoc.* **1985**, *80*, 73–83. [CrossRef]
3. Fahrmeir, L.; Tutz, G.; Hennevogl, W.; Salem, E. *Multivariate Statistical Modelling Based on Generalized Linear Models*, 2nd ed.; Springer Series in Statistics; Springer: Berlin/Heidelberg, Germany, 2001; Volume 425.
4. McKenzie, E. Some simple models for discrete variate time series. *J. Am. Water Resour. Assoc.* **1985**, *21*, 645–650. [CrossRef]
5. Rydberg, T.H.; Shephard, N. *BIN Models for Trade-by-Trade Data. Modelling the Number of Trades in a Fixed Interval of Time*; Technical Report; Nuffield College: Oxford, UK, 1999.
6. Weiß, C.H. *An Introduction to Discrete-Valued Time Series*; John Wiley & Sons: Hoboken, NJ, USA, 2018.
7. Liu, M.; Zhu, F.; Li, J.; Sun, C. A Systematic Review of INGARCH Models for Integer-Valued Time Series. *Entropy* **2023**, *25*, 922. [CrossRef] [PubMed]
8. Kailath, T.; Sayed, A.; Hassibi, B. *Linear Estimation*; Prentice Hall: Hoboken, NY, USA, 2000.
9. Galka, A.; Ozaki, T.; Muhle, H.; Stephani, U.; Siniatchkin, M. A data-driven model of the generation of human EEG based on a spatially distributed stochastic wave equation. *Cogn. Neurodyn.* **2008**, *2*, 101–113. [CrossRef]
10. Hamilton, J.D. *Time Series Analysis*; Princeton University Press: Princeton, NY, USA, 1994.
11. Grewal, M.S.; Andrews, A.P. *Kalman Filtering: Theory and Practice with MATLAB*, 4th ed.; John Wiley & Sons: Hoboken, NJ, USA, 2015.
12. Skoglund, M.; Hendeby, G.; Axehill, D. Extended Kalman filter modifications based on an optimization view point. In Proceedings of the 18th International Conference on Information Fusion, Washington, DC, USA, 6–9 July 2015; pp. 1856–1861.
13. Wang, L.; Libert, G.; Manneback, P. Kalman filter algorithm based on singular value decomposition. In Proceedings of the 31st IEEE Conference on Decision and Control, Tucson, AZ, USA, 16–18 December 1992; pp. 1224–1229.
14. Zeger, S.L.; Irizarry, R.; Peng, R.D. On time series analysis of public health and biomedical data. *Annu. Rev. Public Health* **2006**, *27*, 57–79. [CrossRef]

15. Zeger, S.L. A regression model for time series of counts. *Biometrika* **1988**, *75*, 621–629. [CrossRef]
16. Allard, R. Use of time-series analysis in infectious disease surveillance. *Bull. World Health Organ.* **1998**, *76*, 327.
17. Albert, P.S. A two-state Markov mixture model for a time series of epileptic seizure counts. *Biometrics* **1991**, *47*, 1371–1381. [CrossRef]
18. Balish, M.; Albert, P.S.; Theodore, W.H. Seizure frequency in intractable partial epilepsy: A statistical analysis. *Epilepsia* **1991**, *32*, 642–649. [CrossRef]
19. Krauss, G.L.; Sperling, M.R. Treating patients with medically resistant epilepsy. *Neurol. Clin. Pract.* **2011**, *1*, 14–23. [CrossRef] [PubMed]
20. Kanner, A.M.; Ashman, E.; Gloss, D.; Harden, C.; Bourgeois, B.; Bautista, J.F.; Abou-Khalil, B.; Burakgazi-Dalkilic, E.; Park, E.L.; Stern, J.; et al. Practice guideline update summary: Efficacy and tolerability of the new antiepileptic drugs I: Treatment of new-onset epilepsy: Report of the Guideline Development, Dissemination, and Implementation Subcommittee of the American Academy of Neurology and the American Epilepsy Society. *Neurology* **2018**, *91*, 74–81. [PubMed]
21. Tharayil, J.J.; Chiang, S.; Moss, R.; Stern, J.M.; Theodore, W.H.; Goldenholz, D.M. A big data approach to the development of mixed-effects models for seizure count data. *Epilepsia* **2017**, *58*, 835–844. [CrossRef] [PubMed]
22. Chiang, S.; Vannucci, M.; Goldenholz, D.M.; Moss, R.; Stern, J.M. Epilepsy as a dynamic disease: A Bayesian model for differentiating seizure risk from natural variability. *Epilepsia Open* **2018**, *3*, 236–246. [CrossRef]
23. Wang, E.; Chiang, S.; Cleboski, S.; Rao, V.; Vannucci, M.; Haneef, Z. Seizure count forecasting to aid diagnostic testing in epilepsy. *Epilepsia* **2022**, *63*, 3156–3167. [CrossRef]
24. Galka, A.; Boor, R.; Doege, C.; von Spiczak, S.; Stephani, U.; Siniatchkin, M. Analysis of epileptic seizure count time series by ensemble state space modelling. In Proceedings of the 37th Annual International Conference of the IEEE Engineering in Medicine and Biology Society (EMBC), Milan, Italy, 25–29 August 2015; pp. 5601–5605.
25. Moontaha, S.; Galka, A.; Meurer, T.; Siniatchkin, M. Analysis of the effects of medication for the treatment of epilepsy by ensemble Iterative Extended Kalman filtering. In Proceedings of the 40th Annual International Conference of the IEEE Engineering in Medicine and Biology Society (EMBC), Honolulu, HI, USA, 18–21 July 2018; pp. 187–190.
26. Moontaha, S.; Galka, A.; Siniatchkin, M.; Scharlach, S.; von Spiczak, S.; Stephani, U.; May, T.; Meurer, T. SVD Square-root Iterated Extended Kalman Filter for Modeling of Epileptic Seizure Count Time Series with External Inputs. In Proceedings of the 41st Annual International Conference of the IEEE Engineering in Medicine and Biology Society (EMBC), Berlin, Germany, 23–27 July 2019; pp. 616–619.
27. Galka, A.; Wong, K.F.K.; Ozaki, T.; Muhle, H.; Stephani, U.; Siniatchkin, M. Decomposition of neurological multivariate time series by state space modelling. *Bull. Math. Biol.* **2011**, *73*, 285–324. [CrossRef]
28. Weiß, C.H.; Zhu, F.; Hoshiyar, A. Softplus INGARCH models. *Stat. Sin.* **2022**, *32*, 1099–1120. [CrossRef]
29. Kalman, R.E.; Bucy, R.S. New results in linear filtering and prediction theory. *J. Basic Eng.* **1961**, *83*, 95–108. [CrossRef]
30. Jazwinski, A. *Stochastic Processes and Filtering Theory*; Academic Press: Cambridge, MA, USA, 1974.
31. Björck, Å. *Numerical Methods in Matrix Computations*; Texts in Applied Mathematics; Springer: Berlin/Heidelberg, Germany, 2015; Volume 59.
32. Kulikova, M.V.; Tsyganova, J.V. Improved discrete-time Kalman filtering within singular value decomposition. *IET Control Theory Appl.* **2017**, *11*, 2412–2418. [CrossRef]
33. Nocedal, J.; Wright, S. *Numerical Optimization*; Springer: Berlin/Heidelberg, Germany, 1999.
34. Kulikova, M.V.; Kulikov, G.Y. SVD-based factored-form Cubature Kalman Filtering for continuous-time stochastic systems with discrete measurements. *Automatica* **2020**, *120*, 109110. [CrossRef]
35. Godolphin, E.; Triantafyllopoulos, K. Decomposition of time series models in state-space form. *Comput. Stat. Data Anal.* **2006**, *50*, 2232–2246. [CrossRef]

Disclaimer/Publisher's Note: The statements, opinions and data contained in all publications are solely those of the individual author(s) and contributor(s) and not of MDPI and/or the editor(s). MDPI and/or the editor(s) disclaim responsibility for any injury to people or property resulting from any ideas, methods, instructions or products referred to in the content.

Article

Partial Autocorrelation Diagnostics for Count Time Series

Christian H. Weiß [1,*], Boris Aleksandrov [1], Maxime Faymonville [2] and Carsten Jentsch [2]

[1] Department of Mathematics and Statistics, Helmut Schmidt University, 22043 Hamburg, Germany
[2] Department of Statistics, TU Dortmund University, 44221 Dortmund, Germany
* Correspondence: weissc@hsu-hh.de; Tel.: +49-40-6541-2779

Abstract: In a time series context, the study of the partial autocorrelation function (PACF) is helpful for model identification. Especially in the case of autoregressive (AR) models, it is widely used for order selection. During the last decades, the use of AR-type count processes, i.e., which also fulfil the Yule–Walker equations and thus provide the same PACF characterization as AR models, increased a lot. This motivates the use of the PACF test also for such count processes. By computing the sample PACF based on the raw data or the Pearson residuals, respectively, findings are usually evaluated based on well-known asymptotic results. However, the conditions for these asymptotics are generally not fulfilled for AR-type count processes, which deteriorates the performance of the PACF test in such cases. Thus, we present different implementations of the PACF test for AR-type count processes, which rely on several bootstrap schemes for count times series. We compare them in simulations with the asymptotic results, and we illustrate them with an application to a real-world data example.

Keywords: autoregressive model; count time series; INAR bootstrap; partial autocorrelation function; Yule–Walker equations

1. Introduction

Autoregressive (AR) models for time series date back to Walker [1], Yule [2], and they assume the current observation of the considered process to be generated from its own past by a linear scheme. The ordinary pth-order AR-model for a real-valued process $(Z_t)_{t \in \mathbb{Z}=\{\ldots,-1,0,1,\ldots\}}$, abbreviated as AR(p) model, is defined by the recursive scheme

$$Z_t = \alpha_1 \cdot Z_{t-1} + \ldots + \alpha_p \cdot Z_{t-p} + \varepsilon_t \qquad (\alpha_p \neq 0), \tag{1}$$

where the innovations $(\varepsilon_t)_\mathbb{Z}$ are independent and identically distributed (i.i.d.) real-valued random variables (rv), which are also assumed to be square-integrable ("white noise"). To ensure a (weakly) stationary and causal solution for the AR(p) recursion (1), the AR-parameters $\alpha_1, \ldots, \alpha_p \in \mathbb{R}$ have to be chosen such that the roots of the characteristic polynomial $\alpha(z) = 1 - \alpha_1 z - \ldots - \alpha_p z^p$ are outside the unit circle. Then, if the innovations $(\varepsilon_t)_\mathbb{Z}$ follow a normal distribution, also the observations $(Z_t)_\mathbb{Z}$ are normal, leading to the Gaussian AR(p) process.

A characteristic property of the AR(p) process is given by the fact that its autocorrelation function (ACF), $\rho(h) = Corr[Z_t, Z_{t-h}]$ with $h \in \mathbb{N} = \{1, 2, \ldots\}$ and $\rho(0) = 1$, satisfies the following set of linear equations:

$$\rho(h) = \sum_{i=1}^{p} \alpha_i \rho(|h-i|) \qquad \text{for } h = 1, 2, \ldots \tag{2}$$

These *Yule–Walker (YW) equations*, in turn, give rise to define the *partial autocorrelation function* (PACF), $\rho_{part}(h)$ with time lags $h \in \mathbb{N}$, in the following way (see Appendix A for further details): if $\mathbf{R}_k := \big(\rho(|i-j|)\big)_{i,j=1,\ldots,k}$ and $\mathbf{r}_k := \big(\rho(1), \ldots, \rho(k)\big)^\top \in \mathbb{R}^k$ for $k = 1, 2, \ldots$, and if $\mathbf{a}_k \in \mathbb{R}^k$ denotes the solution of the equation $\mathbf{R}_k \mathbf{a}_k = \mathbf{r}_k$, then the

Citation: Weiß, C.H.; Aleksandrov, B.; Faymonville, M.; Jentsch, C. Partial Autocorrelation Diagnostics for Count Time Series. *Entropy* 2023, 25, 105. https://doi.org/10.3390/e25010105

Academic Editor: Geert Verdoolaege

Received: 6 December 2022
Revised: 23 December 2022
Accepted: 28 December 2022
Published: 4 January 2023

Copyright: © 2023 by the authors. Licensee MDPI, Basel, Switzerland. This article is an open access article distributed under the terms and conditions of the Creative Commons Attribution (CC BY) license (https://creativecommons.org/licenses/by/4.0/).

PACF at lag k is defined as the last component of a_k, i.e., $\rho_{\text{part}}(k) := a_{k,k}$. Hence, if the YW-equations (2) hold, it follows that

$$\rho_{\text{part}}(p) = \alpha_p, \qquad \rho_{\text{part}}(h) = 0 \quad \text{for all } h > p. \tag{3}$$

This characteristic abrupt drop-down of the PACF towards zero after lag $h = p$ is commonly used for model identification in practice, namely by inspecting the sample PACF for such a pattern, see the Box–Jenkins program dating back to Box & Jenkins [3]. Details on the PACF's computation are summarized in Appendix A. There, we also provide a brief discussion on some equivalences between ACF, PACF, and the AR-coefficients, in the sense that the AR(p) model (1) is characterized equivalently by either $\alpha_1, \ldots, \alpha_p$, or $\rho(1), \ldots, \rho(p)$, or $\rho_{\text{part}}(1), \ldots, \rho_{\text{part}}(p)$.

Since the introduction of the ordinary AR(p) model, several other AR-type models have been proposed in the literature, not only for real-valued processes, but also for different types of quantitative processes such as count processes (and even for categorical processes), see the surveys by Holan et al. [4], Weiß [5]. In the present work, the focus is on (stationary and square-integrable) AR-type count processes $(X_t)_\mathbb{Z}$, i.e., where the X_t have a quantitative range contained in $\mathbb{N}_0 = \{0, 1, \ldots\}$. Here, the AR(p) structure is implied by requiring the conditional mean at each time t to be linear in the last p observations [6], i.e.,

$$E[X_t \mid X_{t-1}, \ldots] = \alpha_0 + \alpha_1 X_{t-1} + \ldots + \alpha_p X_{t-p}, \tag{4}$$

because then, the YW-equations (2) immediately follow by using the law of total covariance. Note that one also has to require $\alpha_0 > 0$ and $\alpha_1, \ldots, \alpha_p \geq 0$, as the counts X_t are non-negative rvs having a truly positive mean, computed as $\mu = \alpha_0/(1 - \alpha_1 - \ldots - \alpha_p)$. The considered class of count processes satisfying (4) covers many popular special cases, such as the INAR(p) model (integer-valued AR) by Du & Li [7], the INARCH(p) model ('CH' = conditional heteroscedasticity) by Ferland et al. [8], or their bounded-counts counterparts discussed in Kim et al. [9]; see Section 2 for further details. These count processes satisfying (4), however, are not truly linear processes: by contrast to (1), there is no linear relation between their observations.

As all these AR(p)-like count processes satisfy the YW-equations (2) and, thus, the PACF characterization (3), it is common practice to employ the sample PACF (SPACF) for model identification given a count time series $X_1, \ldots, X_n$. More precisely, one commonly computes the SPACF from $X_1, \ldots, X_n$, $\hat{\rho}_{\text{part}}(h)$ for $h = 1, 2, \ldots$, and checks for the pattern (3) among those SPACF values that are classified as being significantly different from zero. An analogous procedure is common during a later step of the Box–Jenkins program. After having fitted a model to the data, one commonly computes the Pearson residuals to check the model adequacy; see Weiß [5], Jung & Tremayne [10] as well as Section 2. While, for an adequate model fit, the Pearson residuals are expected to be uncorrelated, significant SPACF values computed thereof would indicate that the fitted model does not adequately capture the true dependence structure. In both cases, practitioners usually evaluate the significance of $\hat{\rho}_{\text{part}}(h)$ based on the following asymptotic result (see [11] (Theorem 8.1.2)):

$$\sqrt{n}\,\hat{\rho}_{\text{part}}(h) \stackrel{a}{\sim} N(0,1) \quad \text{for lags } h \geq p, \tag{5}$$

i.e., the value $\hat{\rho}_{\text{part}}(h)$ is compared to the critical values $\pm z_{1-\alpha/2}/\sqrt{n}$ to test the null hypothesis of an AR(h − 1) process on level α. Here, $N(\mu, \sigma^2)$ denotes the normal distribution with mean μ and variance σ^2, and z_γ abbreviates the γ-quantile of $N(0,1)$. The aforementioned critical values are automatically plotted in SPACF plots by common statistical software, e.g., if one uses the command pacf in R. However, Theorem 8.1.2 in Brockwell & Davis [11] assumes that the SPACF is computed from a truly linear AR(p) process as in (1), which is neither the case for the aforementioned AR-type count processes, nor for the Pearson residuals computed thereof. Thus, it is not clear if the approximation (5) is asymptotically correct and sufficiently precise in finite samples. In fact, some special asymptotic results in Kim & Weiß [12], Mills & Seneta [13], see Section 3 for further details, as well as some simulation results for Pearson residuals in Weiß et al. [14] indicate that this is generally indeed *not* the case.

Therefore, several alternative ways of implementing the PACF-test are presented in Section 4, namely relying on different types of bootstrap schemes for count time series. The performance of these bootstrap implementations compared to the asymptotic ones is analyzed in a comprehensive simulation study. In Section 5, we start with the case where the SPACF is applied to the original count time series (X_t) with the aim of identifying the AR model order. Afterwards in Section 6, we consider the case of applying the SPACF to the (non-integer) Pearson residuals computed based on a model fit, i.e., the SPACF is used for checking the model adequacy. Our findings are also illustrated by a real-data example on claims counts in Section 7. Here, the computations and simulations in Sections 5–7 have been performed with the software R, and the documented R-code for Section 7 is provided in the Supplementary Materials to this article. Further R-codes can be obtained from the corresponding author upon request. We conclude the article in Section 8.

2. On AR-Type Count Time Series and Pearson Residuals

Several (stationary and square-integrable) AR-type count processes $(X_t)_\mathbb{Z}$, which also have a conditional linear mean according to (4), have been discussed in the literature. Most of these processes either follow a model recursion using so-called thinning operators (typically referred to as INAR models), or they are defined by specifying the conditional distribution of $X_t|X_{t-1},\ldots$ together with condition (4), leading to INARCH models, see Weiß [5] for a survey. For this research, we focus on the most popular instance of these two classes, namely the INAR(p) model of Du & Li [7] on the one hand, and the INARCH(p) model of Ferland et al. [8] on the other hand.

The INAR(p) model of Du & Li [7] makes use of the binomial thinning operator "$\circ$" introduced by Steutel & van Harn [15]. Having the parameter $\alpha \in (0;1)$ and being applied to a count rv X, it is defined by the conditional binomial distribution $\alpha \circ X|X \sim \text{Bin}(X,\alpha)$, where the boundary cases are included as $0 \circ X = 0$ and $1 \circ X = X$. Let $(\epsilon_t)_\mathbb{Z}$ be square-integrable i.i.d. count rv, denote $\mu_\epsilon = E[\epsilon_t]$ and $\sigma_\epsilon^2 = V[\epsilon_t]$. Then, the INAR(p) process $(X_t)_\mathbb{Z}$ is defined by the recursion

$$X_t = \alpha_1 \circ X_{t-1} + \ldots + \alpha_p \circ X_{t-p} + \epsilon_t, \tag{6}$$

where all thinnings are executed independently of each other, and where $\alpha_\bullet := \sum_{j=1}^p \alpha_j < 1$ is assumed to ensure a stationary solution. The INAR(p) process (6) constitutes a pth-order Markov process, the transition probabilities of which are a convolution between the p binomial distributions $\text{Bin}(X_{t-1},\alpha_1),\ldots,\text{Bin}(X_{t-p},\alpha_p)$ and the innovations' distribution [16] (p. 469). The conditional mean satisfies (4) with $\alpha_0 = \mu_\epsilon$, and the conditional variance is given by

$$V[X_t \mid X_{t-1},\ldots] = \sigma_\epsilon^2 + \sum_{j=1}^p \alpha_j(1-\alpha_j)\, X_{t-j}, \tag{7}$$

see Drost et al. [16] (p. 469). The default choice for ϵ_t in the literature is a Poisson (Poi) distribution (which is the integer counterpart to the normal distribution), leading to the Poi-INAR(p) process. However, any other (non-degenerate) count distribution for ϵ_t might be used as well, such as the negative-binomial (NB) distribution for increased dispersion, leading to the NB-INAR(p) process. In the case of such a parametric specification for ϵ_t, ones computes the moments $\mu_\epsilon, \sigma_\epsilon^2$ according to this model, and then the conditional mean and variance according to (4) and (7), respectively.

The INARCH(p) model of Ferland et al. [8] directly assumes the conditional linear mean (4) to hold, and then specifies the conditional distribution of $X_t|X_{t-1},\ldots$ In Ferland et al. [8], the case of a conditional Poi-distribution is assumed, i.e., altogether

$$X_t|X_{t-1},\ldots \sim \text{Poi}(\alpha_0 + \alpha_1 X_{t-1} + \ldots + \alpha_p X_{t-p}), \tag{8}$$

such that the conditional variance of this Poi-INARCH(p) process equals $V[X_t \mid X_{t-1}, \ldots] = E[X_t \mid X_{t-1}, \ldots]$. However, other choices for the conditional distribution of $X_t \mid X_{t-1}, \ldots$ have been investigated in the literature; see [5] (Section 4.2).

For parameter estimation, one commonly uses either simple method-of-moment (MM) estimators (i.e., derived from marginal sample moments and the sample ACF, also see Appendix A), or the more advanced conditional maximum likelihood (CML) estimators, which are computed by using a numerical optimization routine (see [5] (Section 2.2)). It should be noted that for the INAR(p) model, also a semi-parametric specification exists (where the innovations' distribution is left unspecified). The corresponding semi-parametric CML estimator was analyzed by Drost et al. [16]; see also the small-sample refinement by Faymonville et al. [17]. It leads to non-parametric estimates for the probabilities $p_{\epsilon,k} = P(\epsilon_t = k)$ for k between some finite bounds $0 \le l < u < \infty$ (and $p_{\epsilon,k} = 0$ for $k \notin \{l, \ldots, u\}$), which can then be used for computing $\mu_\epsilon, \sigma_\epsilon^2$ as required for the conditional moments (4) and (7). More precisely, the rth moment, $r \in \mathbb{N}$, is given by $E[\epsilon_t^r] = \sum_{k=l}^{u} k^r p_{\epsilon,k}$.

After having fitted a model to the count time series $X_1, \ldots, X_n$, a widely used approach for checking the model adequacy is to investigate the corresponding (standardized) Pearson residuals [5,10,14,18,19]. Let the parameters of the considered AR(p)-type model be comprised in the vector θ, and let $\hat{\theta}$ denote the estimated parameters of the fitted model. Furthermore, let us write the conditional mean as $E[X_t \mid X_{t-1}, \ldots; \theta]$ and the conditional variance as $V[X_t \mid X_{t-1}, \ldots; \theta]$ to express their dependence on the actual parameter values. Then, the Pearson residuals are defined as

$$R_t := R_t(\hat{\theta}) = \frac{X_t - E[X_t \mid X_{t-1}, \ldots; \hat{\theta}]}{\sqrt{V[X_t \mid X_{t-1}, \ldots; \hat{\theta}]}} \quad \text{for } t = p+1, \ldots, n. \tag{9}$$

If the fitted AR(p)-type model is adequate for $X_1, \ldots, X_n$, then $R_{p+1}, \ldots, R_n$ should have a sample mean (variance) close to zero (one), and they should be uncorrelated. These necessary criteria are then used as adequacy checks. In the present research, our focus is on the SPACF computed from $R_{p+1}, \ldots, R_n$, which, for an adequate model fit, should not have values being significantly different from zero.

3. Some Asymptotic Results for the Sample PACF

The basic asymptotic result (5), which has been shown for the SPACF being computed from a true AR(p) process, has been extended in several directions. First, some refinements have been derived by Anderson [20,21] and further investigated by Kwan [22], who, however, assume the data-generating process (DGP) to be i.i.d. Gaussian, i.e., neither AR dependence nor count rvs are covered by their results. More precisely, Anderson [20] complements the asymptotic variance $1/n$ in (5) by the following $O(n^{-2})$ approximation of the mean:

$$E[\hat{\rho}_{\text{part}}(h)] \stackrel{a}{=} \begin{cases} -1/n + O(n^{-2}) & \text{if } h \text{ odd,} \\ -2/n + O(n^{-2}) & \text{if } h \text{ even.} \end{cases} \tag{10}$$

While the Gaussian assumption is weakened by the statement that the result (10) "seems likely to have some validity for many non-Gaussian distributions" [20] (p. 406), the i.i.d.-assumption is not relaxed.

The $O(n^{-2})$ approximation in (10) is extended to a corresponding $O(n^{-3})$ approximation in Anderson [21] (pp. 565–566).

$$E[\hat{\rho}_{\text{part}}(h)] \stackrel{a}{=} \begin{cases} -\dfrac{1}{n} - \dfrac{h-1}{n^2} + O(n^{-3}) & \text{if } h \text{ odd,} \\ -\dfrac{2}{n} - \dfrac{h/2 - 2}{n^2} + O(n^{-3}) & \text{if } h \text{ even,} \end{cases} \tag{11}$$

$$V[\hat{\rho}_{\text{part}}(h)] \stackrel{a}{=} \frac{1}{n} - \frac{h+2}{n^2} + O(n^{-3}).$$

While the $O(n^{-3})$ extension in (11) seems relevant only for very small sample sizes n, the alternating pattern for the mean in (10) might affect the performance of the normal approximation also for larger n.

Another extension of the basic asymptotic result (5) is due to Kim & Weiß [12], Mills & Seneta [13]. These authors consider two particular types of AR(1) count process, namely a Poi-INAR(1) and a binomial AR(1) process, respectively, and derive an $O(n^{-2})$ approximation of $V[\hat{\rho}_{part}(h)]$ for $h \geq 2$. While their exact formulae are not relevant for the present research, the crucial point is as follows: In both cases, the approximate variance is of the form $(1+c)/n$, where c is inverse proportional to the mean μ, and also depends on the value of $\rho(1)$. Especially for low means μ, the numerator $1+c$ deviates notably from 1. Hence, the basic asymptotics (5) do not hold for these types of count process. An analogous conclusion can be drawn from the simulation results in Weiß et al. [14] (Table 1), where the rejection rate for the SPACF of the Pearson residuals (with CML-fitted Poi-INAR(p) model) under the basic asymptotic critical values (5) is analyzed. These rejection rates are often below the intended level, which indicates that (5) does not hold here.

These possible drawbacks of existing asymptotic results are illustrated by Figure 1. The upper panel refers to the mean of SPACF(h), which is either computed from 10^4 simulated Poi-INAR(1) time series (black and dark grey bars), or according to the refined asymptotic result (11) (light grey bars). Note that the sample size $n = 1000$ was chosen rather large such that sample properties and (true) asymptotic properties should agree reasonably well. In Figure 1a, where the SPACF is computed from the raw counts (X_t), we omit plotting the mean at $h = 1$ as this would violate the graphic's Y-range (recall that $\rho_{part}(1) = \alpha$). From (a) and (b), we recognize that the simple asymptotics (5), where the mean of SPACF is approximated by zero, would be misleading in practice, because a negative bias with an oscillating pattern (odd vs. even lags) is observed. As a consequence, if testing the PACF based on (5) and thus ignoring the bias, we may get unreliable sizes, which is also later observed in our simulation studies. The alternating pattern of the bias in (a) and (b) is similar to the refined asymptotics (11). However, we do not observe an exact agreement to (11), as the simulated means seem to depend on the actual value of the AR-parameter α. The effect of α gets much stronger in (c), where even positive bias values for low h are observed, contradicting (11). This is caused by the use of the MM estimator, which is known to be increasingly biased with increasing α [23]; a possible solution for practice could be to use a bias-corrected version of the MM estimator. The lower panel in Figure 1 shows the corresponding standard deviations (SDs). The strongest deviation between simulated and asymptotic results is observed for lag $h = 1$, followed by lag $h = 2$. In particular, for both types of Pearson residuals and both $h = 1, 2$, the asymptotic SD from (11) is too large (and the asymptotic SD according to (5) would even be larger) such that a corresponding PACF-test is expected to be conservative (which is later confirmed by our simulation study). Therefore, it seems advisable to look for other ways of implementing the PACF-test, neither relying on (5) nor (11). An approximation based on asymptotic results does not look promising in general, as we expect the asymptotics to highly depend on the actual DGP, recall the aforementioned results by Kim & Weiß [12], Mills & Seneta [13]. Thus, in what follows, our idea is to try out different types of bootstrap implementations, i.e., the true distribution of the SPACF is approximated by appropriate resampling schemes. This might also allow to account for the effect of the selected estimator when computing the Pearson residuals.

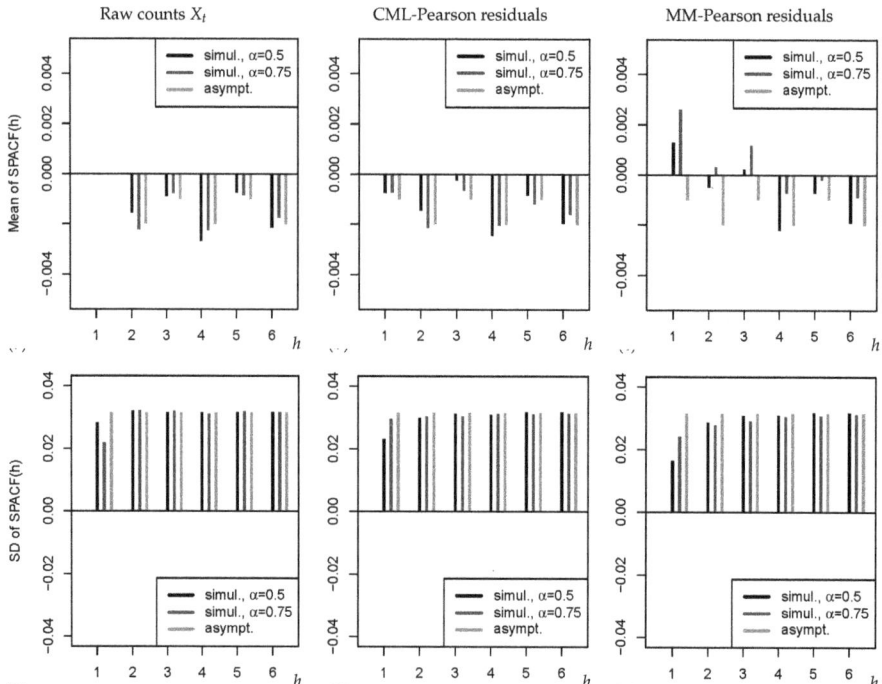

Figure 1. Means in (**a**–**c**) and SDs in (**d**–**f**) of SPACF(h) for sample size $n = 1000$, either simulated values for Poi-INAR(1) DGP with $\mu = 5$ and AR-parameter α, or asymptotic values from (11). SPACF computed from raw counts, and from Pearson residuals with CML or MM estimation.

4. Bootstrap Approaches for the Sample PACF

Let ϑ denote the parameter of interest for the actual DGP (Y_t), and let $\hat{\vartheta} = T(Y_1, \ldots, Y_n)$ denote an estimate of it (in the present research, this parameter is the (S)PACF at some lag $h \in \mathbb{N}$). Analogously, let (Y_t^*) denote a corresponding bootstrap DGP, and let $\hat{\vartheta}^* = T(Y_1^*, \ldots, Y_n^*)$ be the estimator obtained from a bootstrap sample. If $E^*[\cdot]$ denotes the expectation operator of the bootstrap DGP, that is, conditional on the data $X_1, \ldots, X_n$, then the centered bootstrap estimate is given by $\hat{\vartheta}_{\text{cent}}^* := \hat{\vartheta}^* - E^*[\hat{\vartheta}^*]$. A common approach for constructing a two-sided bootstrap confidence interval (CI) for ϑ with confidence level $1 - \alpha \in (0; 1)$ is given by

$$\left[\hat{\vartheta} - q_{1-\alpha/2}(\hat{\vartheta}_{\text{cent}}^*); \; \hat{\vartheta} - q_{\alpha/2}(\hat{\vartheta}_{\text{cent}}^*)\right], \tag{12}$$

where $q_\gamma(\cdot)$ denotes the γ-quantile, see Hall [24]. The bootstrap CI (12) is used for testing the null hypothesis "$H_0 : \vartheta = \vartheta_0$" on level α by applying the following decision rule: reject H_0 if ϑ_0 is not contained in the CI (12). This implies the equivalent decision rule to reject H_0 if

$$\hat{\vartheta} - \vartheta_0 < q_{\alpha/2}(\hat{\vartheta}_{\text{cent}}^*) \quad \text{or} \quad \hat{\vartheta} - \vartheta_0 > q_{1-\alpha/2}(\hat{\vartheta}_{\text{cent}}^*). \tag{13}$$

In the present article, ϑ refers to the PACF at lag h, computed from either the original count process (X_t), or from the Pearson residuals (R_t) obtained after model fitting. In both cases, the PACF at lag h is tested against the hypothetical value $\vartheta_0 = 0$, as it would be the case for an AR-type process of order $< h$.

If we apply the PACF to the original count time series $X_1, \ldots, X_n$, then the following setups are considered:

- fully parametric setup: a fully parametric count AR(p) model with $p \leq 2$ is fitted to the data and then used as the bootstrap DGP; the PACF at certain lags $h > p$ is tested against zero. Here, we focus on the Poi-INAR(p) model, and we use the parametric INAR-bootstrap of Jentsch & Weiß [25].
- semi-parametric setup: a semi-parametric count AR(p) model is fitted to the data [16] and then used as the bootstrap DGP; the PACF at lags $h > p$ is tested against zero. Here, we focus on the INAR(p) model with unspecified innovations, and we use the semi-parametric INAR-bootstrap of Jentsch & Weiß [25].
- non-parametric setup: we use the circular block bootstrap as considered by Politis & White [26], where an automatic block-length selection might be done by using the function b.star in R-package "np" (https://CRAN.R-project.org/package=np, accessed on 31 March 2022).

In case of an INAR(p) bootstrap DGP, the centering at lag h is done by the lag-h PACF corresponding to the fitted model, i.e., which satisfies the YW-equations (2) under estimated parameters, see Appendix A for computational details. In case of the non-parametric block bootstrap, the sample PACF at lag h is used for centering the bootstrap values.

If we apply the PACF to the Pearson residuals (R_t), then again (semi-)parametric setups are considered, where also model fitting is replicated based on the bootstrap time series, as well as the subsequent computation of Pearson residuals based on the bootstrap model fit. This time, a centering is not necessary. Non-parametric bootstrap schemes can be directly applied to the original Pearson residuals (without the need for model fitting during bootstrap replication). Under the null of model adequacy, we expect the available Pearson residuals to be uncorrelated. Thus, a first idea is to apply the classical Efron bootstrap [27], although this bootstrap scheme actually requires i.i.d. data. Therefore, as a second idea, we also apply the aforementioned block bootstrap to (R_t) to account for possible non-linear dependencies.

Remark 1. *For implementing the (semi-)parametric INAR bootstraps, or for computing the Pearson residuals with respect to an INAR model, the model parameters have to be estimated. The following approaches are used for this purpose:*

- *If the fully parametric Poi-INAR(p) model is fitted, we use either the MM estimator of $\theta = (\alpha_1, \ldots, \alpha_p, \mu_\epsilon)$, which is obtained by solving the mean equation $\mu = \mu_\epsilon/(1 - \alpha_1 - \ldots - \alpha_p)$ as well as the YW-equations (2) for $h = 1, \ldots, p$ in $\mu_\epsilon, \alpha_1, \ldots, \alpha_p$ and by plugging-in the sample counterparts for $\mu, \rho(1), \ldots, \rho(p)$, or the CML estimator of θ. The latter is obtained by numerically maximizing the conditional log-likelihood function $\ell(\theta \mid x_p, \ldots, x_1) = \sum_{t=p+1}^{T} \ln p(x_t \mid x_{t-1}, \ldots, x_{t-p}, \theta)$, where the transition probabilities $p(x_t \mid x_{t-1}, \ldots, x_{t-p})$ are computed by evaluating the convolution of the p thinnings' binomial distributions and the innovations' Poisson distribution, i.e., $\text{Bin}(x_{t-1}, \alpha_1) \star \ldots \star \text{Bin}(x_{t-p}, \alpha_p) \star \text{Poi}(\mu_\epsilon)$.*
- *If the semi-parametric Poi-INAR(p) model is fitted, then the innovations' distribution is not specified. As a result, the parameter vector now equals $\theta_{sp} = (\alpha_1, \ldots, \alpha_p, p_{\epsilon,0}, p_{\epsilon,1}, \ldots)$, and we use the semi-parametric CML approach of Drost et al. [16] for estimation. In this case, the transition probabilities for the log-likelihood function $\ell(\theta_{sp} \mid x_p, \ldots, x_1)$ are obtained from the convolution $\text{Bin}(x_{t-1}, \alpha_1) \star \ldots \star \text{Bin}(x_{t-p}, \alpha_p) \star G_\epsilon$, where G_ϵ expresses the unspecified innovations' distribution with probability masses $p_{\epsilon,0}, p_{\epsilon,1}, \ldots$*

5. PACF Diagnostics for Raw Counts

In the first part of our simulation study, we analyze the performance of the asymptotic and (semi-)parametric implementations of PACF-tests if these are applied to the raw counts (X_t) (the results of the non-parametric bootstrap schemes are discussed separately in Remark 2). We consider 1st- and 2nd-order AR-type DGPs, where the aim of applying the PACF-tests (nominal level 0.05) is the identification of the correct AR-order p. As the bootstrap versions of these tests are computationally very demanding (especially the semi-parametric INAR bootstrap), we use the warp-speed method of Giacomini et al. [28] for executing the simulations. This, in turn, allows us to use 10^4 replications throughout our simulation study. We also cross-checked that the achieved rejection rates are close to those obtained by a traditional bootstrap implementation with $B = 500$ bootstrap replications per simulation run. All simulations have been done with the software R, and R-codes can be obtained from the corresponding author upon request.

Table 1 shows the rejection rates of the PACF-tests for different types of AR(1)-like count DGP, recall Section 2. There, the PACFs are computed from a simulated count time series $x_1, \ldots, x_n$ of length n, where the choice $n = 100$ ($n = 1000$) represents the small (large) sample behaviour. The results refer to the medium autocorrelation case $\rho(1) = 0.5$, but further results for $\rho(1) \in \{0.25, 0.75\}$ are summarized in Appendix B, see Table A1. Five implementations of the PACF-test are considered: using the simple asymptotic approximation (5) or the refined one (11) (recall Section 3), using the parametric Poi-INAR(1) bootstrap with either MM or CML estimates, and using the semi-parametric INAR(1) bootstrap with CML estimates (recall Section 4). If first looking at the block "Poi-INAR(1) DGP" in Table 1, we recognize that all implementations perform roughly the same, i.e., the rejection rate at lag $h = 1$ (expressing the power of the PACF-test) is close to 1, and the rejection rates at lags $h \geq 2$ (expressing the size) are close to the 0.05-level. It should be noted, however, that for $\rho(1) = 0.25$, see Table A1, the asymptotic implementations have notably less power at lag $h = 1$. An analogous conclusion holds for the NB-INAR(1) block in Table 1, although now, the model behind the parametric Poi-INAR(1) bootstrap is misspecified. So the parametric bootstrap exhibits robustness properties in finite samples. In the third block, "Poi-INARCH(1)", also the semi-parametric bootstrap is misspecified, but again the rejection rates are robust for $\rho(1) = 0.5$. For $\rho(1) = 0.75$ in Table A1, however, we observe size exceedences for lags $h \geq 2$, i.e., the misspecification of Poi-INARCH(1) as Poi-INAR(1) is not negligible anymore for this DGP. This is plausible in view of Remark 4.1.7 in Weiß [5], where it is shown that these models lead to different sample paths for high autocorrelation. Much more surprising, also both asymptotic implementations deteriorate (even more severely) for a Poi-INARCH(1) DGP with $\rho(1) = 0.75$, see Table A1, i.e., we get too many false rejections in any case. Thus, if one anticipates that the data are generated by an INARCH process, a tailor-made parametric bootstrap implementation of the PACF-tests should be used.

Table 1. Rejection rates of PACF-tests applied to DGP with $\mu = 5$ and $\rho(1) = 0.5$, where semi-parametric (parametric) bootstrap relies on null of (Poi-)INAR(1) process.

True DGP:		Poi-INAR(1) PACF at Lag $h =$				NB-INAR(1), $\frac{\sigma^2}{\mu} = 1.5$ PACF at Lag $h =$				Poi-INARCH(1) PACF at Lag $h =$			
Method	n	1	2	3	4	1	2	3	4	1	2	3	4
asym. (5)	100	0.998	0.053	0.040	0.044	0.998	0.054	0.043	0.046	0.995	0.054	0.046	0.044
	1000	1.000	0.056	0.047	0.049	1.000	0.057	0.051	0.053	1.000	0.056	0.053	0.050
asym. (11)	100	0.998	0.054	0.047	0.048	0.997	0.052	0.050	0.051	0.997	0.047	0.049	0.048
	1000	1.000	0.050	0.049	0.051	1.000	0.061	0.053	0.049	1.000	0.060	0.051	0.048
param. MM	100	1.000	0.052	0.055	0.056	0.999	0.053	0.052	0.055	1.000	0.048	0.053	0.049
	1000	1.000	0.055	0.052	0.049	1.000	0.047	0.052	0.046	1.000	0.046	0.056	0.046
param. CML	100	1.000	0.054	0.051	0.054	0.999	0.049	0.053	0.050	0.999	0.059	0.046	0.049
	1000	1.000	0.048	0.049	0.057	1.000	0.050	0.055	0.051	1.000	0.052	0.052	0.047
semi-p. CML	100	1.000	0.053	0.053	0.051	1.000	0.054	0.051	0.049	0.999	0.044	0.048	0.054
	1000	1.000	0.047	0.054	0.054	1.000	0.051	0.049	0.057	1.000	0.052	0.054	0.052

Let us continue our performance analyses by turning to 2nd-order DGPs. In Table 2, the (semi-)parametric bootstrap schemes are still executed by (erroneously) assuming a 1st-order INAR DGP (like in Table 1), i.e., they are affected by a (further) source of model misspecification. But as seen from the rejection rates in Table 2, we still have good size ($h \geq 3$) and power values ($h = 1, 2$), comparable to those of the refined asymptotic implementation (11). By contrast, the simple asymptotic (5) leads to a clearly reduced power at lag $h = 2$. Finally, in Table 3, the bootstrap schemes now correctly assume a 2nd-order

INAR DGP, i.e., we only have the following misspecifications left: parametric Poi-INAR(2) bootstrap applied to NB-INAR(2) or Poi-INARCH(2) DGP, and semi-parametric INAR(2) bootstrap applied to Poi-INARCH(2) DGP. It can be seen that the parametric bootstrap using MM estimates as well as the semi-parametric bootstrap lead to improved power at lag $h = 2$, whereas the parametric CML-setup even deteriorates (especially under Poi-INARCH(2) misspecification). The latter observation fits well to later results in Section 6, where the parametric bootstrap with CML estimates does again worse than its MM- or semi-CML-counterparts. This can be explained by the fact that for a fully parametric CML approach, model misspecification affects the estimates of all parameters simultaneously, while for the MM approach, for example, the estimation of mean and dependence parameters coincide across all three types of DGP. So it does not seem advisable to use a fully parametric bootstrap in combination with CML estimation for PACF diagnostics.

Table 2. Like Table 1, but 2nd-order DGPs with $\alpha_2 = 0.2$.

True DGP:		Poi-INAR(2)				NB-INAR(2), $\frac{\sigma^2}{\mu} = 1.5$				Poi-INARCH(2)			
		PACF at Lag $h=$				PACF at Lag $h=$				PACF at Lag $h=$			
Method	n	1	2	3	4	1	2	3	4	1	2	3	4
asym. (5)	100	0.984	0.383	0.048	0.047	0.983	0.390	0.047	0.048	0.983	0.384	0.049	0.048
	1000	1.000	1.000	0.053	0.053	1.000	1.000	0.056	0.052	1.000	1.000	0.055	0.053
asym. (11)	100	0.990	0.478	0.048	0.049	0.987	0.480	0.053	0.047	0.986	0.480	0.053	0.054
	1000	1.000	1.000	0.052	0.051	1.000	1.000	0.056	0.051	1.000	1.000	0.056	0.062
param. MM	100	0.996	0.470	0.055	0.058	0.995	0.465	0.053	0.054	0.995	0.482	0.048	0.054
	1000	1.000	1.000	0.056	0.058	1.000	1.000	0.053	0.051	1.000	1.000	0.055	0.054
param. CML	100	0.995	0.470	0.044	0.050	0.994	0.465	0.054	0.051	0.994	0.471	0.052	0.054
	1000	1.000	1.000	0.052	0.053	1.000	1.000	0.051	0.057	1.000	1.000	0.055	0.058
semi-p. CML	100	0.995	0.475	0.055	0.054	0.996	0.472	0.047	0.053	0.994	0.485	0.049	0.058
	1000	1.000	1.000	0.054	0.052	1.000	1.000	0.054	0.056	1.000	1.000	0.054	0.053

Table 3. Rejection rates of PACF-tests applied to DGP with $\mu = 5$, $\rho(1) = 0.5$ and $\alpha_2 = 0.2$, where semi-parametric (parametric) bootstrap relies on null of (Poi-)INAR(2) process.

True DGP:		Poi-INAR(2)				NB-INAR(2), $\frac{\sigma^2}{\mu} = 1.5$				Poi-INARCH(2)			
		PACF at Lag $h=$				PACF at Lag $h=$				PACF at Lag $h=$			
Method	n	1	2	3	4	1	2	3	4	1	2	3	4
asym. (5)	100	0.984	0.383	0.048	0.047	0.983	0.390	0.047	0.048	0.983	0.384	0.049	0.048
	1000	1.000	1.000	0.053	0.053	1.000	1.000	0.056	0.052	1.000	1.000	0.055	0.053
asym. (11)	100	0.990	0.478	0.048	0.049	0.987	0.480	0.053	0.047	0.986	0.480	0.053	0.054
	1000	1.000	1.000	0.052	0.051	1.000	1.000	0.056	0.051	1.000	1.000	0.056	0.062
param. MM	100	0.992	0.510	0.044	0.050	0.992	0.516	0.049	0.048	0.992	0.531	0.054	0.056
	1000	1.000	1.000	0.046	0.051	1.000	1.000	0.055	0.054	1.000	1.000	0.058	0.052
param. CML	100	0.977	0.447	0.057	0.048	0.994	0.478	0.052	0.050	0.991	0.446	0.055	0.056
	1000	1.000	1.000	0.050	0.053	1.000	1.000	0.053	0.051	1.000	1.000	0.054	0.051
semi-p. CML	100	0.993	0.548	0.053	0.047	0.990	0.521	0.053	0.054	0.992	0.498	0.051	0.049
	1000	1.000	1.000	0.055	0.047	1.000	1.000	0.049	0.047	1.000	1.000	0.051	0.051

To sum up, if computing the SPACF from the raw counts (X_t), with the aim of identifying the AR-order of the given count DGP, the overally best performance is shown by the MM-based parametric and CML-based semi-parametric bootstrap implementation of the PACF-test, but also the refined asymptotic implementation relying on (11) does reasonably well. The latter is remarkable as these asymptotics are not the correct ones regarding the considered count DGPs (also recall the discussion of Figure 1), but it appears that their approximation quality is sufficient anyway. The simple asymptotic implementation (5), by contrast, as it is used by statistical software packages by default, leads to reduced power in some cases. From a practical point of view, as the additional benefit of the (semi-)parametric bootstrap schemes compared to the refined asymptotic implementation (11) is not that large, especially in view of the necessary computational effort, it seems advisable for practice to use (11) for doing the PACF-test. Recall that this recommendation refers to the case, where the SPACF is computed from the raw counts (X_t) to identify the DGP's AR-order. The case of applying the PACF-test to Pearson residuals for checking the model adequacy is analyzed in the following Section 6.

6. PACF Diagnostics for Pearson Residuals

While the raw counts' SPACF is typically computed before model fitting (namely for identifying appropriate candidate models), the PACF-analysis of the Pearson residuals is relevant after model fitting, namely for checking the fitted model's adequacy. Thus, the main difference of the simulations in the present section, compared to those of Section 5, is given by the fact that this time, we first fit a (Poi-)INAR model to the data, and then we apply the SPACF to the Pearson residuals computed thereof. For Poi-INAR model fitting, we again use either MM- or CML-estimation, and then we apply the asymptotic or corresponding parametric bootstrap implementations (like before, we use the warp-speed method). An exception is given by the semi-parametric CML estimation, as in this case, also the semi-parametric bootstrap is used for methodological consistency (and Pearson residuals are computed with respect to an unspecified INAR model). We also consider the same scenarios of model orders as before, i.e., 1st-order DGPs and INAR(1)-fit (Tables 4 and 7), 2nd-order DGPs but still INAR(1)-fit (Tables 5 and 8), and 2nd-order DGPs and INAR(2)-fit (Tables 6 and 9). Recall that the fitted model is now used for both the computation of the Pearson residuals and the implementation of (semi-)parametric bootstrap schemes.

Table 4. Rejection rates of PACF-tests applied to Pearson residuals using MM estimates (DGPs with $\mu = 5$ and $\rho(1) = 0.5$), where both residuals and parametric bootstrap rely on null of Poi-INAR(1) process.

| True DGP: | | Poi-INAR(1) | | | | NB-INAR(1), $\frac{\sigma^2}{\mu} = 1.5$ | | | | Poi-INARCH(1) | | | |
| | | PACF at Lag $h =$ | | | | PACF at Lag $h =$ | | | | PACF at Lag $h =$ | | | |
Method	n	1	2	3	4	1	2	3	4	1	2	3	4
asym. (5)	100	0.000	0.026	0.040	0.043	0.001	0.026	0.038	0.044	0.000	0.031	0.041	0.044
	1000	0.000	0.029	0.044	0.053	0.000	0.030	0.043	0.052	0.001	0.033	0.044	0.050
asym. (11)	100	0.001	0.030	0.046	0.050	0.001	0.032	0.042	0.049	0.001	0.037	0.049	0.049
	1000	0.000	0.030	0.047	0.048	0.000	0.030	0.047	0.051	0.000	0.032	0.047	0.050
param. MM	100	0.056	0.051	0.052	0.048	0.050	0.049	0.049	0.045	0.066	0.050	0.047	0.051
	1000	0.051	0.053	0.049	0.045	0.050	0.051	0.053	0.051	0.060	0.046	0.050	0.044

Let us start with the case of fitting a Poi-INAR model by MM estimation, see Tables 4–6. In Table 4 (1st-order models and DGPs; also see Table A3 in the Appendix B), we recognize that both asymptotic implementations lead to undersizing at lags $h = 1, 2$ (particularly severe at $h = 1$). This is in close agreement to our conclusions drawn from Figure 1 as well as to the findings of Weiß et al. [14]. An analogous observation can be done in Table 6 (2nd-order models and DGPs), but now for lags $h = 1, 2, 3$ (particularly severe at $h = 1, 2$). In both cases, however, the MM-based parametric bootstrap holds the nominal 0.05-level reasonably well. The drawback resulting from this undersizing gets clear in Table 5, where the wrong AR-order was selected during model fitting: the asymptotic implementations lead to a very low power for sample size $n = 100$, implying that one will hardly recognize the inadequate model choice. Thus, if model assumptions are used anyway for computing the Pearson residuals, the asymptotic implementation should be avoided, but the model assumptions should also be utilized for executing the PACF-test by using the parametric bootstrap scheme. As a final remark, strictly speaking, we are always concerned with model misspecification if having an NB-INAR or Poi-INARCH DGP. However, all three DGPs per table have the same conditional mean and, thus, the same autocorrelation structure, only their conditional variances differ. Also the MM-estimates required for computing the conditional mean are identical across all models. Thus, it is not surprising that the rejection rates of the PACF-tests do not differ much among these three types of DGP (but again with slight oversizing for the Poi-INARCH DGPs).

Finally, we did the same simulations again, but using CML instead of MM estimation. Table 7 (as well as Table A5 in the Appendix B) refer to the case of both 1st-order models and 1st-order DGPs. In the first block, where the parametric Pearson residuals are computed by correctly assuming a Poi-INAR(1) DGP, we have again strong undersizing at lag 1 for the asymptotic implementation, but a close agreement to the nominal 0.05-level for the parametric bootstrap. The remaining blocks with NB-INAR(1) and Poi-INARCH(1) DGP, however, differ notably from the corresponding blocks in Tables 7 and A5, respectively. This is plausible as the parametric CML approach for a misspecified model leads to misleading estimates for all parameters. While MM estimation leads to the same estimates for the dependence parameters across the three 1st-order models, these differ for parametric CML estimation. Therefore, we have high rejection rates especially at lag 1 (especially if using the parametric bootstrap), which is desirable on the one hand as the fitted model is indeed not adequate. On the other hand, we did not misspecify the (P)ACF structure (a 1st-order model is correct for all DGPs) but the actual data-generating mechanism, i.e., a user might draw the wrong conclusion from this rejection based on the lag-1 PACF. At this point, it is interesting to look at the semi-parametric model fit and bootstrap in Table 7. For both INAR(1) DGPs, the rejection rates are close to the 0.05-level, which is the desirable result as we are concerned with an adequate model fit. For the Poi-INARCH(1) DGP, by contrast, we get moderately increased rejection rates at lag 1, which again has to be assessed ambivalently: on the one hand, the fitted INAR(1) model is indeed not adequate, but on the other hand, the inadequacy does not refer to the autocorrelation structure.

Table 5. Like Table 4, but 2nd-order DGPs with $\alpha_2 = 0.2$.

| True DGP: | | Poi-INAR(2) | | | | NB-INAR(2), $\frac{\sigma^2}{\mu} = 1.5$ | | | | Poi-INARCH(2) | | | |
| | | PACF at Lag $h =$ | | | | PACF at Lag $h =$ | | | | PACF at Lag $h =$ | | | |
Method	n	1	2	3	4	1	2	3	4	1	2	3	4
asym. (5)	100	0.016	0.264	0.079	0.043	0.014	0.266	0.077	0.043	0.018	0.275	0.080	0.045
	1000	0.975	0.998	0.648	0.191	0.958	0.998	0.625	0.192	0.966	0.999	0.642	0.191
asym. (11)	100	0.013	0.359	0.104	0.061	0.010	0.356	0.100	0.059	0.014	0.369	0.105	0.060
	1000	0.972	0.998	0.662	0.212	0.954	0.998	0.640	0.210	0.962	0.999	0.655	0.210
param. MM	100	0.395	0.365	0.102	0.063	0.369	0.356	0.094	0.064	0.396	0.378	0.092	0.069
	1000	1.000	0.999	0.664	0.201	1.000	0.999	0.646	0.204	1.000	0.999	0.662	0.211

Table 6. Rejection rates of PACF-tests applied to Pearson residuals using MM estimates (DGPs with $\mu = 5$, $\rho(1) = 0.5$, and $\alpha_2 = 0.2$), where both residuals and parametric bootstrap rely on null of Poi-INAR(2) process.

| True DGP: | | Poi-INAR(2) | | | | NB-INAR(2), $\frac{\sigma^2}{\mu} = 1.5$ | | | | Poi-INARCH(2) | | | |
| | | PACF at Lag $h =$ | | | | PACF at Lag $h =$ | | | | PACF at Lag $h =$ | | | |
Method	n	1	2	3	4	1	2	3	4	1	2	3	4
asym. (5)	100	0.000	0.001	0.033	0.034	0.000	0.002	0.035	0.038	0.000	0.002	0.034	0.036
	1000	0.000	0.001	0.036	0.042	0.000	0.001	0.033	0.042	0.000	0.001	0.036	0.040
asym. (11)	100	0.000	0.003	0.041	0.043	0.000	0.003	0.040	0.044	0.000	0.002	0.042	0.046
	1000	0.000	0.001	0.036	0.042	0.000	0.001	0.035	0.044	0.000	0.001	0.036	0.045
param. MM	100	0.050	0.037	0.044	0.048	0.053	0.038	0.047	0.050	0.052	0.038	0.059	0.050
	1000	0.049	0.049	0.046	0.052	0.059	0.049	0.052	0.052	0.067	0.054	0.055	0.053

Table 7. Rejection rates of PACF-tests applied to Pearson residuals using CML estimates (DGPs with $\mu = 5$ and $\rho(1) = 0.5$), where both residuals and bootstrap rely on null of Poi-INAR(1) process (parametric bootstrap) or unspecified INAR(1) process (semi-parametric bootstrap), respectively.

True DGP:		Poi-INAR(1) PACF at Lag $h =$				NB-INAR(1), $\frac{\sigma^2}{\mu} = 1.5$ PACF at Lag $h =$				Poi-INARCH(1) PACF at Lag $h =$			
Method	n	1	2	3	4	1	2	3	4	1	2	3	4
asym. (5)	100	0.009	0.035	0.041	0.046	0.028	0.036	0.041	0.039	0.023	0.038	0.045	0.042
	1000	0.008	0.035	0.045	0.048	0.902	0.182	0.069	0.053	0.745	0.148	0.072	0.051
asym. (11)	100	0.009	0.041	0.046	0.048	0.043	0.055	0.047	0.050	0.034	0.053	0.048	0.051
	1000	0.009	0.039	0.046	0.053	0.909	0.198	0.077	0.058	0.753	0.167	0.072	0.055
param. CML	100	0.052	0.049	0.049	0.045	0.238	0.062	0.048	0.049	0.209	0.064	0.053	0.050
	1000	0.049	0.053	0.049	0.053	0.993	0.226	0.075	0.051	0.963	0.188	0.082	0.048
semi-p. CML	100	0.050	0.051	0.054	0.053	0.057	0.048	0.052	0.044	0.070	0.052	0.049	0.051
	1000	0.039	0.053	0.056	0.048	0.052	0.053	0.055	0.049	0.225	0.067	0.058	0.050

Table 8. Like Table 7, but 2nd-order DGPs with $\alpha_2 = 0.2$.

True DGP:		Poi-INAR(2) PACF at Lag $h =$				NB-INAR(2), $\frac{\sigma^2}{\mu} = 1.5$ PACF at Lag $h =$				Poi-INARCH(2) PACF at Lag $h =$			
Method	n	1	2	3	4	1	2	3	4	1	2	3	4
asym. (5)	100	0.026	0.301	0.084	0.043	0.001	0.403	0.090	0.044	0.001	0.404	0.092	0.045
	1000	0.522	0.999	0.696	0.192	0.001	1.000	0.718	0.178	0.000	1.000	0.726	0.174
asym. (11)	100	0.020	0.391	0.110	0.061	0.002	0.502	0.114	0.059	0.001	0.492	0.125	0.064
	1000	0.508	0.999	0.709	0.212	0.001	1.000	0.733	0.193	0.000	1.000	0.727	0.193
param. CML	100	0.099	0.399	0.114	0.059	0.031	0.514	0.105	0.062	0.028	0.495	0.131	0.063
	1000	0.840	1.000	0.710	0.235	0.041	1.000	0.755	0.183	0.023	1.000	0.737	0.194
semi-p. CML	100	0.222	0.350	0.106	0.063	0.172	0.384	0.098	0.054	0.134	0.410	0.109	0.066
	1000	0.999	0.999	0.664	0.222	0.976	0.999	0.665	0.204	0.917	1.000	0.720	0.208

Table 9. Rejection rates of PACF-tests applied to Pearson residuals using CML estimates (DGPs with $\mu = 5$, $\rho(1) = 0.5$, and $\alpha_2 = 0.2$), where both residuals and bootstrap rely on null of Poi-INAR(2) process (parametric bootstrap) or unspecified INAR(2) process (semi-parametric bootstrap), respectively.

True DGP:		Poi-INAR(2) PACF at Lag $h =$				NB-INAR(2), $\frac{\sigma^2}{\mu} = 1.5$ PACF at Lag $h =$				Poi-INARCH(2) PACF at Lag $h =$			
Method	n	1	2	3	4	1	2	3	4	1	2	3	4
asym. (5)	100	0.002	0.002	0.033	0.036	0.003	0.001	0.037	0.038	0.001	0.001	0.037	0.036
	1000	0.000	0.000	0.037	0.044	0.303	0.021	0.082	0.063	0.158	0.005	0.063	0.047
asym. (11)	100	0.002	0.002	0.039	0.041	0.005	0.004	0.047	0.049	0.002	0.002	0.042	0.046
	1000	0.000	0.001	0.038	0.044	0.332	0.028	0.089	0.070	0.174	0.006	0.074	0.055
param. CML	100	0.002	0.002	0.037	0.043	0.003	0.003	0.051	0.047	0.002	0.002	0.048	0.051
	1000	0.000	0.001	0.038	0.044	0.344	0.023	0.086	0.067	0.186	0.007	0.065	0.058
semi-p. CML	100	0.043	0.038	0.050	0.044	0.044	0.036	0.047	0.043	0.038	0.039	0.052	0.050
	1000	0.035	0.045	0.053	0.053	0.051	0.054	0.055	0.051	0.149	0.056	0.054	0.053

Essentially analogous conclusions can be drawn from Table 9, where we are concerned with both 2nd-order models and 2nd-order DGPs. So let us turn to Table 8, where 1st-order models are fitted to 2nd-order DGPs. Thus, we are concerned with at least an inadequate autocorrelation structure (and sometimes also further model misspecification) such that high rejection rates are desirable. Let us start with the first block about the Poi-INAR(2) DGP. As a consequence of the strong undersizing at lag 1, the parametric bootstrap, and especially the asymptotic implementations, show relatively low power values, especially for the small sample size $n = 100$. The semi-parametric bootstrap, by contrast, has substantially higer power values at lag 1. For lags $h \geq 2$, the rejection rates are similar between the different implementations, with a slight advantage for the refined asymptotics as well as the parametric bootstrap. The discrepancy at lag 1 gets even more extreme for the NB-INAR(2) and Poi-INARCH(2) DGP, then all other implementations than the semi-parametric one lead to power close to zero. For lags 2 and 3, by contrast, the refined asymptotics as well as the parametric bootstrap are again more powerful. However, looking back to Table 5, it seems that the overall most appealing power is shown by the MM-based parametric bootstrap. This type of bootstrap also has the advantage that the necessary computational effort is much less than for the CML-based bootstraps. Thus, altogether, while we recommended to use the refined asymptotics (11) if testing the PACF computed from the raw counts, the PACF analysis of Pearson residuals should be done by the MM-based parametric bootstrap: if computing the Pearson residuals from an MM-fitted Poi-INAR model, and if using this model fit for parametric bootstrap, one has good size properties and an appealing power performance at the same time. Certainly, this recommendation does not exclude to do CML-fitting in a second step, once the correct AR-order has been identified. But during the phase of model diagnostics, at least if n is not particularly large, the parametric-MM solution seems to be best suited.

Remark 2. As mentioned in Section 4, we also tried out fully non-parametric bootstrap schemes. For the case where the PACF-tests are applied to the raw counts (X_t), as discussed in Section 5, the circular block bootstrap was used as a fully non-parametric setup, see Table A2 in the Appendix B for the obtained results. While these implementations lead to an appealing power at lag $h = 1$, strong size deteriorations are observed for $h \geq 2$. The strongest deviations are observed for the fixed block length $b = 5$. Increasing b, first the low-lag rejection rates stabilize at 0.05, while we have undersizing for large lags. For $b = 20, 25$, we have good sizes for $h = 5, 6$, but now the low lags lead to exceedances of 0.05. Thus, tailor-made block lengths would be required for different lags h. The automatic block-length selection via b.star typically leads to block lengths between 5 and 10 (depending on the actual extent of $\rho(1)$), but this causes undersizing throughout, getting more severe for increasing h. The reason why b.star tends to pick block lengths that are too small to capture dependence at larger lags is given by the fact that it is designed to select a block length suitable for inference about the sample mean, but not for the sample PACF. In view of the aforementioned size problems and the unclear choice of block lengths, we discourage from using block-bootstrap implementations of the PACF-test for analyzing the raw counts data.

If doing a PACF-analysis of the Pearson residuals, as we investigate it in the present Section 6, then, besides block-bootstrap implementations, also the Efron bootstrap appears reasonable for this task. For the case where the Pearson residuals rely on MM estimates, simulation results are summarized in Table A4 in the Appendix B. If doing an automatic block-length selection via b.star, we often end up with block length 1 (as the Pearson residuals are uncorrelated under model adequacy). Thus, the b.star-block bootstrap shows nearly the same rejection rates as the Efron bootstrap, but these are too low at lags $h = 1, 2$, like for the asymptotic implementations. Increasing the block length to the fixed values $b = 5$ or $b = 10$, we get an even further decrease in size. Therefore, neither Efron nor block bootstrap offer any advantage compared to the asymptotic implementations. Analogous conclusions hold if model fitting is done by CML estimation, see Table A6 in the Appendix B, so we discourage from the use of Efron and block bootstrap also if doing a PACF-test of the Pearson residuals.

7. Real-Data Application

For illustration, we pick up a widely discussed data example from the literature, namely the claims counts data introduced by Freeland [29]. These counts express the monthly number of claims caused by burn-related injuries in the heavy manufacturing industry for the period 1987–1994, i.e., the count time series is of length $n = 96$; see Figure 2. Recall that the R-code used for the subsequent computations is provided in the Supplementary Materials. Freeland [29] suggested to model these data by a Poi-INAR(1) model, but following the discussions of subsequent authors, this model choice is not without controversy. For example, the marginal distribution exhibits moderate overdispersion, as the sample variance 11.357 exceeds the mean 8.604. Therefore, some authors suggested to consider an NB-INAR(1) or Poi-INARCH(1) model instead. Furthermore, one may doubt the 1st-order AR-structure, see Weiß et al. [30], as the SPACF in Figure 2 is only slightly non-significant at lag $h = 2$, where the plotted critical values (dashed lines) refer to the PACF-test on level 0.05 based on the simple asymptotic implementation (5). Thus, altogether, we are concerned with a scenario that fits very well to our simulation study in Sections 5 and 6: the null hypothesis for the data is that of a Poi-INAR(1) model, but this model might be misspecified in terms of marginal distribution, model order, or the actual AR-type data-generating mechanism. Moreover, the sample size $n = 96$ and the lag-1 sample ACF 0.452 are close to the parametrizations considered there. In what follows, we apply the different implementations of the PACF-test to (the Pearson residuals computed from) the claims data. Certainly, as we do not know the true model behind the data, we are not in a position to pass definitive judgement on whether or not a test lead to the correct or wrong decision. But we shall discuss the PACF-tests with respect to our simulation results.

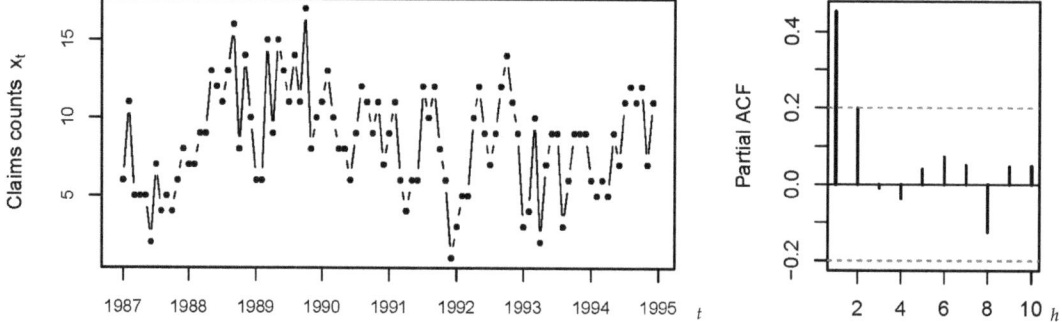

Figure 2. Time series plot and SPACF(h) of claims counts, see Section 7.

Let us start with an analysis of the raw counts' SPACF, in analogy to Section 5. Table 10 summarizes the SPACF(h) values for $h = 1, \ldots, 5$ (bold font) as well as the corresponding critical values (level 0.05). The latter are computed by the five methods considered in Section 5, with the number of bootstrap replications chosen as $B = 1000$. For the simple asymptotic implementation (5), as we have already seen in Figure 2, we get a rejection only at lag 1, whereas the remaining methods reject also at lag 2. Thus, there is indeed evidence that the data might stem from a higher-order model. In addition, the different lag-2 decisions for (5) vs. the remaining implementations appear plausible in view of Table 2, where we found clearly lower power for (5) at $h = 2$. Note that all critical values except (5) are visibly asymmetric, so the SPACF appears rather biased for $n = 96$. Furthermore, all bootstrap implementations lead to quite similar critical values, and the refined asymptotic implementation (11) is also similar to them except for the upper critical value at $h = 1$.

Next, we fit either a Poi-INAR(1) model to the claims counts (via MM or CML), or an unspecified INAR(1) model by the semi-parametric CML approach. Using the resulting model fits, we first compute a set of Pearson residuals for each model, and then the SPACF thereof, like in Section 6. The critical values are determined by both asymptotic approaches as well as by the bootstrap approach corresponding to the respective estimation method. Results are summarized in Table 11. We get only a few rejections anymore, namely for the CML-fit of the Poi-INAR(1) model at lag $h = 2$, both for the refined asymptotics and the parametric bootstrap. The remaining model fits do not lead to a rejection, and one might ask, why? The reason seems to be the respective estimate of the AR(1)-parameter $\alpha_1 = \rho(1)$, which equals only 0.396 for CML, but 0.452 for MM and 0.434 for semi-CML. So the CML-fit explains less of the dependency in the data. The deeper reason for this ambiguous outcome seems to be the low sample size $n = 96$; according to Section 6, we can generally expect only mild power values. It is again interesting to compare the different critical values.

For the Poi-INAR(1) CML-fit, bootstrap and refined asymptotics lead to rather similar critical values, in agreement with our simulation results in Section 6, where a similar performance of both methods was observed. For the remaining estimation approaches, the bootstrap critical values tend to be more narrow than the asymptotic ones, especially at lags 1 and 2. The strongest "shrinkage" of the critical values is observed for $h = 1$, which goes along with our findings in Section 6, where the asymptotic implementation lead to severe undersizing at lag 1, whereas the bootstrap approaches held the nominal level quite well. Furthermore, due to the narrower critical values, the MM and semi-CML bootstraps are also more powerful at lags 1 and 2.

Table 10. SPACF(h) of claims counts (bold font), lower and upper critical values (level 0.05) by different methods, where italic font indicates that critical value is violated.

	Lag h:	1	2	3	4	5
Upper critical value by method ...	asym. (5)	*0.200*	*0.200*	0.200	0.200	0.200
	asym. (11)	*0.186*	*0.175*	0.184	0.173	0.182
	param., MM	*0.134*	*0.175*	0.184	0.181	0.186
	param., CML	*0.148*	*0.167*	0.179	0.182	0.185
	semi-p., CML	*0.138*	*0.171*	0.166	0.162	0.192
	SPACF(h)	**0.452**	**0.198**	**−0.010**	**−0.038**	**0.040**
Lower critical value by method ...	asym. (5)	−0.200	−0.200	−0.200	−0.200	−0.200
	asym. (11)	−0.207	−0.217	−0.205	−0.215	−0.203
	param., MM	−0.211	−0.224	−0.206	−0.202	−0.199
	param., CML	−0.224	−0.223	−0.218	−0.204	−0.197
	semi-p., CML	−0.213	−0.213	−0.198	−0.220	−0.199

Table 11. SPACF(h) of Pearson residuals after fitting a (Poi-)INAR(1) model to the claims counts (bold font). Lower and upper critical values (level 0.05) by different methods, where italic font indicates that critical value is violated.

Poi-INAR(1), MM	Lag h:	1	2	3	4	5
Upper critical value by method ...	asym. (5)	0.201	0.201	0.201	0.201	0.201
	asym. (11)	0.187	0.176	0.185	0.174	0.183
	param., MM	0.108	0.167	0.195	0.184	0.189
	SPACF(h)	**−0.060**	**0.156**	**0.061**	**−0.032**	**−0.007**
Lower critical value by method ...	asym. (5)	−0.201	−0.201	−0.201	−0.201	−0.201
	asym. (11)	−0.208	−0.218	−0.206	−0.216	−0.205
	param., MM	−0.076	−0.190	−0.195	−0.195	−0.202
Poi-INAR(1), CML	Lag h:	1	2	3	4	5
Upper critical value by method ...	asym. (5)	0.201	0.201	0.201	0.201	0.201
	asym. (11)	0.187	*0.176*	0.185	0.174	0.183
	param., CML	0.172	*0.166*	0.183	0.180	0.193
	SPACF(h)	**0.009**	**0.185**	**0.062**	**−0.031**	**−0.002**
Lower critical value by method ...	asym. (5)	−0.201	−0.201	−0.201	−0.201	−0.201
	asym. (11)	−0.208	−0.218	−0.206	−0.216	−0.205
	param., CML	−0.208	−0.213	−0.219	−0.205	−0.201
INAR(1), semi-CML	Lag h:	1	2	3	4	5
Upper critical value by method ...	asym. (5)	0.201	0.201	0.201	0.201	0.201
	asym. (11)	0.187	0.176	0.185	0.174	0.183
	semi-p., CML	0.158	0.171	0.178	0.162	0.203
	SPACF(h)	**−0.041**	**0.165**	**0.064**	**−0.029**	**−0.006**
Lower critical value by method ...	asym. (5)	−0.201	−0.201	−0.201	−0.201	−0.201
	asym. (11)	−0.208	−0.218	−0.206	−0.216	−0.205
	semi-p., CML	−0.142	−0.204	−0.196	−0.215	−0.210

8. Conclusions

In this paper, we considered PACF model diagnostics for AR-type count processes based on raw data and on Pearson residuals, respectively. At first, we illustrated the limitations of the widely used and well-known asymptotic distribution result (as well as some refinements thereof) for the sample PACF values. Then, we introduced appropriate bootstrap schemes for the approximation of the correct sample PACF distribution. We considered a fully parametric bootstrap combined with MM and CML estimation, a semi-parametric bootstrap combined with CML estimation, and a fully non-parametric bootstrap scheme. We compared the performance of the different procedures for first- and second-order AR-type count processes. In the case where we apply the PACF test directly to the raw count data, the best performance was observed for the MM-based parametric bootstrap, CML-based semi-parametric bootstrap, and the refined asymptotic results, where the latter are preferable for computing time reasons. By contrast, when applying the PACF test to the Pearson residuals, we advise using the MM-based parametric bootstrap procedure which simultaneously provides good size properties and power performance. Finally, we applied our different PACF procedures to a well-known data set on claims counts and found some evidence for a higher-order model.

Supplementary Materials: The following supporting information can be downloaded at: https://www.mdpi.com/article/10.3390/e25010105/s1.

Author Contributions: Conceptualization (all authors); Funding acquisition (C.H.W. and C.J.); Methodology (all authors); Software (B.A.); Supervision (C.H.W. and C.J.); Writing—original draft preparation (C.H.W. and B.A.); Writing—review and editing (C.H.W., M.F. and C.J.). All authors have read and agreed to the published version of the manuscript.

Funding: This research was funded by the Deutsche Forschungsgemeinschaft (DFG, German Research Foundation)–Projektnummer 437270842.

Institutional Review Board Statement: Not applicable.

Data Availability Statement: The data presented in this study are available in the supplementary material.

Acknowledgments: The authors thank the two referees for their useful comments on an earlier draft of this article.

Conflicts of Interest: The authors declare no conflict of interest.

Appendix A. On the Equivalence of ACF, PACF, and AR Coefficients

The first p Yule–Walker (YW) equations in

$$\rho(h) = \sum_{i=1}^{p} \alpha_i \rho(|h-i|) \qquad \text{for } h = 1, 2, \ldots \tag{A1}$$

can be rewritten in vector-matrix notation as follows: For $k \in \mathbb{N}$, let $\boldsymbol{\alpha}_k := (\alpha_1, \ldots, \alpha_k)^\top \in \mathbb{R}^k$ with $\alpha_i = 0$ for $i > p$, let $\boldsymbol{r}_k := (\rho(1), \ldots, \rho(k))^\top \in \mathbb{R}^k$, and

$$\mathbf{R}_k := \big(\rho(|i-j|)\big)_{i,j=1,\ldots,k} = \begin{pmatrix} 1 & \rho(1) & \cdots & \rho(k-1) \\ \rho(1) & 1 & \ddots & \vdots \\ \vdots & \ddots & \ddots & \rho(1) \\ \rho(k-1) & \cdots & \rho(1) & 1 \end{pmatrix} \in \mathbb{R}^{k \times k}. \tag{A2}$$

Then, (A1) implies that the AR(p) process satisfies the linear equation

$$\mathbf{R}_p \, \boldsymbol{\alpha}_p = \boldsymbol{r}_p. \tag{A3}$$

Note that $\mathbf{R}_k$ constitutes a symmetric Toeplitz matrix, i.e., it is characterized by having constant diagonals. This type of matrix structure was first considered by Toeplitz [31,32], and it is crucial for efficiently solving (A3) in $\boldsymbol{\alpha}_p$ (see the details below).

Assume that $\mathbf{R}_k$ is invertible, and let $a_k \in \mathbb{R}^k$ be the unique solution of the equation

$$\mathbf{R}_k\, a_k = r_k, \qquad \text{i.e., } a_k = \mathbf{R}_k^{-1} r_k. \tag{A4}$$

Then, the PACF at lag k is defined by $\rho_{\text{part}}(k) := a_{k,k}$ (last component of a_k); let us denote $\pi_k := (\rho_{\text{part}}(1), \ldots, \rho_{\text{part}}(k))^\top \in \mathbb{R}^k$.

If $(X_t)_\mathbb{Z}$ follows an AR(p) model, then (A3) implies that

$$\rho_{\text{part}}(p) = \alpha_p, \qquad \rho_{\text{part}}(h) = 0 \quad \text{for all } h > p, \tag{A5}$$

holds; in particular, we have $a_p = \alpha_p$. Because of the Toeplitz structure of $\mathbf{R}_k$, the YW-equations (A3) can be solved recursively for $k = 1, 2, \ldots$, which was first recognized by Durbin [33], Levinson [34]. The recursive scheme, which is commonly referred to as the *Durbin–Levinson (DL) algorithm*, can be expressed as

$$a_{k+1,k+1} = \frac{\rho(k+1) - \sum_{i=1}^k a_{k,i}\,\rho(k+1-i)}{1 - \sum_{i=1}^k a_{k,i}\,\rho(i)},$$

$$\begin{pmatrix} a_{k+1,1} \\ \vdots \\ a_{k+1,k} \end{pmatrix} = \begin{pmatrix} a_{k,1} \\ \vdots \\ a_{k,k} \end{pmatrix} - a_{k+1,k+1} \begin{pmatrix} a_{k,k} \\ \vdots \\ a_{k,1} \end{pmatrix}. \tag{A6}$$

Given the (sample) ACF, the DL-algorithm (A6) is used to recursively compute the (sample) PACF for $k = 1, 2, \ldots$, where $\rho_{\text{part}}(1) = a_{1,1} = \rho(1)$.

Furthermore, applying the DL-algorithm to (A3), we can compute the AR parameters $\alpha_1, \ldots, \alpha_p$ corresponding to the ACF values $\rho(1), \ldots, \rho(p)$ (or if using the sample ACF, we end up with moment estimates for the AR parameters, referred to as *YW-estimates*). In R, this is readily implemented via `acf2AR`. Given the AR parameters, in turn, (A1) or (A3) can also be solved in the ACF, see Section 3.3 in Brockwell & Davis [11] as well as the R command `ARMAacf`.

The previous discussion shows that an AR(p) model can be characterized equivalently by either α_p or r_p. According to Barndorff-Nielsen & Schou [35], this type of "equivalent parametrization" can be further extended by the one-to-one relationship between α_p and π_p, i.e., we have one-to-one relations between $r_p \leftrightarrow \alpha_p \leftrightarrow \pi_p$. For computing α_p from π_p, Barndorff-Nielsen & Schou [35] suggest to use the DL-algorithm (A6) together with (A4) as follows:

$$\begin{pmatrix} a_{k+1,1} \\ \vdots \\ a_{k+1,k} \\ a_{k+1,k+1} \end{pmatrix} = \begin{pmatrix} a_{k,1} \\ \vdots \\ a_{k,k} \\ 0 \end{pmatrix} - \rho_{\text{part}}(k+1) \begin{pmatrix} a_{k,k} \\ \vdots \\ a_{k,1} \\ -1 \end{pmatrix} \quad \text{for } k = 1, 2, \ldots, \tag{A7}$$

which is initialized by setting $a_{1,1} = \rho_{\text{part}}(1)$. Then, $\alpha_p = a_p$. Altogether, the application of the DL-algorithm allows the transformations

$$r_p \xrightarrow{(A3)} \alpha_p$$
$$(A6) \searrow \quad \nearrow (A7)$$
$$\pi_p$$

By contrast, recall that $\alpha_p \to r_p$ (and thus $\pi_p \to \alpha_p \to r_p$) is done by solving (A1) or (A3) in the ACF, e.g., by using the "third method" in Brockwell & Davis [11] (Section 3.3) or R's `ARMAacf`.

Appendix B. Further Simulation Results

Table A1. Rejection rates of PACF-test applied to DGP with $\mu = 5$, where semi-parametric (parametric) bootstrap relies on null of (Poi-)INAR(1) process.

True DGP:			Poi-INAR(1) PACF at Lag $h=$				NB-INAR(1), $\frac{\sigma^2}{\mu}=1.5$ PACF at Lag $h=$				Poi-INARCH(1) PACF at Lag $h=$			
Method	$\rho(1)$	n	1	2	3	4	1	2	3	4	1	2	3	4
asym. (1.5)	0.25	100	0.629	0.050	0.047	0.043	0.646	0.050	0.042	0.044	0.639	0.049	0.046	0.045
		1000	1.000	0.049	0.050	0.046	1.000	0.054	0.052	0.049	1.000	0.054	0.047	0.050
	0.5	100	0.998	0.053	0.040	0.044	0.998	0.054	0.043	0.046	0.995	0.054	0.046	0.044
		1000	1.000	0.056	0.047	0.049	1.000	0.057	0.051	0.053	1.000	0.056	0.053	0.050
	0.75	100	1.000	0.050	0.042	0.043	1.000	0.057	0.048	0.045	1.000	0.062	0.053	0.047
		1000	1.000	0.050	0.048	0.054	1.000	0.060	0.056	0.055	1.000	0.071	0.063	0.061
asym. (3.2)	0.25	100	0.692	0.045	0.046	0.054	0.688	0.051	0.048	0.052	0.690	0.054	0.052	0.047
		1000	1.000	0.053	0.050	0.052	1.000	0.053	0.051	0.047	1.000	0.051	0.050	0.049
	0.5	100	0.998	0.054	0.047	0.048	0.997	0.053	0.050	0.051	0.997	0.047	0.049	0.048
		1000	1.000	0.050	0.049	0.051	1.000	0.061	0.053	0.049	1.000	0.060	0.051	0.048
	0.75	100	1.000	0.047	0.051	0.046	1.000	0.054	0.054	0.050	1.000	0.060	0.061	0.053
		1000	1.000	0.055	0.054	0.054	1.000	0.062	0.058	0.056	1.000	0.073	0.066	0.060
param. MM	0.25	100	0.742	0.046	0.046	0.050	0.736	0.045	0.054	0.051	0.747	0.053	0.048	0.048
		1000	1.000	0.054	0.048	0.043	1.000	0.048	0.055	0.054	1.000	0.050	0.048	0.050
	0.5	100	1.000	0.052	0.055	0.056	0.999	0.053	0.052	0.055	1.000	0.048	0.053	0.049
		1000	1.000	0.055	0.052	0.049	1.000	0.047	0.052	0.046	1.000	0.046	0.056	0.046
	0.75	100	1.000	0.050	0.057	0.056	1.000	0.052	0.047	0.047	1.000	0.064	0.067	0.061
		1000	1.000	0.061	0.049	0.049	1.000	0.060	0.059	0.054	1.000	0.067	0.062	0.059
param. CML	0.25	100	0.747	0.049	0.049	0.054	0.735	0.046	0.049	0.051	0.726	0.053	0.052	0.055
		1000	1.000	0.051	0.049	0.047	1.000	0.049	0.058	0.049	1.000	0.044	0.051	0.048
	0.5	100	1.000	0.054	0.051	0.054	0.999	0.049	0.053	0.050	0.999	0.059	0.046	0.049
		1000	1.000	0.048	0.049	0.057	1.000	0.050	0.055	0.051	1.000	0.052	0.052	0.047
	0.75	100	1.000	0.052	0.050	0.051	1.000	0.051	0.051	0.050	1.000	0.061	0.055	0.053
		1000	1.000	0.050	0.050	0.052	1.000	0.052	0.058	0.053	1.000	0.069	0.066	0.066
semi-p. CML	0.25	100	0.736	0.043	0.046	0.048	0.723	0.049	0.052	0.051	0.733	0.053	0.053	0.056
		1000	1.000	0.048	0.052	0.047	1.000	0.046	0.053	0.057	1.000	0.054	0.059	0.046
	0.5	100	1.000	0.053	0.053	0.051	1.000	0.054	0.051	0.049	0.999	0.044	0.048	0.054
		1000	1.000	0.047	0.054	0.054	1.000	0.051	0.049	0.057	1.000	0.052	0.054	0.052
	0.75	100	1.000	0.052	0.054	0.054	1.000	0.051	0.051	0.041	1.000	0.064	0.063	0.057
		1000	1.000	0.048	0.046	0.051	1.000	0.051	0.048	0.048	1.000	0.060	0.057	0.060

Table A2. Rejection rates of PACF-test applied to Poi-INAR(1) DGP with $\mu = 5$, where circular block bootstrap with automatically selected ("b.star") or fixed block length b is used.

			PACF at Lag $h=$					
Method	$\rho(1)$	n	1	2	3	4	5	6
b.star	0.25	100	0.922	0.010	0.006	0.007	0.007	0.007
		1000	1.000	0.037	0.025	0.016	0.011	0.006
	0.5	100	1.000	0.024	0.012	0.008	0.008	0.007
		1000	1.000	0.050	0.041	0.032	0.032	0.023
	0.75	100	1.000	0.041	0.027	0.019	0.015	0.010
		1000	1.000	0.045	0.050	0.042	0.037	0.035
$b=5$	0.25	100	0.857	0.031	0.024	0.011	0.005	0.005
		1000	1.000	0.038	0.023	0.011	0.003	0.004
	0.5	100	1.000	0.036	0.020	0.014	0.004	0.008
		1000	1.000	0.042	0.026	0.013	0.004	0.006
	0.75	100	1.000	0.028	0.021	0.008	0.002	0.008
		1000	1.000	0.125	0.074	0.038	0.016	0.029
$b=10$	0.25	100	0.833	0.052	0.047	0.036	0.026	0.022
		1000	1.000	0.045	0.041	0.036	0.031	0.025
	0.5	100	1.000	0.053	0.042	0.032	0.026	0.020
		1000	1.000	0.047	0.040	0.034	0.027	0.024
	0.75	100	1.000	0.045	0.041	0.035	0.023	0.020
		1000	1.000	0.054	0.040	0.034	0.029	0.019
$b=15$	0.25	100	0.831	0.058	0.050	0.051	0.037	0.031
		1000	1.000	0.043	0.048	0.045	0.037	0.030
	0.5	100	1.000	0.058	0.054	0.051	0.042	0.029
		1000	1.000	0.058	0.045	0.046	0.038	0.031
	0.75	100	1.000	0.054	0.042	0.035	0.036	0.034
		1000	1.000	0.053	0.044	0.039	0.031	0.036
$b=20$	0.25	100	0.813	0.064	0.060	0.055	0.051	0.047
		1000	1.000	0.053	0.046	0.045	0.046	0.041
	0.5	100	1.000	0.067	0.065	0.056	0.050	0.044
		1000	1.000	0.052	0.048	0.038	0.045	0.039
	0.75	100	1.000	0.059	0.064	0.051	0.043	0.041
		1000	1.000	0.049	0.048	0.047	0.034	0.038

Table A2. Cont.

Method	$\rho(1)$	n	PACF at Lag $h=$					
			1	2	3	4	5	6
$b=25$	0.25	100	0.819	0.080	0.072	0.058	0.058	0.056
		1000	1.000	0.054	0.058	0.051	0.047	0.044
	0.5	100	1.000	0.078	0.065	0.065	0.052	0.048
		1000	1.000	0.049	0.060	0.050	0.046	0.052
	0.75	100	1.000	0.070	0.064	0.058	0.048	0.051
		1000	1.000	0.054	0.050	0.050	0.048	0.040

Table A3. Rejection rates of PACF-test applied to Pearson residuals using MM estimates (DGPs with $\mu=5$), where both residuals and parametric bootstrap rely on null of Poi-INAR(1) process.

True DGP:			Poi-INAR(1) PACF at Lag $h=$				NB-INAR(1), $\frac{e^2}{\mu}=1.5$ PACF at Lag $h=$				Poi-INARCH(1) PACF at Lag $h=$			
Method	$\rho(1)$	n	1	2	3	4	1	2	3	4	1	2	3	4
asym. (1.5)	0.25	100	0.000	0.038	0.044	0.047	0.000	0.041	0.040	0.046	0.000	0.042	0.046	0.046
		1000	0.000	0.043	0.050	0.050	0.000	0.043	0.049	0.050	0.000	0.045	0.047	0.051
	0.5	100	0.000	0.026	0.040	0.043	0.001	0.026	0.038	0.044	0.000	0.031	0.041	0.044
		1000	0.000	0.029	0.044	0.053	0.000	0.030	0.043	0.052	0.001	0.033	0.044	0.050
	0.75	100	0.011	0.022	0.034	0.037	0.010	0.022	0.030	0.035	0.013	0.026	0.036	0.041
		1000	0.011	0.026	0.035	0.042	0.009	0.024	0.035	0.044	0.017	0.028	0.039	0.045
asym. (3.2)	0.25	100	0.000	0.039	0.046	0.048	0.000	0.043	0.049	0.048	0.000	0.046	0.049	0.045
		1000	0.000	0.046	0.052	0.055	0.000	0.044	0.048	0.048	0.000	0.047	0.047	0.053
	0.5	100	0.001	0.030	0.046	0.050	0.001	0.032	0.042	0.049	0.001	0.037	0.049	0.049
		1000	0.000	0.030	0.047	0.048	0.000	0.030	0.047	0.051	0.000	0.032	0.047	0.050
	0.75	100	0.014	0.031	0.038	0.047	0.015	0.030	0.037	0.045	0.019	0.036	0.045	0.053
		1000	0.011	0.026	0.036	0.042	0.010	0.024	0.034	0.043	0.018	0.029	0.040	0.045
param. MM	0.25	100	0.029	0.050	0.046	0.052	0.030	0.041	0.052	0.047	0.025	0.050	0.046	0.052
		1000	0.047	0.047	0.049	0.050	0.047	0.053	0.047	0.046	0.057	0.053	0.057	0.049
	0.5	100	0.056	0.051	0.052	0.048	0.050	0.049	0.049	0.045	0.066	0.050	0.047	0.051
		1000	0.051	0.053	0.049	0.045	0.050	0.051	0.053	0.051	0.060	0.046	0.050	0.044
	0.75	100	0.059	0.042	0.050	0.053	0.058	0.044	0.046	0.045	0.071	0.056	0.055	0.050
		1000	0.051	0.051	0.047	0.049	0.043	0.046	0.046	0.045	0.064	0.054	0.061	0.053

Table A4. Rejection rates of PACF-test applied to Pearson residuals using MM estimates (Poi-INAR(1) DGPs with $\mu=5$), where circular block bootstrap with automatically selected ("b.star") or fixed block length b is used.

Method	$\rho(1)$	n	PACF at Lag $h=$					
			1	2	3	4	5	6
Efron	0.25	100	0.000	0.047	0.046	0.047	0.051	0.049
		1000	0.000	0.040	0.047	0.051	0.048	0.046
	0.5	100	0.001	0.033	0.054	0.048	0.044	0.050
		1000	0.000	0.033	0.043	0.050	0.050	0.055
	0.75	100	0.014	0.030	0.041	0.044	0.047	0.053
		1000	0.011	0.028	0.032	0.044	0.043	0.046
b.star	0.25	100	0.000	0.041	0.041	0.048	0.045	0.054
		1000	0.000	0.049	0.052	0.047	0.049	0.053
	0.5	100	0.000	0.034	0.044	0.046	0.045	0.054
		1000	0.001	0.030	0.048	0.048	0.047	0.054
	0.75	100	0.008	0.037	0.040	0.044	0.044	0.049
		1000	0.007	0.027	0.033	0.041	0.044	0.049
$b=5$	0.25	100	0.000	0.024	0.035	0.048	0.044	0.047
		1000	0.000	0.023	0.035	0.044	0.055	0.053
	0.5	100	0.001	0.022	0.034	0.044	0.047	0.042
		1000	0.000	0.017	0.033	0.047	0.046	0.049
	0.75	100	0.003	0.017	0.035	0.042	0.050	0.047
		1000	0.003	0.013	0.028	0.039	0.044	0.051
$b=10$	0.25	100	0.000	0.015	0.019	0.027	0.032	0.035
		1000	0.000	0.010	0.017	0.025	0.028	0.033
	0.5	100	0.000	0.016	0.020	0.030	0.033	0.043
		1000	0.000	0.008	0.017	0.021	0.028	0.037
	0.75	100	0.003	0.012	0.018	0.023	0.030	0.034
		1000	0.002	0.008	0.013	0.020	0.025	0.035

Table A5. Rejection rates of PACF-test applied to Pearson residuals using CML estimates (DGPs with $\mu = 5$), where both residuals and bootstrap rely on null of Poi-INAR(1) process (parametric bootstrap) or unspecified INAR(1) process (semi-parametric bootstrap), respectively.

True DGP:			Poi-INAR(1) PACF at Lag $h =$				NB-INAR(1), $\frac{\sigma^2}{\mu} = 1.5$ PACF at Lag $h =$				Poi-INARCH(1) PACF at Lag $h =$			
Method	$\rho(1)$	n	1	2	3	4	1	2	3	4	1	2	3	4
asym. (1.5)	0.25	100	0.001	0.044	0.043	0.047	0.001	0.042	0.040	0.046	0.001	0.045	0.047	0.047
		1000	0.000	0.045	0.051	0.052	0.236	0.057	0.049	0.050	0.000	0.048	0.048	0.052
	0.5	100	0.009	0.035	0.041	0.046	0.028	0.036	0.041	0.039	0.023	0.038	0.045	0.042
		1000	0.008	0.035	0.045	0.048	0.902	0.182	0.069	0.053	0.745	0.148	0.072	0.051
	0.75	100	0.032	0.042	0.043	0.042	0.057	0.041	0.044	0.040	0.364	0.113	0.070	0.047
		1000	0.033	0.040	0.048	0.050	0.581	0.288	0.162	0.094	1.000	0.913	0.488	0.204
asym. (3.2)	0.25	100	0.001	0.043	0.047	0.049	0.002	0.049	0.050	0.048	0.000	0.046	0.049	0.049
		1000	0.000	0.042	0.049	0.048	0.270	0.060	0.049	0.048	0.000	0.045	0.050	0.050
	0.5	100	0.009	0.041	0.046	0.048	0.043	0.055	0.047	0.050	0.034	0.053	0.048	0.051
		1000	0.009	0.039	0.046	0.053	0.909	0.198	0.077	0.058	0.753	0.167	0.072	0.055
	0.75	100	0.034	0.043	0.046	0.048	0.075	0.062	0.052	0.055	0.425	0.165	0.094	0.067
		1000	0.035	0.040	0.046	0.046	0.609	0.310	0.173	0.112	1.000	0.921	0.502	0.221
param. CML	0.25	100	0.040	0.052	0.049	0.048	0.262	0.050	0.051	0.056	0.046	0.051	0.051	0.056
		1000	0.049	0.046	0.053	0.049	1.000	0.065	0.051	0.054	0.194	0.047	0.050	0.047
	0.5	100	0.052	0.049	0.049	0.045	0.238	0.062	0.048	0.049	0.209	0.064	0.053	0.050
		1000	0.049	0.053	0.049	0.053	0.993	0.226	0.075	0.051	0.963	0.188	0.082	0.048
	0.75	100	0.046	0.046	0.050	0.051	0.123	0.079	0.054	0.052	0.606	0.190	0.097	0.068
		1000	0.048	0.053	0.048	0.047	0.709	0.338	0.169	0.121	1.000	0.936	0.522	0.221
semi-p. CML	0.25	100	0.037	0.045	0.047	0.050	0.049	0.044	0.050	0.054	0.045	0.051	0.052	0.051
		1000	0.037	0.054	0.050	0.047	0.037	0.047	0.052	0.052	0.031	0.051	0.054	0.054
	0.5	100	0.050	0.051	0.054	0.053	0.057	0.048	0.052	0.044	0.070	0.052	0.049	0.051
		1000	0.039	0.053	0.056	0.048	0.052	0.053	0.055	0.049	0.225	0.067	0.058	0.050
	0.75	100	0.051	0.044	0.049	0.050	0.049	0.049	0.046	0.047	0.223	0.104	0.072	0.061
		1000	0.046	0.049	0.049	0.051	0.055	0.053	0.050	0.050	0.377	0.217	0.129	0.080

Table A6. Rejection rates of PACF-test applied to Pearson residuals using CML estimates (Poi-INAR(1) DGPs with $\mu = 5$), where circular block bootstrap with automatically selected ("b.star") or fixed block length b is used.

Method	$\rho(1)$	n	PACF at Lag $h =$ 1	2	3	4	5	6
b.star	0.25	100	0.001	0.046	0.050	0.050	0.052	0.049
		1000	0.000	0.050	0.053	0.050	0.050	0.051
	0.5	100	0.007	0.036	0.046	0.054	0.045	0.052
		1000	0.005	0.034	0.046	0.047	0.050	0.050
	0.75	100	0.023	0.038	0.043	0.045	0.042	0.058
		1000	0.022	0.049	0.046	0.047	0.050	0.052
$b = 5$	0.25	100	0.001	0.024	0.037	0.048	0.046	0.048
		1000	0.000	0.020	0.033	0.048	0.045	0.056
	0.5	100	0.003	0.020	0.038	0.039	0.050	0.050
		1000	0.002	0.018	0.035	0.045	0.051	0.049
	0.75	100	0.010	0.023	0.042	0.047	0.050	0.048
		1000	0.009	0.018	0.032	0.045	0.049	0.044
$b = 10$	0.25	100	0.001	0.016	0.020	0.030	0.033	0.042
		1000	0.000	0.012	0.018	0.028	0.029	0.034
	0.5	100	0.003	0.014	0.019	0.030	0.031	0.036
		1000	0.001	0.010	0.015	0.020	0.027	0.031
	0.75	100	0.010	0.016	0.018	0.025	0.040	0.039
		1000	0.006	0.011	0.015	0.023	0.023	0.032

References

1. Walker, G.T. On periodicity in series of related terms. *Proc. R. Soc. Lond. Ser. A* **1931**, *131*, 518–532. [CrossRef]
2. Yule, G.U. On a method of investigating periodicities in disturbed series, with special reference to Wolfer's sunspot numbers. *Philos. Trans. R. Soc. Lond. Ser. A* **1927**, *226*, 267–298.
3. Box, G.E.P.; Jenkins, G.M. *Time Series Analysis: Forecasting and Control*, 1st ed.; Holden-Day: San Francisco, CA, USA, 1970.
4. Holan, S.H.; Lund, R.; Davis, G. The ARMA alphabet soup: A tour of ARMA model variants. *Stat. Surv.* **2010**, *4*, 232–274. [CrossRef]
5. Weiß, C.H. *An Introduction to Discrete-Valued Time Series*; John Wiley & Sons, Inc.: Chichester, UK, 2018.
6. Grunwald, G.; Hyndman, R.J.; Tedesco, L.; Tweedie, R.L. Non-Gaussian conditional linear AR(1) models. *Aust. N. Z. J. Stat.* **2000**, *42*, 479–495. [CrossRef]
7. Du, J.-G.; Li, Y. The integer-valued autoregressive (INAR(p)) model. *J. Time Ser. Anal.* **1991**, *12*, 129–142.
8. Ferland, R.; Latour, A.; Oraichi, D. Integer-valued GARCH processes. *J. Time Ser. Anal.* **2006**, *27*, 923–942. [CrossRef]

9. Kim, H.-Y.; Weiß, C.H.; Möller, T.A. Models for autoregressive processes of bounded counts: How different are they? *Comput. Stat.* **2020**, *35*, 1715–1736. [CrossRef]
10. Jung, R.C.; Tremayne, A.R. Useful models for time series of counts or simply wrong ones? *AStA Adv. Stat. Anal.* **2011**, *95*, 59–91. [CrossRef]
11. Brockwell, P.J.; Davis, R.A. *Time Series: Theory and Methods*, 2nd ed.; Springer: New York, NY, USA, 1991.
12. Kim, H.-Y.; Weiß, C.H. Goodness-of-fit tests for binomial AR(1) processes. *Statistics* **2015**, *49*, 291–315. [CrossRef]
13. Mills, T.M.; Seneta, E. Independence of partial autocorrelations for a classical immigration branching process. *Stoch. Process. Their Appl.* **1991**, *37*, 275–279. [CrossRef]
14. Weiß, C.H.; Scherer, L.; Aleksandrov, B.; Feld, M.H.-J.M. Checking model adequacy for count time series by using Pearson residuals. *J. Time Ser. Econom.* **2020**, *12*, 20180018. [CrossRef]
15. Steutel, F.W.; van Harn, K. Discrete analogues of self-decomposability and stability. *Ann. Probab.* **1979**, *7*, 893–899. [CrossRef]
16. Drost, F.C.; van den Akker, R.; Werker, B.J.M. Efficient estimation of auto-regression parameters and innovation distributions for semiparametric integer-valued AR(p) models. *J. R. Stat. Soc. Ser. B* **2009**, *71*, 467–485. [CrossRef]
17. Faymonville, M.; Jentsch, C.; Weiß, C.H.; Aleksandrov, B. Semiparametric estimation of INAR models using roughness penalization. *Stat. Methods Appl.* **2022**, 1–36. [CrossRef]
18. Harvey, A.C.; Fernandes, C. Time series models for count or qualitative observations. *J. Bus. Econ. Stat.* **1989**, *7*, 407–417.
19. Zhu, F.; Wang, D. Diagnostic checking integer-valued ARCH(p) models using conditional residual autocorrelations. *Comput. Stat. Data Anal.* **2010**, *54*, 496–508. [CrossRef]
20. Anderson, O.D. Approximate moments to $O(n^{-2})$ for the sampled partial autocorrelations from a white noise process. *Comput. Stat. Data Anal.* **1993**, *16*, 405–421. [CrossRef]
21. Anderson, O.D. Exact general-lag serial correlation moments and approximate low-lag partial correlation moments for Gaussian white noise. *J. Time Ser. Anal.* **1993**, *14*, 551–574. [CrossRef]
22. Kwan, A.C.C. Sample partial autocorrelations and portmanteau tests for randomness. *Appl. Econ. Lett.* **2003**, *10*, 605–609. [CrossRef]
23. Weiß, C.H.; Schweer, S. Bias corrections for moment estimators in Poisson INAR(1) and INARCH(1) processes. *Stat. Probab. Lett.* **2016**, *112*, 124–130. [CrossRef]
24. Hall, P. *The Bootstrap and Edgeworth Expansion*; Springer: New York, NY, USA, 1992.
25. Jentsch, C.; Weiß, C.H. Bootstrapping INAR models. *Bernoulli* **2019**, *25*, 2359–2408. [CrossRef]
26. Politis, D.N.; White, H. Automatic block-length selection for the dependent bootstrap. *Econom. Rev.* **2004**, *23*, 53–70; Correction in *Econom. Rev.* **2009**, *28*, 372–375. [CrossRef]
27. Efron, B. Bootstrap methods: Another look at the jackknife. *Ann. Stat.* **1979**, *73*, 1–26. [CrossRef]
28. Giacomini, R.; Politis, D.N.; White, H. A warp-speed method for conducting Monte Carlo experiments involving bootstrap estimators. *Econom. Theory* **2013**, *29*, 567–589. [CrossRef]
29. Freeland, R.K. Statistical Analysis of Discrete Time Series with Applications to the Analysis of Workers Compensation Claims Data. Ph.D. Thesis, University of British Columbia, Vancouver, BC, Canada, 1998. Available online: https://open.library.ubc.ca/cIRcle/collections/ubctheses/831/items/1.0088709 (accessed on 17 January 2019).
30. Weiß, C.H.; Feld, M.H.-J.M.; Mamode Khan, N.; Sunecher, Y. INARMA modeling of count time series. *Stats* **2019**, *2*, 284–320. [CrossRef]
31. Toeplitz, O. Zur Transformation der Scharen bilinearer Formen von unendlichvielen Veränderlichen. *Nachrichten von der Ges. der Wiss. Göttingen Math.-Phys. Kl.* **1907**, *1907*, 110–115. (In German)
32. Toeplitz, O. Zur Theorie der quadratischen und bilinearen Formen von unendlichvielen Veränderlichen. I. Teil: Theorie der L-Formen. *Mathematische Annalen* **1911**, *70*, 351–376. (In German)
33. Durbin, J. The fitting of time-series models. *Rev. Int. Stat. Inst.* **1960**, *28*, 233–244. [CrossRef]
34. Levinson, N. The Wiener (root mean square) error criterion in filter design and prediction. *J. Math. Phys.* **1946**, *25*, 261–278. [CrossRef]
35. Barndorff-Nielsen, O.; Schou, G. On the parametrization of autoregressive models by partial autocorrelations. *J. Multivar. Anal.* **1973**, *3*, 408–419. [CrossRef]

Disclaimer/Publisher's Note: The statements, opinions and data contained in all publications are solely those of the individual author(s) and contributor(s) and not of MDPI and/or the editor(s). MDPI and/or the editor(s) disclaim responsibility for any injury to people or property resulting from any ideas, methods, instructions or products referred to in the content.

Article

Ruin Analysis on a New Risk Model with Stochastic Premiums and Dependence Based on Time Series for Count Random Variables

Lihong Guan [1,*,†] and Xiaohong Wang [2,†]

1 School of Science, Changchun University, Changchun 130022, China
2 Mathematics and Computer College, Jilin Normal University, Siping 136000, China
* Correspondence: guanlihong14@163.com
† These authors contributed equally to this work.

Abstract: In this paper, we propose a new discrete-time risk model of an insurance portfolio with stochastic premiums, in which the temporal dependence among the premium numbers of consecutive periods is fitted by the first-order integer-valued autoregressive (INAR(1)) process and the temporal dependence among the claim numbers of consecutive periods is described by the integer-valued moving average (INMA(1)) process. To measure the risk of the model quantitatively, we study the explicit expression for a function whose solution is defined as the Lundberg adjustment coefficient and give the Lundberg approximation formula for the infinite-time ruin probability. In the case of heavy-tailed claim sizes, we establish the asymptotic formula for the finite-time ruin probability via the large deviations of the aggregate claims. Two numerical examples are provided in order to illustrate our theoretical findings.

Keywords: risk model; stochastic premiums; INAR(1) process; INMA(1) process; ruin probability

MSC: 62P05; 91B30; 97M30

Citation: Guan, L.; Wang, X. Ruin Analysis on a New Risk Model with Stochastic Premiums and Dependence Based on Time Series for Count Random Variables. *Entropy* 2023, 25, 698. https://doi.org/10.3390/e25040698

Academic Editors: Nikolay Kolev Vitanov and Christian H. Weiss

Received: 23 February 2023
Revised: 13 April 2023
Accepted: 19 April 2023
Published: 21 April 2023

Copyright: © 2023 by the authors. Licensee MDPI, Basel, Switzerland. This article is an open access article distributed under the terms and conditions of the Creative Commons Attribution (CC BY) license (https://creativecommons.org/licenses/by/4.0/).

1. Introduction

As an absolutely necessary part of the modern financial system, insurance is one of the most effective ways for people to manage risks, such that it plays a significant role in our daily life. A very important task of insurance companies is to quantitatively analyze future claims. Consequently, risk theory has become an active research field of actuarial science. For the classical mathematical risk model, the so-called Lundberg–Cramér surplus process has the following form:

$$U_t^0 = u + ct - \sum_{i=1}^{N_t^0} Y_i, \quad t \geq 0, \tag{1}$$

in which $u \geq 0$ is the initial capital of an insurance portfolio, $c > 0$ is the constant rate of premium income, $\{N_t^0, t \geq 0\}$ is a homogeneous Poisson process with intensity λ, the total claim numbers are denoted up to time t, and Y_i describes the size of the ith claim. In the literature, Asmussen and Albrecher [1] presented excellent reviews about this well-known and important model.

In model (1), independent structures are usually assumed. For example, the claim amount $\{Y_i, i \geq 1\}$ is a sequence of non-negative independent and identically distributed (i.i.d.) random variables, and the claim numbers of different periods are assumed to be a sequence of i.i.d. random variables. However, these are not always true in practice because of the increasing complexity of individual risks. To avoid this restriction, a growing number of actuaries have been paying attention to the model with dependent risks. As stated in Yang and Zhang [2], there are mainly two kinds of correlation in insurance: one is the correlation among lines of businesses, and the other is temporal dependence, such as the correlation between the current claim and the previous claims. For recent works about the first type of correlation, Refs. [3,4] studied the dependence among individual risks, Refs. [5,6] discussed the two-dimensional risk models with dependent surplus processes, and [7,8] examined the risk models that have multiple classes of insurance business with thinning dependence structure. The relevant results have been used in a variety of actuarial areas, including, among others, value at risk, dividend strategies, reinsurance, capital allocation, etc.

In this paper, we focus on the second type. To deal with this problem, the use of a time series is a critical method. Gerber [9] considered the calculation of ruin probabilities in a Gaussian linear risk model; Gourieroux and Jasiak [10] applied the integer-valued time series model to update the premiums in vehicle insurance; and many researchers have extensively revisited the relevant results afterwards. Considering that the compound distributions are the cornerstones of a great number of risk models in risk theory, Cossette et al. [11] proposed some new discrete-time risk models, where the first-order integer-valued moving average (INMA(1)) and first-order integer-valued autoregressive (INAR(1)) processes are used to describe the dependence structures among the number of claims for each period. The authors derived expressions for the functions that allow people to find the Lundberg adjustment coefficients and discussed the Lundberg approximation formulas for infinite-time ruin probabilities. Along the same line, Cossette et al. [12] determined the distributions of aggregate claim amount and provided an effective way to measure some related risk quantities, including VaR, TVaR, and the stop-loss premium. Shi and Wang [13] gave an approximation method for the risk model with the Poisson INAR(1) claim number process in order to obtain the upper bound of the infinite-time ruin probability. Zhang et al. [14] solved the problem of optimal reinsurance strategy for the risk model with the INMA(1) claim number process. Afterwards, Hu et al. [15] and Chen and Hu [16] further generalized this kind of model by replacing the Poisson innovations with compound Poisson innovations in the INAR(1) and INMA(1) claim number processes, respectively. Guan and Hu [17] even utilized an INAR(1) process with an arbitrary innovations' distribution to specify the temporal dependence among the claim numbers.

In the papers mentioned above, it should be noted that the incomes of all the risk models are linear functions of time t, because the premiums are collected continuously with positive deterministic constant rate c, providing great convenience for risk analysis. However, this assumption is obviously lacking in terms of describing the real situation of insurance portfolios; for example, it cannot capture the uncertainty of the customers' arrivals. As an alternative to a fixed premium rate, Boikov [18] supposesed that the premium income also follows a compound Poisson process and calculates the ruin probability. From then on, the risk models with stochastic premiums have been extensively improved by many actuaries. Wang et al. [19] studied the investment problem of such models. Labbé and Sendova [20] discussed the Gerber–Shiu function. Zhao and Yin [21] proposed a renewal risk model with stochastic incomes. Recently, Su et al. [22] provided a statistical method for estimating the Gerber–Shiu function; Ragulina [23] investigated the De Vylder approximation for the ruin probability and a constant dividend strategy in the risk model with stochastic premiums; and Dibu and Jacob [24] focused on a double barrier hybrid dividend strategy. Wang et al. [25] quantitatively assessed the impact of the stochastic income process on some ruin quantities in detail.

Similar to the classical risk model, the premium numbers of different periods are commonly set to be a sequence of i.i.d. random variables in the aforementioned papers. To better characterize the uncertainty and capture the variability of an insurer's income process, Guan and Wang [26] proposed modeling the temporal dependence among the premium numbers of each period by a Poisson INAR(1) process. In this paper, we follow this trend of research. We also aim to study a new dependent risk model with stochastic premiums based on time series for count random variables, in which the INAR(1) process and INMA(1) process are applied to fit the temporal dependence among the premium numbers and the temporal dependence among the claim numbers of consecutive periods, respectively. Our goal is to approximate the infinite-time ruin probability of the proposed surplus process by the Lundberg adjustment coefficient and discuss the asymptotic formula for the finite-time ruin probability when the claim sizes follow distributions with heavy tails.

Our model generalizes the classical discrete-time surplus process of an insurance portfolio with stochastic premiums to a new dependent risk model, and our results extend what has been studied in the existing literature. The contributions of our paper mainly include the following two aspects:

- In contrast to the assumption that either claim numbers or premium numbers have a temporally dependent structure, we propose a new risk model of an insurance portfolio with both claim numbers and premium numbers being dependent within the integer-valued time series framework, which is more flexible in insurance practice.
- In addition to studying the distribution of the aggregate claims, the Lundberg adjustment coefficient, and the Lundberg approximation formula for the infinite-time ruin probability in the case of light-tailed claim sizes, we also explore the large deviations of the aggregate claims and the asymptotic formula for the finite-time ruin probability when the claim sizes are heavy-tailed, which enlarges the applicability of the risk model.

The remainder of the paper is organized as follows: Section 2 introduces our concerned risk model and considers some probabilistic properties of the proposed model. Section 3 defines the Lundberg adjustment coefficient via the solution of an explicit equation. Section 4 establishes an exponential asymptotic estimation for the infinite-time ruin probability. Section 5 studies the large deviations of the aggregate claims when the claim sizes follow a class of heavy-tailed distributions and presents an asymptotic formula for the finite-time ruin probability. Section 6 illustrates the main results by numerical simulations. Section 7 finally concludes this paper.

2. Risk Model and Basic Properties

In this section, we first describe the new dependent risk model, and then, provide some moment results of the premiums and claims. Let U_t be the surplus of an insurance portfolio at the end of period t, and we define the surplus process by the dynamic equation

$$U_t = U_{t-1} + P_t - L_t = U_{t-1} + \sum_{k=1}^{M_t} X_{t,k} - \sum_{j=1}^{N_t} Y_{t,j}, \quad t = 1, 2, \cdots, \quad (2)$$

where $U_0 = u \geq 0$ is the initial surplus level; $P_t = \sum_{k=1}^{M_t} X_{t,k}$ aggregates the premiums during period t, in which M_t counts the number of individual income and $X_{t,k}$ represents the amount of the kth premium income for the insurance portfolio during period t; $L_t = \sum_{j=1}^{N_t} Y_{t,j}$ is the aggregate claims during period t, in which N_t denotes the number of claims and $Y_{t,j}$ is the size of the jth payment to the insured in period t. For mathematical tractability, the following assumptions are made:

(1) Both $\{X_{t,k}, t = 1, 2, \cdots, k = 1, 2, \cdots\}$ and $\{Y_{t,j}, j = 1, 2, \cdots, k = 1, 2, \cdots\}$ are arrays of i.i.d. random variables, which have the same distributions as non-negative X and Y, respectively.

(2) $\{X_{t,k}, t = 1, 2, \cdots, k = 1, 2, \cdots\}$, $\{Y_{t,j}, j = 1, 2, \cdots, k = 1, 2, \cdots\}$, $\{M_t, t = 1, 2, \cdots\}$, and $\{N_t, t = 1, 2, \cdots\}$ are mutually independent.

The dependence structures of the model are constructed in the following ways:

(i) $\{M_t, t = 1, 2, \cdots\}$ constitutes a Poisson INAR(1) process that satisfies

$$M_t = \alpha \circ M_{t-1} + \varepsilon_t, \quad t = 2, 3, \cdots, \quad (3)$$

where the so-called binomial thinning operator "∘" is given by

$$\alpha \circ M_{t-1} = \sum_{m=1}^{M_{t-1}} B_{t,m}^{(1)}, \quad t = 2, 3, \cdots, \quad (4)$$

in which the following statements are true:
- The thinning parameter $\alpha \in [0, 1)$.
- $\{B_{t,m}^{(1)}, t = 2, 3, \cdots, m = 1, 2, \cdots\}$ is an array of i.i.d. Bernoulli random variables with mean α.
- $\{\varepsilon_t, t = 2, 3, \cdots\}$ is a sequence of i.i.d. Poisson random variables with mean λ_1.
- M_1, $\{B_{t,m}^{(1)}, t = 2, 3, \cdots, m = 1, 2, \cdots\}$ and $\{\varepsilon_t, t = 2, 3, \cdots\}$ are independent.

(ii) $\{N_t, t = 1, 2, \cdots\}$ constitutes a Poisson INMA(1) process that satisfies

$$N_t = \beta \circ \eta_{t-1} + \eta_t, \quad t = 1, 2, \cdots, \quad (5)$$

where "∘" is similarly defined by

$$\beta \circ \eta_{t-1} = \sum_{m=1}^{\eta_{t-1}} B_{t,m}^{(2)}, \quad t = 1, 2, \cdots, \quad (6)$$

in which the following are true:
- The thinning parameter $\beta \in [0, 1)$.
- $\{B_{t,m}^{(2)}, t = 1, 2, \cdots, m = 1, 2, \cdots\}$ is an array of i.i.d. Bernoulli random variables with mean β.
- $\{\eta_t, t = 0, 1, \cdots\}$ is a sequence of i.i.d. Poisson random variables with mean λ_2.
- $\{B_{t,m}^{(2)}, t = 1, 2, \cdots, m = 1, 2, \cdots\}$ and $\{\eta_t, t = 1, 2, \cdots\}$ are independent.

Remark 1. Time series analysis is one of the most important methods for dealing with dependent data and has attracted a lot of interest during the last decades. However, the classical real-valued time series models with continuous ranges can not account for discreteness, so they are of limited use for fitting the premium numbers and the claim numbers, which are typical count random variables fairly common in practice. Their poor performances in modeling this class of data mainly include: (1) the data generating mechanism can not be explained; (2) the approximate errors are big; and (3) the forecast results are not integer-valued. Therefore, models and methods for integer-valued time series have been covered by a large number of papers in recent years. Refs. [27–30] present comprehensive surveys on this fascinating research area. As two core models of integer-valued time series, INAR(1) process and INMA(1) process have been extensively applied in the literature of actuarial science, and the relevant results have been widely used in a variety of risk management.

Remark 2. The INAR(1) process (3) shows that the premium number in period t is composed of two parts: ε_t denotes the new incomes arriving between period $t - 1$ and t, and $\alpha \circ M_{t-1}$ presents a random proportion of the premium number in the previous period. This can be reasonably explained for the insurance practice that states: every insured entity could continue to pay a premium with probability α; or withdraw from the contract with probability $1 - \alpha$ in the next period. When $\alpha = 0$, (3) becomes $M_t = \varepsilon_t$, meaning that the premium number in period t could be totally determined by ε_t, and our model (2) will reduce to the classical discrete-time risk model with stochastic premiums, where the premium numbers of different periods are independent (please see Appendix A for details).

Remark 3. The INMA(1) process (5) reveals that the claim number in period t also consists of two parts: η_t is the new claim during period t, and $\beta \circ \eta_{t-1}$ indicates the claims of period $t - 1$ that could produce another accident with probability β in period t. Instead of (3), we use the INMA(1) process (5) to fit the temporal dependence among the claim numbers for each period, considering that the insured parties cannot receive benefits every year for some insurance products. Taking unemployment insurance as an example, every time the claimant is out of work, they could receive the benefits for up to 2 years, if the premiums for at least 1 year have been paid. Another appropriate example might be some medical insurance contracts, which state that no matter how long the patient stays in the hospital, the insurer would pay the benefits for at most (for instance) 2 months. Similarly, if $\beta = 0$, our proposed model will reduce to the classical case, where the claim numbers of different periods are independent.

As stated in Al-Osh and Alzaid [31], under the condition of $0 \leq \alpha < 1$, it follows that the process of premium numbers $\{M_t, t = 1, 2, \cdots\}$ is a stationary and ergodic Markov chain. Furthermore, if we assume $\varepsilon_t \sim P(\lambda_1)$, then M_t is also Poisson distributed with mean $\frac{\lambda_1}{1-\alpha}$. Hence, by the law of iterated expectation and the assumption that $\{X_{t,k}, k = 1, 2, \cdots\}$ and M_t are independent, it is easy to find that

$$E(P_t = E[E(P_t|M_t)] = E\left[E\left(\sum_{k=1}^{M_t} X_{t,k}|M_t\right)\right] = E\left[\sum_{k=1}^{M_t} E(X_{t,k}|M_t)\right] = E\left[\sum_{k=1}^{M_t} E(X_{t,k})\right] = E(M_t)E(X) = \frac{\lambda_1}{1-\alpha}E(X). \quad (7)$$

Meanwhile, by the law of total variance, we can obtain

$$\begin{aligned}
\text{Var}(P_t) &= \text{Var}[E(P_t|M_t)] + E[\text{Var}(P_t|M_t)] \\
&= \text{Var}\left[E\left(\sum_{k=1}^{M_t} X_{t,k}|M_t\right)\right] + E\left[\text{Var}\left(\sum_{k=1}^{M_t} X_{t,k}|M_t\right)\right] \\
&= \text{Var}\left[\sum_{k=1}^{M_t} E(X_{t,k}|M_t)\right] + E\left[\sum_{k=1}^{M_t} \text{Var}(X_{t,k}|M_t)\right] \\
&= \text{Var}\left[\sum_{k=1}^{M_t} E(X_{t,k})\right] + E\left[\sum_{k=1}^{M_t} \text{Var}(X_{t,k})\right] \\
&= \text{Var}[M_t \cdot E(X)] + E[M_t \cdot \text{Var}(X)] \\
&= E(M_t)\text{Var}(X) + \text{Var}(M_t)[E(X)]^2 = \frac{\lambda_1}{1-\alpha}E(X^2). \quad (8)
\end{aligned}$$

Furthermore, Al-Osh and Alzaid [31] show that

$$\text{Cov}(M_t, M_{t+h}) = \alpha^h \text{Var}(M_t) = \frac{\lambda_1 \alpha^h}{1-\alpha},$$

from which we can obtain

$$\begin{aligned}
\text{Cov}(P_t, P_{t+h}) &= E(P_t P_{t+h}) - E(P_t)E(P_{t+h}) \\
&= E[E(P_t P_{t+h}|M_t, M_{t+h})] - E(M_t)E(M_{t+h})[E(X)]^2 \\
&= E[E(P_t|M_t)E(P_{t+h}|M_{t+h})] - E(M_t)E(M_{t+h})[E(X)]^2 \\
&= E(M_t M_{t+h})[E(X)]^2 - E(M_t)E(M_{t+h})[E(X)]^2 \\
&= [E(X)]^2 \text{Cov}(M_t, M_{t+h}) = \frac{\lambda_1 \alpha^h}{1-\alpha}[E(X)]^2, \quad h = 1, 2, \cdots. \quad (9)
\end{aligned}$$

Similarly, for the process of claim numbers $\{N_t, t = 1, 2, \cdots\}$, under the condition of $0 \leq \beta < 1$, its marginal distribution is uniquely determined by the law of $\{\eta_t, t = 0, 1, \cdots\}$. Therefore, the assumption of $\eta_t \sim P(\lambda_2)$ will result in N_t being Poisson distributed with a mean of $(1+\beta)\lambda_2$. Consequently, by the same method to drive (7)–(9), we have

$$E(L_t) = E(N_t)E(Y) = (1+\beta)\lambda_2 E(Y),$$

$$\text{Var}(L_t) = E(N_t)\text{Var}(Y) + \text{Var}(N_t)[E(Y)]^2 = (1+\beta)\lambda_2 E(Y^2).$$

and

$$\text{Cov}(L_t, L_{t+h}) = [E(Y)]^2 \text{Cov}(N_t, N_{t+h}) = \begin{cases} \lambda_2 \beta [E(Y)]^2, & h = 1, \\ 0, & h > 1. \end{cases} \quad (10)$$

These results are consistent with those in [11].

3. Definition of the Lundberg Adjustment Coefficient

In this section, we first consider how to calculate the moment generating functions (m.g.f.) of the aggregate premiums and aggregate claims up to period t, and then, define the Lundberg adjustment coefficient of the proposed dependent risk model with stochastic premiums based on time series for count random variables by means of a equation.

After recursive calculation, we can rewrite model (2) as

$$U_t = U_{t-1} + P_t - L_t = U_{t-1} + \sum_{k=1}^{M_t} X_{t,k} - \sum_{j=1}^{N_t} Y_{t,j}$$

$$= u + \sum_{i=1}^{t}\sum_{k=1}^{M_i} X_{i,k} - \sum_{i=1}^{t}\sum_{j=1}^{N_i} Y_{i,j} = u + W_t - S_t, \quad t = 1, 2, \cdots, \quad (11)$$

in which $W_t = \sum_{i=1}^{t} P_i = \sum_{i=1}^{t}\sum_{k=1}^{M_i} X_{i,k}$ and $S_t = \sum_{i=1}^{t} L_i = \sum_{i=1}^{t}\sum_{j=1}^{N_i} Y_{i,j}$ represent the aggregate premium incomes and aggregate claim payments up to time t, respectively. As for the m.g.f. of W_t and S_t, by the definition, we have that

$$\begin{aligned} M_{W_t}(r) &= E(e^{rW_t}) \\ &= E[e^{r(P_1 + \cdots + P_t)}] \\ &= M_{P_1, \cdots, P_t}(r, \cdots, r) \\ &= P_{M_1, \cdots, M_t}(M_X(r), \cdots, M_X(r)) \\ &= E[M_X(r)^{M_1} \cdots M_X(r)^{M_t}] \\ &= P_{M_1 + \cdots + M_t}(M_X(r)), \end{aligned} \quad (12)$$

where $M_X(\cdot)$ denotes the m.g.f. of X and $P_{M_1+\cdots+M_t}(\cdot)$ presents the probability generating function (p.g.f.) of the total premium number up to period t of the proposed model (2).

Similarly, it holds that

$$M_{S_t}(r) = P_{N_1 + \cdots + N_t}(M_Y(r)), \quad (13)$$

where $M_Y(\cdot)$ denotes the m.g.f. of Y, and $P_{N_1+\cdots+N_t}(\cdot)$ presents the p.m.f. of the total claim number up to period t of the proposed model (2).

In order to compute $M_{W_t}(r)$ and $M_{S_t}(r)$, we find the explicit expressions for $P_{M_1+\cdots+M_t}(\cdot)$ and $P_{N_1+\cdots+N_t}(\cdot)$ in the following two lemmas, respectively.

Lemma 1. *For $t = 1, 2, \cdots$, when $0 \leq s \leq 1$, the p.g.f. of $M_1 + \cdots + M_t$ is given by*

$$P_{M_1+\cdots+M_t}(s) = \exp\left\{\lambda \frac{s-1}{1-\alpha s}\left[t + \frac{1-(\alpha s)^t}{1-\alpha} - \frac{1-(\alpha s)^t}{1-\alpha s}\right]\right\}. \quad (14)$$

Proof. Since $M_1 \sim P(\frac{\lambda_1}{1-\alpha})$, it is obvious that

$$P_{M_1}(s) = \exp\left\{\frac{\lambda_1}{1-\alpha}(s-1)\right\}.$$

When $t \geq 2$, we denote

$$\alpha^{(t)} \circ M_1 = \underbrace{\alpha \circ \cdots \circ \alpha \circ}_{t-\text{fold operation}} M_1.$$

By the property of the binomial thinning operator (see Scotto et al. [28] for example), we can rewrite $M_1 + \cdots + M_t$ as

$$\begin{aligned}
M_1 + \cdots + M_t &= M_1 + \alpha \circ M_1 + \alpha^{(2)} \circ M_1 + \cdots + \alpha^{(t-1)} \circ M_1 \\
&\quad + \varepsilon_2 + \alpha \circ \varepsilon_2 + \cdots + \alpha^{(t-2)} \circ \varepsilon_2 \\
&\quad + \cdots \\
&\quad + \varepsilon_{t-1} + \alpha \circ \varepsilon_{t-1} \\
&\quad + \varepsilon_t.
\end{aligned} \tag{15}$$

For the p.g.f. calculation, we have

$$\begin{aligned}
P_{M_1+\alpha\circ M_1+\alpha^{(2)}\circ M_1+\cdots+\alpha^{(t-1)}\circ M_1}(s) &= E(s^{M_1+\alpha\circ M_1+\alpha^{(2)}\circ M_1+\cdots+\alpha^{(t-1)}\circ M_1}) \\
&= E\left[s^{M_1} s^{\alpha\circ M_1} \cdots s^{\alpha^{(t-2)}\circ M_1} E\left(s^{\alpha^{(t-1)}\circ M_1} | M_1, \cdots, \alpha^{(t-2)}\circ M_1\right)\right] \\
&= E\left[s^{M_1} s^{\alpha\circ M_1} \cdots s^{\alpha^{(t-2)}\circ M_1} (\alpha s + 1 - \alpha)^{\alpha^{(t-2)}\circ M_1}\right] \\
&= E\left[s^{M_1} s^{\alpha\circ M_1} \cdots s^{\alpha^{(t-3)}\circ M_1} (h_2(s))^{\alpha^{(t-2)}\circ M_1}\right] \\
&= E\left[s^{M_1} s^{\alpha\circ M_1} \cdots s^{\alpha^{(t-3)}\circ M_1} (\alpha h_2(s) + 1 - \alpha)^{\alpha^{(t-3)}\circ M_1}\right] \\
&= E\left[s^{M_1} s^{\alpha\circ M_1} \cdots s^{\alpha^{(t-4)}\circ M_1} (h_3(s))^{\alpha^{(t-3)}\circ M_1}\right] \\
&= \cdots \\
&= E\left[s^{M_1} (\alpha h_{t-1}(s) + 1 - \alpha)^{\alpha\circ M_1}\right] \\
&= \exp\left\{\frac{\lambda_1}{1-\alpha}(h_t(s) - 1)\right\},
\end{aligned} \tag{16}$$

in which $h_1(s) = s$ and $h_t(s) = s(\alpha h_{t-1}(s) + 1 - \alpha)$.

Similarly, we can obtain

$$P_{\varepsilon_2+\alpha\circ\varepsilon_2+\cdots+\alpha^{(t-2)}\circ\varepsilon_2}(s) = \exp\{\lambda_1(h_{t-1}(s) - 1)\}, \cdots, P_{\varepsilon_t}(s) = \exp\{\lambda_1(h_1(s) - 1)\}. \tag{17}$$

Combining (15)–(17), it follows that

$$\begin{aligned}
&P_{M_1+\cdots+M_t}(s) \\
&= P_{M_1+\alpha\circ M_1+\alpha^{(2)}\circ M_1+\cdots+\alpha^{(t-1)}\circ M_1}(s) \times P_{\varepsilon_2+\alpha\circ\varepsilon_2+\cdots+\alpha^{(t-2)}\circ\varepsilon_2}(s) \times \cdots \times P_{\varepsilon_{t-1}+\alpha\circ\varepsilon_{t-1}}(s) \times P_{\varepsilon_t}(s) \\
&= \exp\left\{\frac{\lambda_1}{1-\alpha}(h_t(s) - 1)\right\} \prod_{i=1}^{t-1} \exp\{\lambda_1(h_i(s) - 1)\}.
\end{aligned} \tag{18}$$

Moreover, from the definition $h_t(s) = s(\alpha h_{t-1}(s) + 1 - \alpha)$, it is easy to find that

$$h_t(s) - 1 = s - 1 + \alpha s(h_{t-1}(s) - 1).$$

Then, recursive calculation results in

$$h_t(s) - 1 = (s - 1)\frac{1 - (\alpha s)^t}{1 - \alpha s}. \tag{19}$$

Finally, inserting (19) into (18), we can obtain

$$P_{M_1+\cdots+M_t}(s) = \exp\left\{\lambda \frac{s-1}{1-\alpha s}\left[t + \frac{1-(\alpha s)^t}{1-\alpha} - \frac{1-(\alpha s)^t}{1-\alpha s}\right]\right\}.$$

This completes the proof. □

Lemma 2. *For* $t = 1, 2, \cdots$, *when* $s \geq 0$, *the p.g.f. of* $N_1 + \cdots + N_t$ *is given by*

$$P_{N_1+\cdots+N_t}(s) = \exp\left\{\lambda_2(1+\beta)(s-1) + \lambda_2(t-1)[\beta s^2 + (1-\beta)s - 1]\right\}. \tag{20}$$

Proof. By (5), it holds that

$$\begin{aligned}
P_{N_1+\cdots+N_t}(s) &= E(s^{N_1+\cdots+N_t}) \\
&= E(s^{\beta \circ \eta_0 + \eta_1} s^{\beta \circ \eta_1 + \eta_2} \cdots s^{\beta \circ \eta_{t-1} + \eta_t}) \\
&= E(s^{\beta \circ \eta_0} s^{\eta_1 + \beta \circ \eta_1} \cdots s^{\eta_{t-1} + \beta \circ \eta_{t-1}} s^{\eta_t}) \\
&= E(s^{\beta \circ \eta_0}) E(s^{\eta_1 + \beta \circ \eta_1}) \cdots E(s^{\eta_{t-1} + \beta \circ \eta_{t-1}}) E(s^{\eta_t}) \\
&= \exp\left\{\lambda_2(1+\beta)(s-1) + \lambda_2(t-1)[\beta s^2 + (1-\beta)s - 1]\right\},
\end{aligned}$$

which follows from $\eta_t \sim P(\lambda_2)$, $\beta \circ \eta_0 \sim P(\beta \lambda_2)$ and

$$\begin{aligned}
P_{\varepsilon_i + \beta \circ \varepsilon_i}(s) &= E(s^{\varepsilon_i + \beta \circ \varepsilon_i}) \\
&= E[E(s^{\varepsilon_i} s^{\beta \circ \varepsilon_i} | \varepsilon_i)] \\
&= E[s^{\varepsilon_i} E(s^{\beta \circ \varepsilon_i} | \varepsilon_i)] \\
&= E[s^{\varepsilon_i} (\beta s + 1 - \beta)^{\varepsilon_i}] \\
&= \exp\left\{\lambda_2[\beta s^2 + (1-\beta)s - 1]\right\}, \quad i = 1, 2, \cdots, t-1.
\end{aligned}$$

The proof then is completed. □

To further analyze the insurance portfolio, we write

$$c_t(r) = \frac{1}{t}\ln E([e^{r(S_t - W_t)}]), \tag{21}$$

and let

$$c(r) = \lim_{t \to +\infty} c_t(r). \tag{22}$$

Then, the positive solution to the equation $c(r) = 0$ can be defined as the Lundberg adjustment coefficient, which is denoted by R and can be used to approximate the infinite-time ruin probability of the proposed model (2). The following result gives the explicit expression for $c(r)$.

Theorem 1. *For* $r \geq 0$, *we have*

$$c(r) = \lambda_1 \frac{M_X(-r) - 1}{1 - \alpha M_X(-r)} + \lambda_2[\beta M_Y^2(r) + (1-\beta)M_Y(r) - 1]. \tag{23}$$

Proof. Due to the non-negativity of r and X, it follows that

$$0 \leq M_X(-r) \leq 1, \ 0 \leq \alpha M_X(-r) < 1.$$

Then, by Lemma 1, we have

$$\lim_{t\to+\infty} \frac{1}{t} \ln E(e^{-rW_t}) = \lim_{t\to+\infty} \frac{1}{t} \ln P_{M_1+\cdots+M_t}(M_X(-r))$$
$$= \lim_{t\to+\infty} \frac{1}{t} \ln \left(\exp\left\{ \lambda \frac{M_X(-r)-1}{1-\alpha M_X(-r)} \left[t + \frac{1-(\alpha M_X(-r))^t}{1-\alpha} - \frac{1-(\alpha M_X(-r))^t}{1-\alpha M_X(-r)} \right] \right\} \right)$$
$$= \lambda \frac{M_X(-r)-1}{1-\alpha M_X(-r)}. \tag{24}$$

On the other hand, from (13) and (20), we obtain

$$\lim_{t\to+\infty} \frac{1}{t} \ln E(e^{rS_t}) = \lim_{t\to+\infty} \frac{1}{t} \ln P_{N_1+\cdots+N_t}(M_Y(r))$$
$$= \lim_{t\to+\infty} \frac{1}{t} \ln \left(\exp\left\{ \lambda_2(1+\beta)(M_Y(r)-1) + \lambda_2(t-1)[\beta M_Y^2(r) + (1-\beta)M_Y(r) - 1] \right\} \right)$$
$$= \lambda_2(1+\beta) \lim_{t\to+\infty} \frac{1}{t}(M_Y(r)-1) + \lim_{t\to+\infty} \frac{t-1}{t} \lambda_2[\beta M_Y^2(r) + (1-\beta)M_Y(r) - 1]$$
$$= \lambda_2[\beta M_Y^2(r) + (1-\beta)M_Y(r) - 1]. \tag{25}$$

Then, combining (24) and (25) with (21) and (22) yields

$$c(r) = \lim_{t\to+\infty} c_t(r)$$
$$= \lim_{t\to+\infty} \frac{1}{t} \ln E([e^{r(S_t-W_t)}])$$
$$= \lim_{t\to+\infty} \frac{1}{t} \ln E(e^{rS_t}) + \lim_{t\to+\infty} \frac{1}{t} \ln E(e^{-rW_t})$$
$$= \lambda_1 \frac{M_X(-r)-1}{1-\alpha M_X(-r)} + \lambda_2[\beta M_Y^2(r) + (1-\beta)M_Y(r) - 1],$$

This completes the proof. □

Remark 4. When $\alpha = 0$, the proposed model (2) degenerates to the discrete-time risk model based on the Poisson INMA(1) process studied by [11,14], where only the temporal dependence among the claim numbers of consecutive periods is considered. Consequently, (23) becomes

$$c(r) = \lambda_1[M_X(-r) - 1] + \lambda_2[\beta M_Y^2(r) + (1-\beta)M_Y(r) - 1],$$

which corresponds to (7) in [11,14].

Remark 5. When $\beta = 0$, the proposed model (2) reduces to the discrete-time risk model with stochastic premiums and dependence based on the Poisson INAR(1) process studied by [26], where only the temporal dependence among the premium numbers of consecutive periods is considered. As a result, (23) becomes

$$c(r) = \lambda_1 \frac{M_X(-r)-1}{1-\alpha M_X(-r)} + \lambda_2[M_Y(r) - 1],$$

which corresponds to (3.10) in [26].

4. Lundberg Approximation Formula for the Infinite-Time Ruin Probability

Let the ruin time of our proposed surplus process (2) be $T = \inf_{t\in\{0,1,2,\cdots\}} \{t, U_t \leq 0\}$ if U_t goes below 0 at least once; otherwise, take $T = +\infty$. As a consequence, the infinite-time ruin probability $\psi(u)$ is defined by

$$\psi(u) = P(T < +\infty | U_0 = u).$$

Ruin probability $\psi(u)$ is well-known as one of the most common and important quantities used to measure the riskiness of an insurance portfolio in the risk-theoretic context. However, as can be seen from the expression (11), our proposed model releases the condition that P_t and L_t are independent of U_{t-1}, which is a key but defective assumption in the classical risk model with stochastic premiums and allows for the temporal dependence among the premium numbers and claim numbers. Therefore, P_t and L_t are correlated with U_{t-1}, and $\{U_t, t = 1, 2, \cdots\}$ is no longer a Lévy process with stationary independent increments in our model. Consequently, it is not easy to derive the upper bounds and explicit expression for the infinite-time ruin probability such as those in some classical models. As an efficient alternative, the following result gives an asymptotic estimation for $\psi(u)$ of our proposed model (2).

Theorem 2. *In the discrete-time dependent risk model with stochastic premiums based on the Poisson INAR(1) process and Poisson INMA(1) process, if*

$$\frac{\lambda_1}{1-\alpha}E(X) > \lambda_2(1+\beta)E(Y), \tag{26}$$

we can obtain the Lundberg approximation formula for the infinite-time ruin probability $\psi(u)$, which has the following expression:

$$\lim_{u \to +\infty} -\frac{\ln(\psi(u))}{u} = R, \tag{27}$$

where u and R are the initial capital and the Lundberg adjustment coefficient, respectively.

Proof. According to Theorem 2.1 in Müller and Pflug [32], it is sufficient for us to prove that the equation $c(r) = 0$ has a unique positive solution, which can be defined as the Lundberg adjustment coefficient R. To this end, we derive the following four properties of the function $c(r)$.

Firstly, noting that $M_X(0) = M_Y(0) = 1$, we have

$$c(0) = \lambda_1 \frac{M_X(0) - 1}{1 - \alpha M_X(0)} + \lambda_2[\beta M_Y^2(0) + (1-\beta)M_Y(0) - 1] = 0. \tag{28}$$

Secondly, it is easy to calculate that

$$c'(r) = \frac{-\lambda_1(1-\alpha)M_X'(-r)}{[1-\alpha M_X(-r)]^2} + \lambda_2[2\beta M_Y(r)M_Y'(r) + (1-\beta)M_Y'(r)].$$

Together with the fact $M_X'(0) = E(X)$ and $M_Y'(0) = E(Y)$, we obtain

$$c'(0) = \frac{-\lambda_1(1-\alpha)M_X'(0)}{[1-\alpha M_X(0)]^2} + \lambda_2[2\beta M_Y(0)M_Y'(0) + (1-\beta)M_Y'(0)]$$

$$= \lambda_2(1+\beta)E(Y) - \frac{\lambda_1}{1-\alpha}E(X) < 0. \tag{29}$$

Thirdly, it is easy to verify the convexity of $c(r)$, which results from the fact that $c_t(r)$ is convex and the definition of $c(r) = \lim_{t \to +\infty} c_t(r)$.

Finally, when the m.g.f. of Y exists, i.e., there exists some quantity r_0, $0 < r_0 \leq +\infty$, such that $M_Y(r)$ is finite for all $r < r_0$ with

$$\lim_{r \to r_0^-} M_Y(r) = +\infty,$$

then, it holds that

$$\lim_{r \to r_0^-} c(r) = \lim_{r \to r_0^-} \left(\lambda_1 \frac{M_X(-r) - 1}{1 - \alpha M_X(-r)} + \lambda_2 [\beta M_Y^2(r) + (1-\beta) M_Y(r) - 1] \right) = +\infty. \quad (30)$$

Therefore, it can be concluded that there exists a unique positive solution to the equation $c(r) = 0$, and then, (27) follows immediately. □

Remark 6. *In risk and ruin theory, the assumption (26) is the so-called relative safety loading condition, which implies that the expected premium incomes should be more than the expected claim expenses to guarantee that the insurance company can operate normally and profitably.*

Remark 7. *As a result of the approximation formula (27), we can asymptotically estimate the infinite-time ruin probability $\psi(u)$ by*

$$\psi(u) \simeq e^{-Ru}, \quad (31)$$

if the initial surplus u becomes large enough.

From (9) and (10), it can be seen that the thinning parameters α and β could quantitatively measure the degree of the dependence in the risk model (2); hence, it is necessary for us to discuss their impacts on the adjustment coefficient and further on the risk of the insurance portfolio.

Proposition 1. *As a function of the two thinning parameters, the Lundberg adjustment coefficient R of our proposed risk model (2) increases with respect to α and decreases with respect to β.*

Proof. For convenience, we now rewrite $c(r)$ as $c(\alpha, \beta, r)$; the Lundberg adjustment coefficient R is determined by $c(\alpha, \beta, R) = 0$ and can be taken as a function of α and β. By the properties derived in the proof of Theorem 2, we know that

$$\frac{\partial c(\alpha, \beta, R)}{\partial R} > 0.$$

Meanwhile, with $R > 0$ in mind, it follows that $0 \leq M_X(-R) < 1$. Thus, we take the partial derivative of $c(\alpha, \beta, R)$ with respect to variable α and then have

$$\frac{\partial c(\alpha, \beta, R)}{\partial \alpha} = \frac{\partial}{\partial \alpha} \left(\lambda_1 \frac{M_X(-R) - 1}{1 - \alpha M_X(-R)} + \lambda_2 [\beta M_Y^2(R) + (1-\beta) M_Y(R) - 1] \right)$$

$$= \frac{-\lambda_1 M_X(-R)[1 - \alpha M_X(-R)] + \lambda_1 (1-\alpha) M_X(-R) M_X(-R)}{[1 - \alpha M_X(-R)]^2}$$

$$= \frac{\lambda_1 [M_X^2(-R) - M_X(-R)]}{[1 - \alpha M_X(-R)]^2} < 0.$$

As a result, using implicit function theorem, it holds that

$$\frac{\partial R}{\partial \alpha} = -\frac{(\partial/\partial \alpha) c(\alpha, \beta, R)}{(\partial/\partial R) c(\alpha, \beta, R)} > 0,$$

implying that R increases with respect to α.

Similarly, because $M_Y(R) > 1$ for $R > 0$, taking the partial derivative of $c(\alpha, \beta, R)$ with respect to variable β yields

$$\frac{\partial c(\alpha, \beta, R)}{\partial \beta} = \frac{\partial}{\partial \beta} \left(\lambda_1 \frac{M_X(-R) - 1}{1 - \alpha M_X(-R)} + \lambda_2 [\beta M_Y^2(R) + (1-\beta) M_Y(R) - 1] \right)$$

$$= \lambda_2 [M_Y^2(R) - M_Y(R)] > 0,$$

from which we apply implicit function theorem again and obtain

$$\frac{\partial R}{\partial \beta} = -\frac{(\partial/\partial \beta)c(\alpha, \beta, R)}{(\partial/\partial R)c(\alpha, \beta, R)} < 0,$$

meaning that R decreases with respect to β. □

Remark 8. *As shown in Proposition 1, the degree of riskiness can be measured and quantified by the Lundberg adjustment coefficient R, in the sense that it decreases with the thinning parameter α, while it increases with the thinning parameter β. In insurance practice, it can be naturally explained that when α increases, the insured parties would like to renew their insurance contracts with a higher probability in the next period, which would lower the risk of the portfolio. On the contrary, when β increases, a reported claim becomes more likely to produce another insurance accident in the next period, which could make the portfolio much riskier.*

5. Asymptotic Formula for the Finite-Time Ruin Probability

In this section, we turn our focus to the case of heavy-tailed claim sizes, which are frequently used in insurance practice for catastrophe risks, such as earthquakes, hurricanes, floods, financial crises, agricultural disasters, and so on. In these instances, the Lundberg adjustment coefficient and Lundberg approximation estimation for infinite-time ruin probability can no longer be applied because $M_Y(r)$ (the m.g.f. of Y) does not exist for $r > 0$. Therefore, increasing numbers of researchers have increasingly paid close attention to the precise large deviations in the aggregate of claims, as well as the asymptotic formulas for infinite-time and finite-time ruin probabilities. The relevant study was initiated by Klüppelberg and Mikosch [33] and then has been revisited by many researchers afterwards. We refer to Chen et al. [34] and Fu et al. [35] for some recent contributions on this topic. Cheng and Wang [36], Yang et al. [37], and Jing et al. [38] considered the asymptotic ruin probabilities in risk models with dependence among the claim sizes. Xun et al. [39] obtained the uniformly asymptotic result of ruin probability in a general risk model with stochastic premiums. Yu [40] derived the precise large deviations of the aggregate amount of claims for a risk model with the Poisson ARCH claim number process. Along the same line, in this section, we investigate our proposed model (2) when the distribution of claim sizes belongs to a heavy-tailed class.

First, we give some brief notations. Let $a(x)$ and $b(x)$ be two positive functions. We denote $a(x) \sim b(x)$ if $\lim_{x \to +\infty} a(x)/b(x) = 1$; denote $a(x) \lesssim b(x)$ if $\limsup_{x \to +\infty} a(x)/b(x) \leq 1$; denote $a(x) \gtrsim b(x)$ if $\liminf_{x \to +\infty} a(x)/b(x) \geq 1$; and denote $a(x) = o(b(x))$ if $\limsup_{x \to +\infty} a(x)/b(x) = 0$. We denote the common distribution functions of premium amount X and claim size Y with $F_X(x)$ and $F_Y(y)$, respectively.

Then, we recall a class of heavy-tailed distributions and one of its important properties. More detailed discussions can be found in Embrechts et al. [41], Asmussen and Albrecher [1], etc.

A distribution function F on $[0, \infty]$ is said to have a consistently varying tail, denoted by $F \in \mathcal{C}$, if

$$\lim_{y \uparrow 1} \limsup_{x \to +\infty} \frac{\overline{F}(xy)}{\overline{F}(x)} = \lim_{y \downarrow 1} \liminf_{x \to +\infty} \frac{\overline{F}(xy)}{\overline{F}(x)} = 1, \qquad (32)$$

where $\overline{F}(x)$ is the tail probability with $\overline{F}(x) = 1 - F(x)$. The class $\mathcal{C}$ is a wide class of distributions commonly used in actuarial science, including the well-known Pareto, Burr, and loggamma distributions. Ng et al. [42] established a very useful result for the distributions of class $\mathcal{C}$, which is given in the following lemma.

Lemma 3. *Suppose that* $\{Y_j, j = 1, 2, \cdots\}$ *is a sequence of i.i.d. non-negative random variables with common distribution function* $F_Y(y) \in \mathcal{C}$ *and* $E(Y) < +\infty$. *Taking* $Q_t = \sum_{j=1}^{t} Y_j$, *for any fixed* $\gamma > 0$, *it holds uniformly for all* $y > \gamma t$ *that*

$$P(Q_t - tE(Y) > y) \sim t\overline{F}_Y(y), \quad t \to +\infty, \tag{33}$$

in which the uniformity is understood in the following sense:

$$\lim_{t \to +\infty} \sup_{y \geq \gamma t} \left| \frac{P(Q_t - tE(Y) > y)}{t\overline{F}_Y(y)} - 1 \right| = 0.$$

Analogous to the infinite-time ruin probability $\psi(u)$, for any fixed $t = 1, 2, \cdots$, we define the finite-time ruin probability $\psi(u, t)$ of the discrete-time risk model (2) as

$$\psi(u, t) = P(T \leq t | U_0 = u).$$

In order to further study the asymptotic formula of $\psi(u, t)$, which is also a core actuarial quantity, we revise Lemma 3 as follows.

Lemma 4. *Suppose that* $\{Y_j, j = 1, 2, \cdots\}$ *is a sequence of i.i.d. non-negative random variables with the common distribution function* $F_Y(y) \in \mathcal{C}$ *and* $E(Y) < +\infty$. *Define* $Q_t = \sum_{j=1}^{t} Y_j$; *then, for any fixed* $\gamma > 0$ *and* $\delta > 0$, *it holds uniformly for all* $y > \gamma t^{1+\delta}$ *that*

$$P(Q_t > y) \sim t\overline{F}_Y(y), \quad t \to +\infty. \tag{34}$$

Proof. By the definition of class $\mathcal{C}$, it follows for any fixed $\theta > 0$ and sufficiently large y that

$$\frac{\overline{F}_Y((1+\theta)y)}{\overline{F}_Y(y)} \leq \frac{\overline{F}_Y(y + o(y))}{\overline{F}_Y(y)} \leq \frac{\overline{F}_Y((1-\theta)y)}{\overline{F}_Y(y)},$$

from which we can obtain

$$1 = \lim_{\theta \downarrow 0} \liminf_{y \to +\infty} \frac{\overline{F}_Y((1+\theta)y)}{\overline{F}_Y(y)} \leq \liminf_{y \to +\infty} \frac{\overline{F}_Y(y + o(y))}{\overline{F}_Y(y)}$$

$$\leq \limsup_{y \to +\infty} \frac{\overline{F}_Y(y + o(y))}{\overline{F}_Y(y)} \leq \lim_{\theta \downarrow 0} \limsup_{y \to +\infty} \frac{\overline{F}_Y((1-\theta)y)}{\overline{F}_Y(y)} = 1.$$

Hence, it holds that

$$\lim_{y \to +\infty} \frac{\overline{F}_Y(y + o(y))}{\overline{F}_Y(y)} = 1. \tag{35}$$

Furthermore, by Lemma 3 and (35), it follows that

$$\lim_{t\to+\infty}\sup_{y>\gamma t^{1+\delta}}\left|\frac{P(Q_t>y)}{t\overline{F}_Y(y)}-1\right| = \lim_{t\to+\infty}\sup_{y>\gamma t^{1+\delta}}\left|\frac{P(Q_t-tE(Y)>y-tE(Y))}{t\overline{F}_Y(y-tE(Y))}\times\frac{\overline{F}_Y(y-tE(Y))}{\overline{F}_Y(y)}-1\right|$$

$$\leq \lim_{t\to+\infty}\sup_{y>\gamma t^{1+\delta}}\frac{\overline{F}_Y(y-tE(Y))}{\overline{F}_Y(y)}\times\left|\frac{P(Q_t-tE(Y)>y-tE(Y))}{t\overline{F}_Y(y-tE(Y))}-1\right|$$

$$+\lim_{t\to+\infty}\sup_{y>\gamma t^{1+\delta}}\left|\frac{\overline{F}_Y(y-tE(Y))}{\overline{F}_Y(y)}-1\right|$$

$$=\lim_{y\to+\infty}\frac{\overline{F}_Y(y+o(y))}{\overline{F}_Y(y)}\times\lim_{t\to+\infty}\sup_{y>\gamma t^{1+\delta}}\left|\frac{P(Q_t-tE(Y)>y-tE(Y))}{t\overline{F}_Y(y-tE(Y))}-1\right|$$

$$+\lim_{y\to+\infty}\left|\frac{\overline{F}_Y(y+o(y))}{\overline{F}_Y(y)}-1\right|$$

$$=0.$$

The proof is then completed. □

Now, we give the precise large deviations of the aggregate claims, S_t, which is described in model (11).

Theorem 3. *For our proposed model (2), let $F_Y(y)$ and $E(Y)$ be the common distribution function and expectation of the claim sizes, respectively. Assuming $F_Y(y) \in \mathcal{C}$ and $E(Y) < +\infty$, then for any fixed $\gamma > 0$ and $\delta > 0$, it holds uniformly for all $y > \gamma t^{1+\delta}$ that*

$$P(S_t > y) \sim \lambda_2(1+\beta)t\overline{F}_Y(y), \quad t \to +\infty. \tag{36}$$

Proof. Let $\{Y_j, j = 1, 2, \cdots\}$ be a sequence of i.i.d. non-negative random variables, with their common distribution function denoted by $F_Y(y)$. Suppose that $\varphi_{S_t}(r)$ is the characteristic function of $S_t = \sum_{i=1}^{t}\sum_{j=1}^{N_i} Y_{i,j}$. With the same method to derive (12) and (13), we can obtain

$$\varphi_{S_t}(r) = E\left[(\varphi_Y(r))^{N_1+\cdots+N_t}\right], \tag{37}$$

where $\varphi_Y(r)$ is the characteristic function of Y.

On the other hand, direct calculation leads to

$$E\left[\exp\left\{ir\sum_{j=1}^{N_1+\cdots+N_t}Y_j\right\}\right] = \sum_n E\left[\exp\left\{ir\sum_{j=1}^{n}Y_j\right\}\times I_{\{N_1+\cdots+N_t=n\}}\right]$$

$$= \sum_n [E\exp\{irY\}]^n \times P\{N_1+N_2+\cdots+N_t = n\}$$

$$= E[(\varphi_Y(r))^{N_1+\cdots N_t}]. \tag{38}$$

We conclude after checking (37) and (38) that

$$S_t \stackrel{d}{=} \sum_{j=1}^{N_1+\cdots+N_t} Y_j, \tag{39}$$

where "$\stackrel{d}{=}$" means the identical distribution.

For any $0 < \eta < \lambda_2(1+\beta)$, we have

$$P\left(\sum_{j=1}^{N_1+\cdots+N_t} Y_j > y\right)$$
$$= P\left(\sum_{i=1}^{t} N_i < \lfloor(\lambda_2(1+\beta)+\eta)t\rfloor, \sum_{j=1}^{N_1+\cdots+N_t} Y_j > y\right) + P\left(\sum_{i=1}^{t} N_i \geq \lfloor(\lambda_2(1+\beta)+\eta)t\rfloor, \sum_{j=1}^{N_1+\cdots+N_t} Y_j > y\right)$$
$$\leq P\left(\sum_{j=1}^{\lfloor(\lambda_2(1+\beta)+\eta)t\rfloor} Y_j > y\right) + P\left(\sum_{i=1}^{t} N_i \geq \lfloor(\lambda_2(1+\beta)+\eta)t\rfloor\right)$$
$$= \Delta_1 + \Delta_2, \tag{40}$$

in which $\lfloor \cdot \rfloor$ denotes the maximum integer not exceeding $"."$.
From Lemma 4, we know it holds uniformly for all $y > \gamma t^{1+\delta}$ that

$$\Delta_1 \sim \lfloor(\lambda_2(1+\beta)+\eta)t\rfloor \overline{F}_Y(y). \tag{41}$$

As for Δ_2, for $t = 1, 2, \cdots$, we write

$$\sum_{i=1}^{t} N_i = \sum_{i=1}^{\lfloor t/2 \rfloor} N_{2i} + \sum_{i=1}^{\lfloor t/2 \rfloor + p} N_{2i-1},$$

where $p = 0$ if t is a even number, and $p = 1$ if t is an odd number. From the definition, we know that $\{N_i, i = 1, 2, \cdots\}$ is a one-dependent stationary sequence with the common Poisson distribution of mean $\lambda_2(1+\beta)$ and m.g.f $M_N(r) = \exp\{\lambda_2(1+\beta)(e^r - 1)\}$; then, it is easy to see that $\{N_{2i}, 1 \leq i \leq \lfloor t/2 \rfloor\}$ and $\{N_{2i-1}, 1 \leq i \leq \lfloor t/2 \rfloor + p\}$ are two sequences of i.i.d. random variables. Let $a = \lambda_2(1+\beta) + \eta$; by Cramér Theorem (Theorem 2.2.3 in Dembo and Zeitonui [43]), we have

$$\Delta_2 = P\left(\sum_{i=1}^{\lfloor t/2 \rfloor} N_{2i} + \sum_{i=1}^{\lfloor t/2 \rfloor + p} N_{2i-1} \geq \lfloor(\lambda_2(1+\beta)+\eta)t\rfloor\right)$$
$$\leq P\left(\sum_{i=1}^{\lfloor t/2 \rfloor} N_{2i} \geq \lfloor\lambda_2(1+\beta)+\eta\rfloor\lfloor t/2 \rfloor\right) + P\left(\sum_{i=1}^{\lfloor t/2 \rfloor + p} N_{2i-1} \geq \lfloor\lambda_2(1+\beta)+\eta\rfloor(\lfloor t/2 \rfloor + p)\right)$$
$$\sim e^{-\lfloor t/2 \rfloor I(a)} + e^{-(\lfloor t/2 \rfloor + p)I(a)} \to 0, \quad t \to +\infty, \tag{42}$$

in which $I(a) = \sup_{-\infty < r < +\infty} \{ar - \log(M_N(r))\} > 0$.
Combining (40)–(42) gives

$$P\left(\sum_{j=1}^{N_1+\cdots+N_t} Y_j > y\right) \lesssim \lfloor(\lambda_2(1+\beta)+\eta)t\rfloor \overline{F}_Y(y). \tag{43}$$

On the other hand, it holds that

$$P\left(\sum_{j=1}^{N_1+\cdots+N_t} Y_j > y\right) \geq P\left(\sum_{i=1}^{t} N_i \geq \lfloor(\lambda_2(1+\beta)-\eta)t\rfloor, \sum_{j=1}^{N_1+\cdots+N_t} Y_j > y\right)$$
$$\geq P\left(\sum_{i=1}^{t} N_i \geq \lfloor(\lambda_2(1+\beta)-\eta)t\rfloor, \sum_{j=1}^{\lfloor(\lambda_2(1+\beta)-\eta)t\rfloor} Y_j > y\right)$$
$$= P\left(\sum_{i=1}^{t} N_i \geq \lfloor(\lambda_2(1+\beta)-\eta)t\rfloor\right) \times P\left(\sum_{j=1}^{\lfloor(\lambda_2(1+\beta)-\eta)t\rfloor} Y_j > y\right), \tag{44}$$

in which

$$P\left(\sum_{i=1}^{t} N_i \geq \lfloor(\lambda_2(1+\beta)-\eta)t\rfloor\right) = P\left(\sum_{i=1}^{t} N_i - \lambda_2(1+\beta)t \geq \lfloor(\lambda_2(1+\beta)-\eta)t\rfloor - \lambda_2(1+\beta)t\right)$$

$$\geq P\left(\sum_{i=1}^{t} N_i - \lambda_2(1+\beta) \geq -\eta t\right)$$

$$\geq P\left(\left|\sum_{i=1}^{t} N_i - \lambda_2(1+\beta)t\right| \leq \eta t\right) \to 1, \quad t \to +\infty, \quad (45)$$

because of the fact that

$$P\left(\left|\sum_{i=1}^{t} N_i - \lambda_2(1+\beta)t\right| > \eta t\right)$$
$$\leq P\left(\left|\sum_{i=1}^{\lfloor t/2\rfloor} [N_{2i} - \lambda_2(1+\beta)]\right| > \eta\lfloor t/2\rfloor\right) + P\left(\left|\sum_{i=1}^{\lfloor t/2\rfloor+p} [N_{2i-1} - \lambda_2(1+\beta)]\right| > \eta(\lfloor t/2\rfloor+p)\right)$$
$$\to 0, \quad t \to +\infty,$$

obtained from the classical law of large numbers.

Then, combining (44), (45), and Lemma 4 yields

$$P\left(\sum_{j=1}^{N_1+\cdots+N_y} Y_j > y\right) \gtrsim \lfloor(\lambda_2(1+\beta)-\eta)t\rfloor\overline{F}_Y(y). \quad (46)$$

Generally, letting $\eta \downarrow 0$ in (43) and (46) and keeping (39) in mind, we finally conclude that

$$P(S_t > y) = P\left(\sum_{j=1}^{N_1+\cdots+N_t} Y_j > y\right) \sim \lfloor\lambda_2(1+\beta)t\rfloor\overline{F}_Y(y) \sim \lambda_2(1+\beta)t\overline{F}_Y(y).$$

Then, the proof is completed. □

With the help of the above conclusion, we can manifest the asymptotic formula for the finite-time ruin probability in the following theorem.

Theorem 4. *Under the conditions of Theorem 3, for any fixed $\gamma > 0$ and $\delta > 0$, the asymptotic formula*

$$\psi(u,t) \sim \lambda_2(1+\beta)t\overline{F}_Y(u) \quad (47)$$

holds uniformly for all $u > \gamma t^{1+\delta}$ as $t \to +\infty$.

Proof. From the definition of finite-time ruin probability, it is clear that

$$\psi(u,t) = P\left(\sup_{m\in\{0,1,\cdots,t\}} (S_m - W_m) > u\right)$$
$$\geq P(S_t - W_t > u)$$
$$= P(S_t > u + W_t)$$
$$= P\left(S_t > u + \frac{\lambda_1}{1-\alpha}tE(X) + W_t - \frac{\lambda_1}{1-\alpha}tE(X)\right). \quad (48)$$

Noting that $W_t = \sum_{i=1}^{t} P_i = \sum_{i=1}^{t} \sum_{k=1}^{M_i} X_{i,k}$ and keeping (9) in mind, for any $\eta > 0$, we have

$$P\left(\left|W_t - \frac{\lambda_1}{1-\alpha}tE(X)\right| \geq \eta t\right) = P\left(\left|\frac{1}{t}\sum_{i=1}^{t} P_i - \frac{\lambda_1}{1-\alpha}E(X)\right| \geq \eta\right)$$

$$\leq \frac{E\left(\frac{1}{t}\sum_{i=1}^{t} P_i - \frac{\lambda_1}{1-\alpha}E(X)\right)^2}{\eta^2}$$

$$= \frac{1}{(\eta t)^2} \sum_{i,j=1}^{t} Cov(P_i, P_j)$$

$$= \frac{\lambda_1 E(X^2)}{(1-\alpha)(\eta t)^2}\left(t + 2(t-1)\alpha + 2(t-2)\alpha^2 + \cdots + 2\alpha^{t-1}\right)$$

$$\leq \frac{2\lambda_1 E(X^2)(1-\alpha^t)}{t\eta^2(1-\alpha)^2} \to 0, \quad t \to +\infty,$$

from which we can obtain

$$\lim_{t \to +\infty} \sup_{u > \gamma t^{1+\delta}} \frac{1}{u}\left(\frac{\lambda_1}{1-\alpha}tE(X) + W_t - \frac{\lambda_1}{1-\alpha}tE(X)\right) = 0.$$

Then, for any $\theta > 0$, if t is sufficiently large such that u is sufficiently large, it holds that

$$P(S_t > u + \theta u) \leq P\left(S_t > u + \frac{\lambda_1}{1-\alpha}tE(X) + W_t - \frac{\lambda_1}{1-\alpha}tE(X)\right) \leq P(S_t > u - \theta u),$$

Furthermore, by Theorem 3 and let $\theta \downarrow 0$, we have

$$P\left(S_t > u + \frac{\lambda_1}{1-\alpha}tE(X) + W_t - \frac{\lambda_1}{1-\alpha}tE(X)\right) \sim \lambda_2(1+\beta)\overline{F}_Y(u), \text{ uniformly for } u > \gamma t^{1+\delta} \text{ as } t \to +\infty. \quad (49)$$

Plugging (49) into (48) gives

$$\psi(u,t) \gtrsim \lambda_2(1+\beta)t\overline{F}_Y(u). \quad (50)$$

On the other hand, for any fixed $\gamma > 0$ and $\delta > 0$, we have uniformly for all $u > \gamma t^{1+\delta}$ that

$$\psi(u,t) = P\left(\sup_{m \in \{0,1,\cdots,t\}}(S_m - W_m) > u\right) \leq P(S_t > u) \sim \lambda_2(1+\beta)t\overline{F}_Y(u).$$

which implies

$$\psi(u,t) \lesssim \lambda_2(1+\beta)t\overline{F}_Y(u). \quad (51)$$

Therefore, we complete the proof by combining (50) and (51). □

Remark 9. *Applying Lemma 3 instead of Lemma 4 in Theorem 3, it is not difficult to see that the precise large deviation (36) also holds uniformly for all $y > \gamma t$. In this paper, we restrict ourselves to the interval $y > \gamma t^{1+\delta}$ in order to provide convenience for investigating the finite-time ruin probability $\psi(u,t)$. Moreover, we can prove that the asymptotic formula (47) in Theorem 4 holds uniformly for all $u \in \Omega = \{u; t = o(u)\}$, which includes $u > \gamma t^{1+\delta}$ as a special case. In practice, when t is large enough, we can asymptotically estimate $\psi(u,t)$ by $\lambda_2(1+\beta)t\overline{F}_Y(u)$, as the size of claims belong to the distributions of class $\mathcal{C}$ and the insurer's initial surplus is adequate in the sense of $u > \gamma t^{1+\delta}$.*

6. Numerical Examples

In this section, we aim to perform some numerical simulations to demonstrate and assess the Lundberg adjustment coefficient and the Lundberg approximation results for the infinite-time ruin probability $\psi(u)$, as well as the asymptotic formula for the finite-time ruin probability $\psi(u,t)$, of our proposed model.

Example 1. *We suppose that the gain amount X and the claim size Y follow exponential distributions that have means $1/\mu_1$ and $1/\mu_2$, respectively. Therefore, we have the moment generating functions of X and Y as follows:*

$$M_X(-r) = \frac{\mu_1}{\mu_1 + r}, \quad M_Y(r) = \frac{\mu_2}{\mu_2 - r}, \quad r > 0. \tag{52}$$

Then, from Theorem 2, $c(r) = 0$ is equivalent to

$$\lambda_1 \frac{1}{(1-\alpha)\mu_1 + r} = \lambda_2 \frac{(1+\beta)\mu_2 - r}{(\mu_2 - 2)^2}. \tag{53}$$

The unique positive solution to Equation (53) can be found but appears tedious. In what follows, we give some numerical results to show the properties and performance of R and e^{-Ru}.

Without loss of generality, we set $\lambda_1 = 1$, $\lambda_2 = 0.4$, $\mu_1 = 1$, and $\mu_2 = 0.5$, and then, calculate and discuss the Lundberg adjustment coefficient R and the approximated ruin probability e^{-Ru} for different values of α and β. When we consider the impacts of α and β on the main results, it should be noted that the relative safety loading condition (26) has to be satisfied, i.e.,

$$\frac{\lambda_1}{1-\alpha} \cdot \frac{1}{\mu_1} > \lambda_2 \cdot (1+\beta) \cdot \frac{1}{\mu_2}, \tag{54}$$

which, in our parameter scenario, implies

$$\frac{1}{1-\alpha} > 0.8(1+\beta). \tag{55}$$

Table 1 gives the computed values of Lundberg adjustment coefficients corresponding to different values of α and β. We also illustrate these results in Figure 1, from which it can be clearly seen that R increases as α increases, implying that the insurance portfolio would become less and less dangerous because the approximated infinite-time ruin probability e^{-Ru} decreases. In the same sense, when β increases, R will decrease, meaning that there could be higher risks in the insurance portfolio. (In Table 1, the notation " $-$ " means that the values of α and β do not satisfy the relative safety loading condition, and the Lundberg adjustment coefficients are not considered for these situations.)

Table 1. Lundberg adjustment coefficients for different α and β.

	β								
α	0.1	0.2	0.3	0.4	0.5	0.6	0.7	0.8	0.9
0.1	0.0680	0.0414	0.0183	-	-	-	-	-	-
0.2	0.0968	0.0706	0.0481	0.0282	0.0104	-	-	-	-
0.3	0.1256	0.1000	0.0781	0.0588	0.0416	0.0259	0.0115	-	-
0.4	0.1545	0.1295	0.1082	0.0897	0.0731	0.0581	0.0443	0.0316	0.0198
0.5	0.1834	0.1591	0.1386	0.1208	0.1049	0.0906	0.0776	0.0655	0.0544
0.6	0.2124	0.1888	0.1691	0.1522	0.1371	0.1236	0.1113	0.1000	0.0895
0.7	0.2415	0.2187	0.2000	0.1839	0.1698	0.1571	0.1457	0.1351	0.1254
0.8	0.2707	0.2489	0.2312	0.2162	0.2031	0.1913	0.1807	0.1711	0.1622
0.9	0.3000	0.2794	0.2630	0.2491	0.2370	0.2264	0.2167	0.2080	0.2000

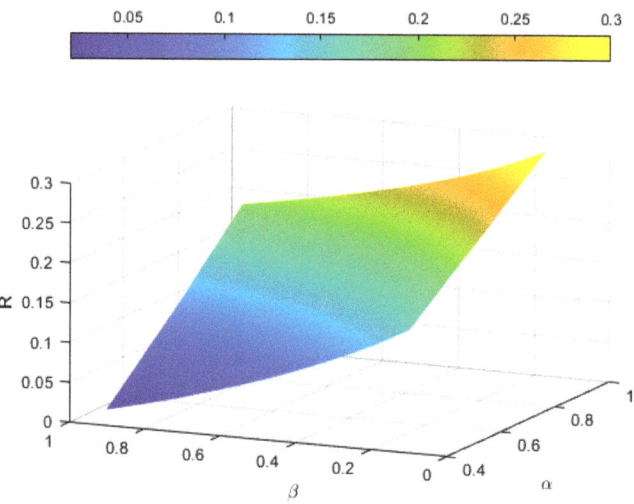

Figure 1. Lundberg adjustment coefficients corresponding to different α and β.

In order to evaluate the performance of the approximated infinite-time ruin probability e^{-Ru}, we fix $\alpha = \beta = 0.5$ in the proposed risk model and compute the true ruin probabilities corresponding to different values of u by the Monte Carlo method used in Albrecher and Kantor [44]. For this purpose, we randomly draw sample paths according to the Poisson INAR(1) process and the Poisson INMA(1) process for the premium arrivals $\{M_t, t = 1, 2, \cdots\}$ and the claim numbers $\{N_t, t = 1, 2, \cdots\}$, respectively. Afterwards, we simulate the surplus process (2) starting at $U_0 = u$, with the premium amounts and claim sizes following the given exponential distributions. These simulations are replicated $n = 3000$ times; then, the trajectories with negative values (i.e., ruin event occurs) are counted, and we denote this number by n_1. Hence, the infinite-time ruin probability $\psi(u)$ can be estimated by

$$\hat{\psi}(u) = \frac{n_1}{n}. \tag{56}$$

In addition, because of the fact that $U_t \to +\infty$ with probability one as $t \to +\infty$ when the relative safety loading condition holds, we know that U_t will never become negative when t is large enough. Therefore, it is necessary for us to choose a suitable T_{st} at which we should stop the simulated surplus process for each sample path if the ruin event does not occur before this time. As a consequence, (56) is actually the estimate of the finite-time ruin probability $\psi(u, T_{st}) = P(T \leq T_{st}|U_0 = u)$. In this paper, we set $T_{st} = 1000$. In practice, we can choose larger values for T_{st} so that the bias of the estimator for $\psi(u)$ is less significant.

In Table 2 and Figure 2, we compare the simulated ruin probability with the approximated ruin probability. As can be seen, when u grows, both of the ruin probabilities approach zero. However, as alternatives to the true ruin probabilities, the approximations do not work well when the values of u are small. We can explain these results with the following three reasons. Firstly, as the limit of $\psi(u)$ as $u \to +\infty$, e^{-Ru} may be very different than $\psi(u)$ at the beginning. Secondly, the simulated infinite-time ruin probabilities are indeed the estimated values for the finite-time ruin probability $\psi(u, T_{st})$, which are smaller than the true values of $\psi(u)$. Thirdly, the total number of simulated trajectories n and the chosen time T_{st} affect the simulated results. We could increase n and T_{st} to improve the performance, but a longer run time is needed.

On the other hand, the values of simulated ruin probability and the approximated ruin probability become closer and closer with the increase in u, implying that the approximation method could work better as u grows. To strengthen this statement, we define $\gamma(u) = \frac{\hat{\psi}(u)}{e^{-Ru}}$, and then, calculate the values of $\gamma(u)$ with respect to different values of u; the results are listed in the last column of

Table 2. It can be seen that $\gamma(u)$ approaches 1 asymptotically, which indicates that it is valid to take e^{-Ru} as the approximated result for $\hat{\psi}(u)$ and, furthermore, for $\psi(u)$ when u is large enough. Figure 3 also illustrates this conclusion visually. In practice, an insurer is always required to hold a huge number of initial surplus to guarantee its solvency under certain regulatory frameworks; therefore, the approximation method is of importance and is applicable in the risk management of insurance.

Table 2. Comparison of the simulated and approximated ruin probability.

u	$\hat{\psi}(u)$	e^{-Ru}	$\gamma(u) = \dfrac{\hat{\psi}(u)}{e^{-Ru}}$
10	0.2280	0.3503	0.6509
15	0.1386	0.2073	0.6686
20	0.0819	0.1227	0.6678
25	0.0497	0.0726	0.6846
30	0.0294	0.0430	0.6835
35	0.0183	0.0254	0.7186
40	0.0112	0.0151	0.7388
45	0.0067	0.0089	0.7575
50	0.0043	0.0053	0.8125

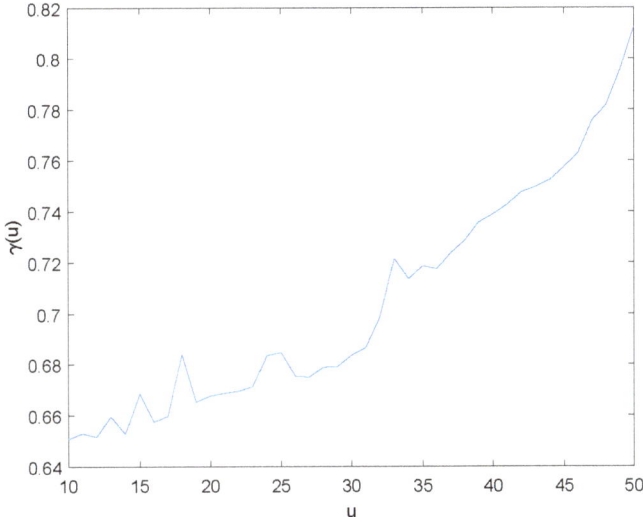

Figure 2. The simulated and approximated ruin probabilities with respect to different values of u.

Example 2. We suppose that the gain amount X is distributed by the exponential distribution with mean of $1/\mu_1$, and the claim size Y follows the Pareto distribution, which has shape parameter τ_1 and scale parameter τ_2, i.e., the distribution function $F_Y(y)$ is given by $F_Y(y) = 1 - \left(\dfrac{\tau_2}{\tau_2 + x}\right)^{\tau_1}$, $y > 0$. To perform the calculations, we set $\lambda_1 = 1$, $\mu_1 = 1$, $\lambda_2 = 0.1$, $\tau_1 = 3$, $\tau_2 = 16$, and $\alpha = \beta = 0.5$. It is not difficult to check that these values satisfy the relative safety loading condition (26). Our goal is to compare the asymptotic result $\lambda_2(1+\beta)t\bar{F}_Y(y)$ (AS for simplification) with the simulated results of the finite-time ruin probabilities obtained using the Monte Carlo method (MC for simplification). As can be seen from Table 3, the ratio of MC to AS becomes closer and closer to one as t increases for different u, indicating that the asymptotic formula stated in Theorem 4 is valid and applicable in practice.

Table 3. Comparison of the simulated results and the asymptotic results for $\psi(u,t)$.

		$u=60$	$u=70$	$u=80$	$u=90$	$u=100$
$t=50$	AS	0.0760	0.0560	0.0437	0.0300	0.0210
	MC	0.0700	0.0483	0.0347	0.0258	0.0197
	AS/MC	1.0860	1.1595	1.2576	1.1628	1.0660
$t=40$	AS	0.0703	0.0487	0.0440	0.0280	0.0200
	MC	0.0560	0.0386	0.0278	0.0206	0.0157
	AS/MC	1.2563	1.2596	1.5840	1.3592	1.2739
$t=30$	AS	0.0517	0.0360	0.0330	0.0170	0.0140
	MC	0.0420	0.0290	0.0208	0.0155	0.0118
	AS/MC	1.2305	1.2423	1.5840	1.0985	1.1856
$t=20$	AS	0.0377	0.0247	0.0223	0.0140	0.0120
	MC	0.0280	0.0193	0.0139	0.0103	0.0079
	AS/MC	1.3456	1.2768	1.6043	1.3570	1.5190
$t=10$	AS	0.0190	0.0130	0.0113	0.0077	0.0063
	MC	0.0140	0.0097	0.0069	0.0052	0.0039
	AS/MC	1.3575	1.3402	1.6377	1.4862	1.6090

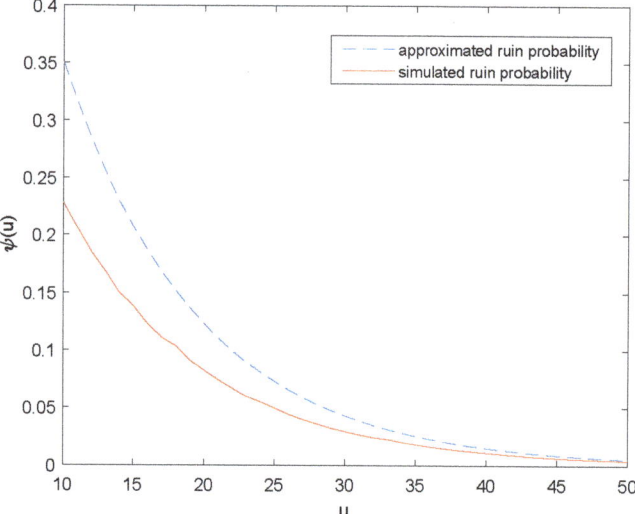

Figure 3. The values of ratio $\gamma(u)$ with respect to different u.

7. Conclusions

In this paper, we examine a generalization of the classical discrete-time risk model of an insurance portfolio with stochastic premiums, using a Poisson INAR(1) process and a Poisson INMA(1) process to fit the temporal dependence among the premium numbers and the temporal dependence among the claim numbers, respectively. We give the explicit expression for the function satisfied by the Lundberg adjustment coefficient and find the Lundberg approximation formula for the infinite-time ruin probability. Furthermore, we discuss and analyze the impact of the two thinning parameters and manifest that the dependence structure in the model has a significant influence on the risk of the surplus process in an insurance company. When the claim sizes follow a class of heavy-tailed distributions, we establish the large deviations of the aggregate claims and investigate the asymptotic formula for the finite-time ruin probability. In the numerical examples, we use MATLAB to randomly draw the sample paths of the proposed surplus process and compute estimates of the true ruin probabilities corresponding to different values of u using the Monte

Carlo method. From the simulated results, it can be seen that the approximation formula and asymptotic formula we obtained are effective. Furthermore, these two formulas are much simpler to use for calculating and estimating the ruin probabilities than the Monte Carlo method.

As for future work, we could implement the same methodology by applying the time series for count data with other distributed innovations or an arbitrary innovations' distribution. Generally, using the same approach as that in Lemma 1 and Lemma 2, we can extend (14) and (20) to

$$P_{M_1+\cdots+M_t}(s) = P_{M_1}(h_t(s)) \prod_{i=1}^{t-1} P_\varepsilon(h_i(s)),$$

and

$$P_{N_1+\cdots+N_t}(s) = P_\eta(s) P_\eta(\beta s + (1-\beta))[P_\eta(\beta s^2 + (1-\beta)s)]^{t-1},$$

respectively. Therefore, if we could derive the explicit expression of $c(r)$, the properties of the solution to the equation $c(r) = 0$ can be discussed, and the adjustment coefficient can be obtained to measure the risk.

Additionally, we could adopt some higher-order processes to make the insurance risk model much more practical and flexible. In this situation, it becomes more challenging to find the expressions of $P_{M_1+\cdots+M_t}(s)$ and $P_{N_1+\cdots+N_t}(s)$. As a consequence, there might be some difficulties in deriving $c(r)$ and defining the adjustment coefficient for an insurance portfolio.

On the other hand, instead of fixing the distributions and the parameters to illustrate the results by simulation, we can use the real dataset to fit the distributions and obtain the statistical estimates of the parameters, so that the ruin problems of the risk model could be analyzed in a more scientific way.

Author Contributions: Conceptualization, L.G. and X.W.; methodology, L.G.; software, X.W.; validation, L.G. and X.W.; formal analysis, L.G.; investigation, L.G.; resources, X.W.; data curation, X.W.; writing—original draft preparation, L.G. and X.W.; writing—review and editing, L.G. and X.W.; visualization, X.W.; supervision, L.G. and X.W.; project administration, L.G.; funding acquisition, L.G. and X.W. All authors have read and agreed to the published version of the manuscript.

Funding: This research was funded by the Natural Science Foundation of Jilin Province, grant number YDZJ202201ZYTS516.

Institutional Review Board Statement: Not applicable.

Informed Consent Statement: Not applicable.

Data Availability Statement: Not applicable.

Conflicts of Interest: The authors declare no conflict of interest.

Appendix A

In this appendix, we explicate the motivation and rationale of our proposed model (2) by the following descriptions, in order to make this paper more accessible to the readers. Let us begin with the classical discrete-time Lundberg–Cramér risk model

$$U_t = U_{t-1} + c - L_t, \ t = 1, 2, \cdots, \tag{A1}$$

where U_t corresponds to the surplus of an insurance portfolio at time t, with $U_0 = u$ being the initial surplus; c being the constant premium income per period, and L_t representing the aggregate claim amount in period t that is defined as

$$L_t = \sum_{j=1}^{N_t} Y_{t,j}, \tag{A2}$$

in which N_t denotes the number of claims and $Y_{t,j}$ is the size of the jth payment to the insured in period t. After recursively calculating, it is easy to see that the risk model (A1) can be rewritten as

$$U_t = u + ct - \sum_{i=1}^{t} L_i = u + ct - \sum_{i=1}^{t} \sum_{j=1}^{N_i} Y_{i,j}, \ t = 1, 2, \cdots,$$

which is equivalent to the model (1) by denoting $N_t^0 = \sum_{i=1}^{t} N_i$.

For simplicity, it is assumed that the claim numbers of different periods are independent in the Lundberg–Cramér risk model, i.e., the claim number process $\{N_t, t = 1, 2, \cdots\}$ is a sequence of i.i.d. random variables, which is certainly not realistic. As a consequence, ref. [11] proposes some new discrete-time risk models, where the Poisson INMA(1) process and Poisson INAR(1) process are used to describe the dependence structures among the numbers of claims. That is to say, the claim number process $\{N_t, t = 1, 2, \cdots\}$ satisfies

$$N_t = \alpha \circ N_{t-1} + \varepsilon_t, \ t = 2, 3, \cdots,$$

or

$$N_t = \beta \circ \eta_{t-1} + \eta_t, \ t = 1, 2, \cdots.$$

On the other hand, both of the above two types of risk models suppose that the premiums are collected with positive deterministic constant rate c, which is also lacks the ability of describing the real situation of insurance portfolio. As an alternative to this case, ref. [18] proposes the risk model with stochastic premiums that can be expressed as

$$U_t = U_{t-1} + P_t - L_t, \ t = 1, 2, \cdots, \tag{A3}$$

where L_t is defined by (A2), and P_t aggregates the premiums in period t that is defined as

$$P_t = \sum_{k=1}^{M_t} X_{t,k}, \tag{A4}$$

in which M_t counts the number of individual income, and $X_{t,k}$ represents the amount of the kth premium income for the insurance portfolio in period t.

In the risk model (A3), it should be noted that both the premium number process $\{M_t, t = 1, 2, \cdots\}$ and the claim number process $\{N_t, t = 1, 2, \cdots\}$ are supposed to be sequences of i.i.d. random variables. Therefore, the goal of this paper is to introduce the idea of [11] into the risk model with stochastic premiums by using time series for count random variables to fit the temporal dependence among $\{M_t, t = 1, 2, \cdots\}$ and $\{N_t, t = 1, 2, \cdots\}$, respectively. Furthermore, considering some insurance practices (please see Remarks 2 and 3), we assume that $\{M_t, t = 1, 2, \cdots\}$ constitutes a Poisson INAR(1) process that satisfies

$$M_t = \alpha \circ M_{t-1} + \varepsilon_t, \ t = 2, 3, \cdots,$$

and $\{N_t, t = 1, 2, \cdots\}$ constitutes a Poisson INMA(1) process that satisfies

$$N_t = \beta \circ \eta_{t-1} + \eta_t, \ t = 1, 2, \cdots.$$

Our proposed risk model can also be generalized in several aspects to make itself more flexible and applicable, as discussed in the Conclusions.

References

1. Asmussen, S.; Albrecher, H. *Ruin Probabilities*, 2nd ed.; World Scientiffic: Singapore, 2010.
2. Yang, H.L.; Zhang, L.H. Martingale method for ruin probability in an autoregressive model with constant interest rate. *Probab. Eng. Inform. Sci.* **2003**, *17*, 183–198. [CrossRef]

3. Chen, Y.Y.; Lin, L.Y.; Wang, R.D. Risk aggregation under dependence uncertainty and an order constraint. *Insur. Math. Econ.* **2022**, *102*, 169–187. [CrossRef]
4. Hanbali, H.; Linders, D.; Dhaene, J. Value-at-Risk, Tail Value-at-Risk and upper tail transform of the sum of two counter-monotonic random variables. *Scand. Actuar. J.* **2022**, *2022*, 219–243. [CrossRef]
5. Albrecher, H.; Cheung, E.C.K.; Liu H.; Woo, J-K. A bivariate Laguerre expansions approach for joint ruin probabilities in a two-dimensional insurance risk process. *Insur. Math. Econ.* **2022**, *103*, 96–118. [CrossRef]
6. Strietzel, P.L.; Heinrich, H.E. Optimal dividends for a two-dimensional risk model with simultaneous ruin of both branches. *Risk* **2022**, *10*, 116. [CrossRef]
7. Chen, M.; Zhou, M.; Liu, H.Y.; Yuen, K.C. Optimal dividends and reinsurance with capital injection under thinning dependence. *Commun. Stat. Theory Methods* **2022**, *51*, 5728–5749. [CrossRef]
8. Wang, F.D.; Liang, Z.B. Optimal per-loss reinsurance for a risk model with a thinning-dependence structure. *Mathematics* **2022**, *10*, 4621. [CrossRef]
9. Gerber, H.U. Ruin theory in the linear model. *Insur. Math. Econ.* **1982**, *1*, 177–184. [CrossRef]
10. Gourieroux, C.; Jasiak, J. Heterogeneous INAR(1) model with application to car insurance. *Insur. Math. Econ.* **2004**, *34*, 177–192. [CrossRef]
11. Cossette, H.; Marceau, E.; Maume-Deschamps, V. Discrete-time risk models based on time series for count random variables. *ASTIN Bull.* **2010**, *40*, 123–150. [CrossRef]
12. Cossette, H.; Marceau, E.; Toureille, F. Risk models based on time series for count random variables. *Insur. Math. Econ.* **2011**, *48*, 19–28. [CrossRef]
13. Shi, H.F.; Wang, D.H. An approximation model of the collective risk model with INAR(1) claim process. *Commun. Stat. Theory Methods* **2014**, *43*, 15305–15317. [CrossRef]
14. Zhang, L.Z.; Hu, X.; Duan, B.G. Optimal reinsurance under adjustment coefficient measure in a discrete risk model based on Poisson MA(1) process. *Scand. Actuar. J.* **2015**, *2015*, 455–467. [CrossRef]
15. Hu, X.; Zhang, L.Z.; Sun, W.W. Risk model based on the first-order integer-valued moving average process with compound Poisson distributed innovations. *Scand. Actuar. J.* **2018**, *2018*, 412–425. [CrossRef]
16. Chen, M.; Hu, X. Risk aggregation with dependence and overdispersion based on the compound Poisson INAR(1) process. *Commun. Stat. Theory Methods* **2020**, *49*, 3985–4001. [CrossRef]
17. Guan, G.H.; Hu, X. On the analysis of a discrete-time risk model with INAR(1) processes. *Scand. Actuar. J.* **2022**, *2022*, 115–138. [CrossRef]
18. Boikov, A.V. The Cramér-Lundberg model with stochastic premium process. *Theory Probab. Its Appl.* **2003**, *47*, 489–493. [CrossRef]
19. Wang, R.M.; Xu, L.; Yao, D.J. Ruin problems with stochastic premium stochastic return on investments. *Front. Math. China* **2017**, *2*, 467–490. [CrossRef]
20. Labbé, C.; Sendova, K.P. The expected discounted penalty function under a risk model with stochastic income. *Appl. Math. Comput.* **2019**, *215*, 1852–1867. [CrossRef]
21. Zhao, Y.X.; Yin, C.C. The expected discounted penalty function under a renewal risk model with stochastic income. *Appl. Math. Comput.* **2012**, *217*, 6144–6154. [CrossRef]
22. Su, W.; Shi, B.X.; Wang, Y.Y. Estimating the Gerber-Shiu function under a risk model with stochastic income by Laguerre series expansion. *Commun. Stat. Theory Methods* **2020**, *49*, 5686–5708. [CrossRef]
23. Ragulina, O. Simple approximations for the ruin probability in the risk model with stochastic premiums and a constant dividend strategy. *Mod. Stochastics Theory Appl.* **2020**, *7*, 489–493. [CrossRef]
24. Dibu, A.S.; Jacob, M.J. On a double barrier hybrid dividend strategy in a compound Poisson risk model with stochastic income. *Ann. Oper. Res.* **2022**, *315*, 969–984. [CrossRef]
25. Wang, Z.J.; Landriault, D.; Li, S. An insurance risk process with a generalized income process: A solvency analysis. *Insur. Math. Econ.* **2021**, *98*, 133–146. [CrossRef]
26. Guan, L.H.; Wang, X.H. A discrete-time dual risk model with dependence based on a Poisson INAR(1) process. *AIMS Math.* **2022**, *7*, 20823–20837. [CrossRef]
27. Weiß, C.H. Thinning operations for modeling time series of counts—A survey. *AStA Adv. Stat. Anal.* **2008**, *92*, 319–343. [CrossRef]
28. Scotto, M.G.; Weiß, C.H.; Gouveia, S. Thinning-based models in the analysis of integer-valued time series: A review. *Stat. Model.* **2015**, *15*, 590–618. [CrossRef]
29. Weiß, C.H. *An Introduction to Discrete-Valued Time Series*; John Wiley and Sons Ltd.: Hoboken, NJ, USA, 2018.
30. Weiß, C.H. Stationary count time series models. *WIREs Comput. Stat.* **2021**, *11*, e1502. [CrossRef]
31. Al-Osh, M.A.; Alzaid, A.A. First-order integer-valued autoregressive (INAR(1)) process. *J. Time Ser. Anal.* **1987**, *8*, 261–275. [CrossRef]
32. Müller, A.; Pflug, G. Asymptotic ruin probabilities for risk processes with dependent increments. *Insur. Math. Econ.* **2001**, *28*, 381–392. [CrossRef]
33. Klüppelberg, C.; Mikosch, T. Large deviations of heavy-tailed random sums with applications in insurance and finance. *J. Appl. Probab.* **1997**, *24*, 293–308. [CrossRef]
34. Chen, Y.Q.; White, T.; Yuen, K.C. Precise large deviations of aggregate claims with arbitrary dependence between claim sizes and waiting times. *Insur. Math. Econ.* **2021**, *97*, 1–6. [CrossRef]

35. Fu, K.A.; Liu, Y.; Wang, J.F. Precise large deviations in a bidimensional risk model with arbitrary dependence between claim-size vectors and waiting times. *Stat. Probab. Lett.* **2022**, *184*, 109365. [CrossRef]
36. Cheng, J.H.; Wang, D.H. Ruin problems for an autoregressive risk model with dependent rates of interest. *Appl. Math. Comput.* **2011**, *218*, 3822–3833. [CrossRef]
37. Yang, Y.; Wang, K.Y.; Konstantinides, D.G. Uniform asymptotics for discounted aggregate claims in dependent risk models. *J. Appl. Probab.* **2014**, *51*, 669–684. [CrossRef]
38. Jing, H.J.; Peng, J.Y.; Jiang, Z.Q.; Bao, Q. Asymptotic estimates for finite-time ruin probability in a discrete-time risk model with dependence structures and CMC simulations. *Commun. Stat. Theory Methods* **2022**, *51*, 3761–3786. [CrossRef]
39. Xun, B.Y.; Wang, K.Y.; Yuen, Y.C. The finite-time ruin probability of a risk model with a general counting process and stochastic return. *J. Ind. Manag. Optim.* **2022**, *18*, 1541–1556. [CrossRef]
40. Yu, S.H. Precise large deviations of aggregate claims in a discrete-time risk model with Poisson ARCH claim-number process. *J. Inequalities Appl.* **2016**, *2016*, 140. [CrossRef]
41. Embrechts, P.; Klüppelberg, C.; Mikosch, T. *Modelling Extremal Events: For Insurance and Finance*; Springer Science & Business Media: Berlin, Germany, 1997.
42. Ng, K.W.; Tang, Q.H.; Yan, J.A.; Yang, H.L. Precise large deviations for sums of random variables with consistently varying tails. *J. Appl. Probab.* **2004**, *41*, 93–107. [CrossRef]
43. Dembo, A.; Zeitouni, O. *Large Deviations Techniques and Applications*, 2nd ed.; Springer: Berlin, Germany, 1998.
44. Albrecher, H.; Kantor, J. Simulation of ruin probabilities for risk processes of Markovian type. *Monte Carlo Methods Appl.* **2002**, *8*, 111–127. [CrossRef]

Disclaimer/Publisher's Note: The statements, opinions and data contained in all publications are solely those of the individual author(s) and contributor(s) and not of MDPI and/or the editor(s). MDPI and/or the editor(s) disclaim responsibility for any injury to people or property resulting from any ideas, methods, instructions or products referred to in the content.

Article

Two Features of the GINAR(1) Process and Their Impact on the Run-Length Performance of Geometric Control Charts

Manuel Cabral Morais

Department of Mathematics & CEMAT (Center for Computational and Stochastic Mathematics), Instituto Superior Técnico, Universidade de Lisboa, 1049-001 Lisbon, Portugal; maj@math.ist.utl.pt

Abstract: The geometric first-order integer-valued autoregressive process (GINAR(1)) can be particularly useful to model relevant discrete-valued time series, namely in statistical process control. We resort to stochastic ordering to prove that the GINAR(1) process is a discrete-time Markov chain governed by a totally positive order 2 (TP_2) transition matrix. Stochastic ordering is also used to compare transition matrices referring to pairs of GINAR(1) processes with different values of the marginal mean. We assess and illustrate the implications of these two stochastic ordering results, namely on the properties of the run length of geometric charts for monitoring GINAR(1) counts.

Keywords: discrete-time Markov chain; TP_2 transition probability matrix; Kalmykov order; statistical process control; run length

MSC: 60J10; 60E15; 62P30

1. Introduction

The INAR(1) and GINAR(1) processes were originally proposed by McKenzie [1,2]; the latter model was soon after discussed in more detail by Alzaid and Al-Osh [3]. They rely on the binomial thinning operation due to Steutel and van Harn [4] which is defined below.

Definition 1. *Let X be a non-negative integer-valued r.v. with range $\mathbb{N}_0 = \{0, 1, \dots\}$ and ρ a scalar in $(0, 1)$. Then the binomial thinning operation on X results in the r.v.*

$$\rho \circ X = \sum_{t=1}^{X} Y_t, \tag{1}$$

where $\circ$ represents the binomial thinning operator; $\{Y_t : t \in \mathbb{N}\}$ is a sequence of i.i.d. Bernoulli r.v. with parameter ρ; $\{Y_t : t \in \mathbb{N}\}$ is independent of X.

We usually refer to $\rho \circ X$ as the r.v. that arises from X by binomial thinning. Furthermore, we define $0 \circ X = 0$ and $1 \circ X = X$.

Now that we have defined the binomial thinning operation, a sort of scalar multiplication counterpart in the integer-valued setting, the reader is reminded of the definition of McKenzie's GINAR(1) process and its main properties.

Definition 2. *Let $\rho, p \in (0, 1)$. Then $\{X_t : t \in \mathbb{N}_0\}$ is said to be a GINAR(1) process if X_t is written in the form*

$$X_t = \rho \circ X_{t-1} + B_t \times G_t, \tag{2}$$

where $\{B_i : i \in \mathbb{N}\}$ and $\{G_i : i \in \mathbb{N}\}$ are independent sequences of i.i.d. Bernoulli r.v. with parameter $(1 - \rho)$ and of i.i.d. geometric r.v. with parameter p, respectively; the sequence of innovations $\{\varepsilon_t = B_t \times G_t : t \in t \in \mathbb{N}\}$ and $\{X_0, \dots, X_{t-2}, X_{t-1}\}$ are independent; all thinning operations are performed independently of each other and of $\{\varepsilon_t : t \in \mathbb{N}\}$; and all the thinning operations at time t are independent of $\{X_0, \dots, X_{t-2}, X_{t-1}\}$.

According to McKenzie [2] and Alzaid and Al-Osh [3], if $X_0 \sim$ geometric(p) then $\{X_t : t \in \mathbb{N}_0\}$ is a stationary AR(1) process with geometric(p) marginal distribution.

McKenzie [2] also adds that $\{X_t : t \in \mathbb{N}_0\}$ is a DTMC with TPM, $\mathbf{P}(p,\rho) = [p_{ij}(p,\rho)]_{i,j\in\mathbb{N}_0}$ = $[P(X_t = j \mid X_{t-1} = i)]_{i,j\in\mathbb{N}_0}$, where

$$p_{ij}(p,\rho) = \sum_{m=0}^{\min\{i,j\}} \binom{i}{m} \rho^m (1-\rho)^{i-m} \times (1-p)(1-p)^{j-m} p$$

$$+ \binom{i}{j} \rho^j (1-\rho)^{i-j} \times p \times I_{\mathbb{N}_0}(i-j), \quad i,j \in \mathbb{N}_0, \quad (3)$$

where $I_{\mathbb{N}_0}$ represents the indicator function of the set of non-negative integers. These entries can be obtained by taking advantage of a few facts: $(\rho \circ X_{t-1} \mid X_{t-1} = 0) = 0$ with probability 1; $(\rho \circ X_{t-1} \mid X_{t-1} = i) \sim$ binomial(i,ρ), for $i \in \mathbb{N}$; the p.f. of the innovations, $\varepsilon_t = B_t \times G_t$, is equal to

$$P(\varepsilon_t = j) = \begin{cases} P(B_t = 0 \text{ or } G_t = 0) = p(1-\rho) + \rho, & j = 0 \\ P(B_t = 1, G_t = j) = (1-\rho)(1-p)^j p, & j \in \mathbb{N}. \end{cases} \quad (4)$$

The autocorrelation function of the GINAR(1) process is equal to

$$corr(X_t, X_{t+k}) = \rho^k, \quad k, t \in \mathbb{N}_0. \quad (5)$$

We ought to point out that the GINAR(1) process is a particular case of the generalized geometric INAR(1) or GGINAR(1) process, introduced by (Al-Osh and Aly [5], Section 3). Moreover, autocorrelated geometric counts can also be modeled by the new geometric INAR(1) or NGINAR(1) process, proposed by Ristić et al. [6] and relying on the negative binomial thinning operator. Finally, the NGINAR(1) process is a special instance of the ZMGINAR(1) process, the zero-modified geometric first-order integer-valued autoregressive, introduced and thoroughly described by Barreto-Souza [7].

The remainder of the paper is organized as follows. In Section 2, we shall prove that $\mathbf{P}$ has two important features stated in the two following theorems.

Theorem 1. *The TPM $\mathbf{P}(p,\rho)$ of a GINAR(1) process is totally positive of order 2,*

$$\mathbf{P}(p,\rho) \in TP_2, \quad (6)$$

i.e., all the 2×2 minors of the $\mathbf{P}(p,\rho)$ are non-negative.

Theorem 2. *Let: $\{X_t(p,\rho) : t \in \mathbb{N}_0\}$ and $\{X_t(p',\rho) : t \in \mathbb{N}_0\}$ be two independent GINAR(1) processes, with parameters (p,ρ) and (p',ρ); $\mathbf{P}(p,\rho) = [p_{ij}(p,\rho)]_{i,j\in\{0,1,\dots,n\}}$ and $\mathbf{P}(p',\rho) = [p_{ij}(p',\rho)]_{i,j\in\{0,1,\dots,n\}}$ be their corresponding TPM. Then $\mathbf{P}(p',\rho)$ is stochastically smaller than $\mathbf{P}(p,\rho)$ in the usual (or in the Kalmykov order) sense,*

$$\mathbf{P}(p',\rho) \leq_{st} \mathbf{P}(p,\rho), \quad (7)$$

if $0 \leq \sqrt{\rho}/(\sqrt{\rho}+1) < p \leq p' < 1$, that is,

$$\sum_{j=l}^{n} p_{ij}(p,\rho) \leq \sum_{j=l}^{n} p_{mj}(p',\rho), \quad i,l,m \in \{0,1,\dots,n\}, \quad i \leq m,$$

in case $1/\sqrt{\rho} > E[X_t(p,\rho)] \geq E[X_t(p',\rho)] > 0$.

In Section 3, we discuss and illustrate the impact of (6) and (7) on the run length of an upper one-sided geometric chart for monitoring GINAR(1) processes. In Section 4, we sum up our findings and briefly refer to related and future work.

2. Proving the Two Features of the GINAR(1) Process

Demonstrating that the 2 × 2 minors of the TPM of a GINAR(1) process are all non-negative is not simple, due to the aspect of the transition probabilities defined in (3). However, by adopting the reasoning of (Morais and Pacheco [8] Section 2) and resorting to some auxiliary definitions and lemmas in Appendix A.1, we can prove (6).

Proof of Theorem 1. Note that

$$(X_{t+1}(p,\rho) \mid X_t(p,\rho) = i) = \rho \circ X_{t-1}(p,\rho) + \varepsilon_t \stackrel{st}{=} B(i,\rho) + BG(p,\rho),$$

where: $B(0,\rho) \stackrel{st}{=} 0$; $B(i,\rho) \sim \text{binomial}(i,\rho)$, $i \in \mathbb{N}$; $BG(p,\rho)$ a r.v. with p.f. given by (4); $B(i,\rho)$ and $BG(p,\rho)$ are two independent r.v.

In accordance to Lemmas A1 and A3, $B(i)$ stochastically increases with i in the likelihood ratio sense and $B(i,\rho), BG(p,\rho) \in \text{PF}_2$. Hence, we can invoke the closure of the stochastic order $\leq_{lr}$ (see Definition A1) under the sum of independent PF_2 r.v. (see Shaked and Shanthikumar [9] p. 46, Theorem 1.C.9) or Karlin and Proschan [10]) to conclude that

$$B(i,\rho) + BG(p,\rho) \leq_{lr} B(i+1,\rho) + BG(p,\rho)$$
$$(X_{t+1}(p,\rho) \mid X_t(p,\rho) = i) \leq_{lr} (X_{t+1}(p,\rho) \mid X_t(p,\rho) = i+1),$$

for $i \in \mathbb{N}_0$, i.e., $\mathbf{P} \in \text{TP}_2$ or $\mathbf{P}$ is a stochastically monotone TPM in the likelihood ratio sense ($\mathbf{P} \in \mathcal{M}_{lr}$), according to Definition A2. □

The next proof refers to a stochastic ordering between the TPM that govern two DTMC with the same state space, thus associated with what Kulkarni [11] (pp. 148–149) terms the Kalmykov-dominance or Kalmykov order(see Kalmykov [12] Theorem 2).

Proof of Theorem 2. Result (7) can be shown to hold by successively capitalizing on: Lemmas A1 and A2; the closure of $\leq_{lr}$ under the sum of independent PF_2 r.v.; $\mathbf{P}(p',\rho) \in \text{TP}_2$; and $X \leq_{lr} Y$ implies that the r.v. X is stochastically smaller than the r.v. Y in the usual sense, in short $X \leq_{st} Y$ (see Shaked and Shanthikumar [9] p. 42, Theorem 1.C.1). Then, for $i, m \in \{0, 1, \ldots, n\}$, $i \leq m$, and $0 \leq \sqrt{\rho}/(\sqrt{\rho}+1) < p \leq p' < 1$:

$$\begin{aligned}
(X_{t+1}(p',\rho) \mid X_t(p',\rho) = i) &\stackrel{st}{=} B(i,\rho) + BG(p',\rho) \\
&\leq_{lr} B(i+1,\rho) + BG(p,\rho) \\
&\stackrel{st}{=} (X_{t+1}(p,\rho) \mid X_t(p,\rho) = i) \\
&\leq_{lr} (X_{t+1}(p,\rho) \mid X_t(p,\rho) = m) \\
(X_{t+1}(p',\rho) \mid X_t(p',\rho) = i) &\leq_{st} (X_{t+1}(p,\rho) \mid X_t(p,\rho) = m) \\
\sum_{j=l}^{n} p_{ij}(p',\rho) &\leq \sum_{j=l}^{n} p_{mj}(p,\rho), \quad l \in \{0, 1, \ldots, n\},
\end{aligned}$$

i.e., $\mathbf{P}(p',\rho) \leq_{st} \mathbf{P}(p,\rho)$ if $1/\sqrt{\rho} > 1/p - 1 = E[X_t(p,\rho)] \geq E[X_t(p',\rho)] = 1/p' - 1 > 0$. □

3. Practical Implications in Statistical Process Control

Time series of counts arise naturally in several applications, namely the manufacturing industry, health care, service industry, insurance, and network analysis. Using control charts for monitoring the underlying count processes is essential to swiftly detect changes in such processes and start preventive or corrective actions (see Weiß [13]). For an overview of control charts for count processes, we refer the reader to Weiß [14].

As noted by Ristić et al. [6], counts with geometric marginal distributions play a *major role* in several areas, for instance reliability, medicine, and precipitation modeling. These counts may refer to the number of *machines waiting for maintenance, congenital malformations,* or *thunderstorms in a day.*

In statistical process control, the GINAR(1) process can be used to model, for example, the cumulative counts of conforming items between two nonconforming items when these successive counts are no longer independent, say because the observations are generated by automated high-frequency sampling.

The literature review reveals that no charts have been proposed for monitoring GINAR(1) or GGINAR(1) counts. However, Li et al. [15] proposed a combined jumps chart, a cumulative sum (CUSUM) chart, and a combined exponentially weighted moving average (EWMA) chart for monitoring the NGINAR(1) counts. Furthermore, Li et al. [16] described upper and lower one-sided CUSUM charts for monitoring the mean of ZMGINAR(1) counts.

Let us consider that the following quality control chart is being used to detect decreases in the parameter p of the GINAR(1) process.

Definition 3. *Let $\{X_t : t \in \mathbb{N}_0\}$ be a GINAR(1) process. The upper one-sided geometric chart makes use of the set of control statistics $\{X_t : t \in \mathbb{N}\}$ and triggers a signal at time t ($t \in \mathbb{N}$) if $X_t > U$, where U is a fixed upper control limit (UCL) in $\mathbb{N}_0$.*

We should bear in mind that the control statistic X_t becomes stochastically smaller in the usual sense as p increases (see Lemma A4). Consequently and as suggested by (Xie et al. [17] p. 42), it is clear that when an observed value of X_t exceeds the UCL of the chart, this should be taken as a sign that the p has decreased, that is, an indication of a potential increase in the process mean $(1-p)/p$.

The performance of the upper one-sided geometric chart is about to be assessed in terms of the run length (RL), the random number of samples collected before a signal is triggered by this control chart. Consequently, the following first passage time of the stochastic process $\{X_t : t \in \mathbb{N}_0\}$, under the condition that $X_0 = u \in \{0,1,\ldots,U\}$, is a vital performance measure of this chart for monitoring a GINAR(1) process:

$$RL^u \equiv RL^u(U) = \min\{t \in \mathbb{N} : X_t > U \mid X_0 = u\}, \tag{8}$$

where u is a fixed initial value in the set $\{0,1,\ldots,U\}$.

U is chosen in such a way that false alarms are rather infrequent and increases in the process mean $(1-p)/p$ (i.e., decreases in p) are detected as quickly as possible. Hence, we should be dealing with a large in-control RL and smaller out-of-control run lengths.

3.1. Significance of $\mathbf{P} \in TP_2$

By invoking the first part of Theorem 3.1 of Assaf et al. [18], we can state that the TP_2 character of the TPM of the GINAR(1) process leads to the following result.

Corollary 1. *Let $\{X_t : t \in \mathbb{N}_0\}$ be a GINAR(1) process. Then*

$$RL^0 = \min\{t \in \mathbb{N} : X_t > U \mid X_0 = 0\} \in PF_2, \tag{9}$$

i.e., $[P_{RL^0}(x+1)]^2 \geq P_{RL^0}(x) \times P_{RL^0}(x+2)$, for $x \in \mathbb{N}_0$.

Corollary 1 implies that RL^0 has an increasing hazard rate ($RL^0 \in \text{IHR}$), that is, $\lambda_{RL^0}(m) = P(RL^0 = m)/P(RL^0 \geq m)$ is a nondecreasing function of $m \in \mathbb{N}$ (see Kijima [19] p. 118, Theorem 3.7(ii)). $RL^0 \in \text{IHR}$ means that signaling, given that no observation has previously exceeded the UCL, becomes more likely as we proceed with the collection of observations provided that $X_0 = 0$.

Note, however, that RL^u may not be IHR, for $u \in \{1,\ldots,U\}$. In fact, the second part of Theorem 3.1 of Assaf et al. [18] allows us to state that the p.f. $P_{RL^u}(l+n)$ is TP_2 in l and n ($l, n \in \mathbb{N}_0$), i.e., $P_{RL^u}(l+n) \times P_{RL^u}(l'+n') \geq P_{RL^u}(l'+n) \times P_{RL^u}(l+n')$, for $l, n \in \mathbb{N}_0$ ($l < l', n < n'$). As a consequence, $[P_{RL^u}(x+1)]^2 \leq P_{RL^u}(x) \times P_{RL^u}(x+2)$, for $x \in \mathbb{N}_0$, thus we can add that RL^u has an decreasing hazard rate ($RL^u \in \text{DHR}$).

The next corollary translates the stochastic influence of an increase in the initial value u and can be shown to be valid by capitalizing on (Karlin [20] pp. 42–43, Theorem 2.1).

Corollary 2. *Let* $\{X_t : t \in \mathbb{N}_0\}$ *be a GINAR(1) process. Then, for* $u, u' \in \{0, 1, \ldots, U\}$,

$$RL^{u'} \leq_{lr} RL^u, \quad u \leq u'. \tag{10}$$

Let us denote the upper one-sided geometric chart with $X_0 = u'$ (resp. $X_0 = u$) by Scheme 1 (resp. Scheme 2). Then (10) can be interpreted as follows: the odds of Scheme 1 signaling at sample m against Scheme 2 triggering a signal at the same sample decreases as m increases (see [21] p. 5).

Result (10) seems *quite evident*; nevertheless, it would not be valid if the GINAR(1) process was not governed by a TP$_2$ TPM.

3.2. Other Comparisons of Run Lengths

The stochastic inequality $\mathbf{P}(p', \rho) \leq_{st} \mathbf{P}(p, \rho)$, for $0 \leq \sqrt{\rho}/(\sqrt{\rho} + 1) < p \leq p' < 1$, allows us to stochastically compare two GINAR(1) processes. As a matter of fact, by invoking Lemma A4 and Theorem 6.B.32 of (Shaked and Shanthikumar [9] p. 282), we can state the next result.

Corollary 3. *Let* $\{X_t(p', \rho) : t \in \mathbb{N}_0\}$ *and* $\{X_t(p, \rho) : t \in \mathbb{N}_0\}$ *two GINAR(1) processes. If* $0 \leq \sqrt{\rho}/(\sqrt{\rho}+1) < p \leq p' < 1$ *and the initial states are deterministic* $X_0 = u' \leq X_0(p') = u$ *or random, say* $X_0(p', \rho) \sim \text{geometric}(p') \leq_{st} X_0(p, \rho) \sim \text{geometric}(p)$, *then*

$$\{X_t(p', \rho) : t \in \mathbb{N}_0\} \leq_{st} \{X_t(p, \rho) : t \in \mathbb{N}_0\}. \tag{11}$$

From (11) we can infer from (11) that $X_1(p', \rho) \leq_{st} X_1(p, \rho)$.

The next lemma plays a vital role in the comparison of run lengths and is taken from (Shaked and Shanthikumar [9] p. 283).

Lemma 1. *If two stochastic processes* $\{X_t : t \in \mathcal{T}\}$ *and* $\{Y_t : t \in \mathcal{T}\}$ *satisfy* $\{X_t : t \in \mathcal{T}\} \leq_{st} \{Y_t : t \in \mathcal{T}\}$ *then*

$$\inf\{t \in \mathcal{T} : Y_t > U\} \leq_{st} \inf\{t \in \mathcal{T} : X_t > U\}. \tag{12}$$

Lemma 1 states what could be considered obvious: if we are dealing with two ordered stochastic processes in the usual sense, the larger stochastic process in the usual sense exceeds the critical level U stochastically sooner also in the usual sense.

By combining Corollary 3 and Lemma 1, we can provide a stochastic flavor to the influence of an increase in p not only on RL^u but also on another important RL:

$$RL^{X_1} = \min\{t \in \mathbb{N} : X_t > U \mid X_1\}, \tag{13}$$

which we coin as *overall run length*, following (Weiß [22] Section 20.2.2). RL^{X_1} refers to a first passage time of the stochastic process $\{X_t : t \in \mathbb{N}\}$ under the condition that the initial state coincides with the r.v. X_1. In point of fact, it is reasonable to resort to this performance measure because in practice we do not know X_0, hence it is plausible to rely, for example, on $X_1 \sim \text{geometric}(p)$.

Corollary 4. *The following stochastic ordering results hold for the run lengths of the upper one-sided geometric chart for monitoring GINAR(1) processes:*

$$RL^u(p, \rho) \leq_{st} RL^{u'}(p', \rho) \tag{14}$$
$$RL^{X_1(p,\rho)}(p, \rho) \leq_{st} RL^{X_1(p',\rho)}(p', \rho), \tag{15}$$

for $u' \leq u$ *and* $0 \leq \sqrt{\rho}/(\sqrt{\rho}+1) < p \leq p' < 1$.

Note that we could have also invoked (14) and the closure of the usual stochastic order $\leq_{st}$ under mixtures (see Shaked and Shanthikumar [9] p. 6, Theorem 1.A.3.(d)) to prove (15).

Results (14) and (15) mean that the upper one-sided geometric chart for the GINAR(1) process stochastically increases its detection speed (in the usual sense) as the downward shift in p becomes more extreme. This stochastic ordering result parallels with the notion of a sequentially repeated uniformly powerful test.

3.3. An Illustration

Ristić et al. [6] found that an NGINAR(1) model with estimated parameters $\hat{p}_0 = 1/(1 + 0.5872) = 0.63$ and $\hat{\rho}_0 = 0.1650$ adequately described the monthly counts of sex offenses reported in the 21st police car beat in Pittsburgh. This data set comprises 144 observations, starting in January 1990 and ending in December 2001.

Note that the GINAR(1) and NGINAR(1) processes share the same geometric marginal distribution; and, as far as the offense data set is concerned, the value of the Akaike information criterion (AIC) for the NGINAR(1) and GINAR(1) models are very close, namely 302.67 and 303.74, respectively, as (Ristić et al. [6] Table 2) attest. Hence, we are going to consider the upper one-sided geometric chart from Definition 3 with $p_0 = 0.63$ and $\rho_0 = 0.1650$ for monitoring such counts.

An UCL equal to $U = 5$ and an initial state $u = 0$ (resp. $u = U$) yield an in-control ARL of $E[RL^0(p_0, \rho_0)] \simeq 393.7$ (resp. $E[RL^U(p_0, \rho_0)] \simeq 391.4$). These and other RL-related performance measures used in this subsection are described in Appendix A.2.

The plots of the hazard rate function in Figure 1 give additional insights into the RL performance of the geometric chart as we proceed with the sampling and to the impact of the adoption of a head start. Indeed, it illustrates two results that follow from Corollary 1: $RL^0(p_0, \rho_0) \in IHR$ and $RL^U(p_0, \rho_0) \in DHR$. This last result suggests that the false-alarm rate conveniently decreases in the first samples when we adopt a head start ($u = U > 0$).

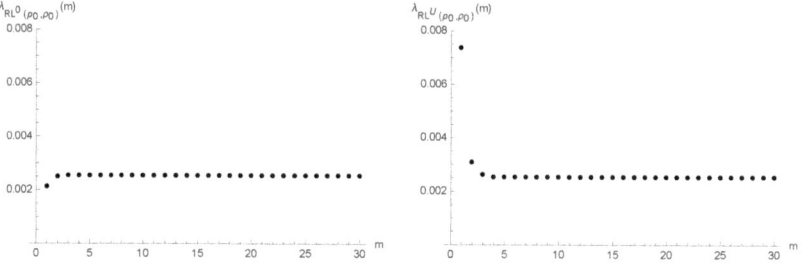

Figure 1. Hazard rate functions of $RL^0(p_0, \rho_0)$ and $RL^U(p_0, \rho_0)$.

According to Brook and Evans [23], the limiting form of the p.f. of the RL is geometric-like with parameter $1 - \xi(p, \rho)$, where $\xi(p, \rho)$ is the maximum real eigenvalue of $\mathbf{Q}(p, \rho) = [p_{ij}(p, \rho)]_{i,j \in \{0,1,...,U\}}$, regardless of the initial value u of the control statistic X_t. Therefore, it comes as no surprise that the values of the hazard rate functions of $RL^0(p_0, \rho_0)$ and $RL^U(p_0, \rho_0)$ converge to

$$\lim_{m \to +\infty} \lambda_{RL^0(p_0,\rho_0)}(m) = \lim_{m \to +\infty} \lambda_{RL^U(p_0,\rho_0)}(m) = 1 - \xi(p, \rho) \simeq 0.002541, \quad (16)$$

as suggested by Figure 1.

Furthermore, the hazard rate function of $RL^0(p_0, \rho_0)$ is pointwise below the one of $RL^U(p_0, \rho_0)$ because Corollary 2 establishes that $RL^U(p_0, \rho_0) \leq_{lr} RL^0(p_0, \rho_0)$ and this result in turn implies $RL^U(p_0, \rho_0) \leq_{hr} RL^0(p_0, \rho_0)$, that is, $\lambda_{RL^U(p_0,\rho_0)}(m) \geq \lambda_{RL^0(p_0,\rho_0)}(m)$, for $m \in \mathbb{N}$ (see Definition A4).

We now illustrate the first result of Corollary 4 and also of a consequence of its second result: $RL^0(p,\rho) \leq_{st} RL^0(p',\rho)$, for $0 \leq \sqrt{\rho}/(\sqrt{\rho}+1) < p \leq p' < 1$; $E[RL^{X_1(p,\rho)}(p,\rho)]$ is an increasing function of p in the interval, $(\sqrt{\rho}/(\sqrt{\rho}+1),1)$.

In the left panel of Figure 2, we plotted the survival functions of $RL^0(0.9p_0,\rho_0)$ and $RL^0(p_0,\rho_0)$.

Since $RL^0(0.9p_0,\rho) \leq_{st} RL^0(p_0,\rho)$, the plot of survival function of $RL^0(0.9p_0,\rho)$ is pointwise below the one of $RL^0(p_0,\rho)$, as Figure 2 plainly demonstrates. Hence, the number of samples taken until the detection of a 10% decrease in p by the upper one-sided geometric chart is indeed stochastically smaller than the number of samples we collect until this chart emits a false alarm.

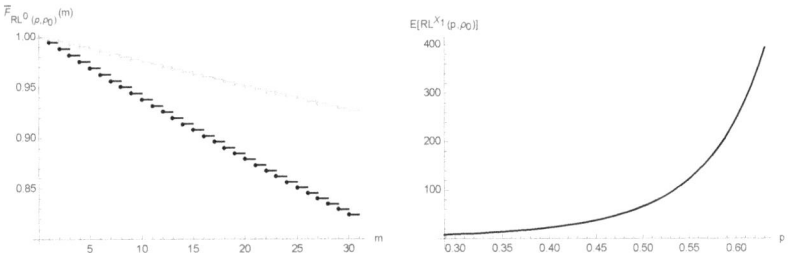

Figure 2. Survival function of $RL^0(p,\rho_0)$, for $p = 0.9p_0$ and $p = p_0$ (black and gray solid lines); overall ARL function, $E[RL^{X_1}(p,\rho_0)]$, for $\sqrt{\rho}/(\sqrt{\rho}+1) < p \leq p_0$.

The right panel of Figure 2 refers to the overall ARL function, $E[RL^{X_1}(p,\rho_0)]$, for $\sqrt{\rho}/(\sqrt{\rho}+1) < p \leq p_0$. It increases with p in this particular interval from $E[RL^{X_1}(\sqrt{\rho_0}/(\sqrt{\rho_0}+1),\rho_0)] \simeq 8.3$ to $E[RL^{X_1}(p_0,\rho_0)] \simeq 393.5$. We ought to note that it increases further when we take $p \in (p_0,1)$, therefore the upper one-sided geometric chart cannot detect increases in p in an expedient manner, as we have anticipated.

We wrote a program for Mathematica 10.3 (Wolfram [24]) to produce all the graphs and results in this subsection.

4. Concluding Remarks

As expertly put by Montgomery and Mastrangelo [25], the independence assumption is often violated in practice. As a consequence, we often deal with discrete-valued time series, namely when we are dealing with very high sampling rates, as suggested by Weiß and Testik [26], and Rakitzis et al. [27].

In this paper, we considered the GINAR(1) count process, resorted to stochastic ordering to prove two features of its TPM, and discussed the implications of these two traits on RL-related performance measures of an upper one-sided geometric control chart that accounts for the autocorrelated character of such process.

For example: the TP_2 character of the TPM of the GINAR(1) process implies an IHR behaviour of the run length RL^0 of that same chart; the run length RL^u and the overall run length RL^{X_1} stochastically increase in the usual sense in the interval $(\sqrt{\rho_0}/(\sqrt{\rho_0}+1),1)$.

These features of the GINAR(1) process and the associated results are comparable to the ones derived by (Morais [21] Section 3.2) and Morais and Pacheco [8,28].

It is important to note that the notion of stochastically monotone matrices in the usual sense was introduced by Daley [29] for real-valued discrete-time Markov chains. Moreover, Karlin [20] implicitly states that a TP_2 TPM possesses a monotone likelihood ratio property and, thus, virtually defines stochastically monotone Markov chains in the likelihood ratio sense. Furthermore, the comparison of counting processes and queues in the usual sense can be traced back, for instance, to Whitt [30] and the multivariate likelihood ratio order of random vectors (or TP_2 order) is discussed, for example, by (Shaked and Shanthikumar [9] pp. 298–305).

Coincidentally, the stochastic order in the likelihood sense for stochastic processes or TPM has not been defined up to now, as far as we have investigated. For this reason and the fact that the $\leq_{lr}$ order is not closed under mixtures (see Shaked and Shanthikumar [31] p. 33), we did not state or prove the $\leq_{lr}$ analogue of the two results in Corollary 4.

We also failed to prove that $\mathbf{P}(p,\rho') \leq_{st} \mathbf{P}(p,\rho)$, for $0 \leq \rho \leq \rho' < 1$, because of two opposing stochastic behaviors of the summands $(\rho \circ X_t \mid X_t = i)$ and ε_{t+1}: the r.v. binomial(i,ρ) (resp. $BG(p,\rho)$) stochastically increases (resp. decreases) with ρ in the likelihood ratio sense. Had we proven that result, we could have concluded that the larger the upward shifts in the autocorrelation parameter, the longer it takes the upper one-sided geometric chart to detect such a change in ρ.

It would be pertinent to investigate the stochastic properties of the RL and overall RL of lower one-sided geometric charts for detecting increases in the parameter p of a GINAR(1) process.

Another possibility of further work which certainly deserves some consideration is to investigate the extension of Theorems 1 and 2 to the NGINAR(1) process, the novel geometric INAR(1) process proposed by Guerrero et al. [32], or the new INAR(1) process with Poisson binomial-exponential 2 innovations studied by Zhang et al. [33], and assess the impact of these two results in the RL performance of upper one-sided charts for monitoring such autocorrelated geometric counts.

We ought to mention that deriving results similar to (6) and (7) seems to be very unlikely for the mixed generalized Poisson INAR process [34]. This follows from the fact that the generalized Poisson distribution has not a PF$_2$ p.f.

Funding: The author acknowledges the financial support of the Portuguese FCT—Fundação para a Ciência e a Tecnologia, through the projects UIDB/04621/2020 and UIDP/04621/2020 of CEMAT/IST-ID (Center for Computational and Stochastic Mathematics), Instituto Superior Técnico, Universidade de Lisboa.

Acknowledgments: We are grateful to the three reviewers who selflessly devoted their time to scrutinizing this work and offered pertinent comments that led to an improved version of the original manuscript. The author would also like to thank Christian H. Weiß for the opportunity to celebrate the vital role of "Discrete-valued Time Series" in this special issue of "Entropy".

Conflicts of Interest: The author declares no conflict of interest.

Abbreviations

The following abbreviations are used in this manuscript:

c.d.f.	cumulative distribution function
DHR	decreasing hazard rate
DTMC	discrete-time Markov chain
GGINAR(1)	generalized geometric first-order integer-valued autoregressive process
GINAR(1)	geometric first-order integer-valued autoregressive process
i.i.d.	independent and identically distributed
IHR	increasing hazard rate
INAR(1)	first-order integer-valued autoregressive process
NGINAR(1)	new geometric first-order integer-valued autoregressive process
p.f.	probability function
PF$_2$	Pólya frequency of order 2
RL	run length
r.v.	random variable
TP$_2$	totally positive of order 2
TPM	transition probability matrix
UCL	upper control limit
ZMGINAR(1)	zero-modified geometric first-order integer-valued autoregressive process

Appendix A

Appendix A.1. Auxiliary Definitions and Lemmas

This appendix has the sole purpose of providing a few notions and results that are crucial to prove (6), (7), and some of the implications of these two stochastic ordering results.

The notions of *stochastically smaller in the likelihood ratio sense, stochastically monotone in the likelihood ratio sense*, and *Pólya frequency of order 2 p.f.* are taken or follow from (Shaked and Shanthikumar [9] p. 42), (Kijima [19] pp. 129–131), and (Kijima [19] p. 106) (respectively).

Definition A1. *Let X and Y be two non-negative integer r.v., with p.f. P_X and P_Y. Then X is said to be stochastically smaller than Y in the likelihood ratio sense if*

$$\frac{P_X(x)}{P_Y(x)} \text{ is a nonincreasing function of } x, \tag{A1}$$

over the union of the supports of the r.v. X and Y. In shorthand notation, $X \leq_{lr} Y$.

Lemma A1. *Let $B(i,\rho) \sim \text{binomial}(i,\rho)$. Then*

$$B(i,\rho) \leq_{lr} B(i+1,\rho), \quad i \in \mathbb{N}_0. \tag{A2}$$

$B(i,\rho)$ stochastically increases with i in the likelihood ratio sense because the ratio

$$\frac{P_{B(i,\rho)}(x)}{P_{B(i+1,\rho)}(x)} = \frac{i+1-x}{(1-p)(i+1)}$$

is a nonincreasing function of $x \in \{0, 1, \ldots, i+1\}$.

Lemma A2. *Let $BG(p,\rho)$ be a r.v. with p.f. given by (4). Then*

$$BG(p',\rho) \leq_{lr} BG(p,\rho), \quad \frac{\sqrt{\rho}}{\sqrt{\rho}+1} < p \leq p' < 1. \tag{A3}$$

Note that

$$r_{BG}(x) = \frac{P_{BG(p',\rho)}(x)}{P_{BG(p,\rho)}(x)} = \begin{cases} \frac{p'(1-\rho)+\rho}{p'(1-\rho)+\rho'}, & x = 0 \\ \frac{p'}{p} \times \left(\frac{1-p'}{1-p}\right)^x, & x \in \mathbb{N}. \end{cases}$$

Since $(1-p')/(1-p) \leq 1$, $r_{BG}(x)$ is a nonincreasing function of $x \in \mathbb{N}$. We still have to verify that $r_{BG}(0) \geq r_{BG}(1)$: this inequality is valid if $f(p) = [p(1-\rho)+\rho]/[p(1-p)] = 1/(1-p) + \rho/p$ has a positive derivative, i.e., if $1/(1-p)^2 - \rho/p^2 > 0$ or, equivalently, $p > \sqrt{\rho}/(\sqrt{\rho}+1)$. Hence $BG(p,\rho)$ stochastically decreases with p in the likelihood ratio sense as long as $0 \leq \sqrt{\rho}/(\sqrt{\rho}+1) < p \leq p' < 1$.

Definition A2. *Let $\{X_t : t \in \mathbb{N}_0\}$ be an irreducible DTMC with TPM $\mathbf{P}$. Then $\{X_t : t \in \mathbb{N}_0\}$ is said to be stochastically monotone in the likelihood ratio sense if*

$$(X_{t+1} \mid X_t = i) \leq_{lr} (X_{t+1} \mid X_t = i+1), \tag{A4}$$

for all i. In this case, we write $\mathbf{P} \in \mathcal{M}_{lr}$ or $\mathbf{P} \in TP_2$.

Definition A3. *Let X be a non-negative r.v. with probability function (p.f.) $P_X(x)$. If*

$$\frac{P_X(x)}{P_X(x+1)} \text{ is nondecreasing in } x \in \mathbb{N}_0, \tag{A5}$$

i.e., $[P_X(x+1)]^2 \geq P_X(x) \times P_X(x+2)$, $x \in \mathbb{N}_0$, then X is said to have a Pólya frequency of order 2 (PF$_2$) p.f. and we write $X \in PF_2$.

Lemma A3. *If $B(i, \rho) \sim binomial(i, \rho)$ and $BG(p, \rho)$ is a r.v. with p.f. given by (4) then $B(i, \rho)$, $BG(p, \rho) \in PF_2$.*

We have

$$\frac{P_{B(i,\rho)}(x)}{P_{B(i,\rho)}(x+1)} = \frac{1-p}{p} \times \frac{x+1}{i-x}, \quad x = 0, 1, \ldots, i,$$

$$\frac{P_{BG(p,\rho)}(x)}{P_{BG(p,\rho)}(x+1)} = \begin{cases} \frac{p(1-\rho)+\rho}{(1-\rho)(1-p)p}, & x = 0 \\ \frac{1}{1-p}, & x \in \mathbb{N}. \end{cases}$$

Since $[p(1-\rho)+\rho]/[(1-\rho)(1-p)p] \leq 1/(1-p) \Leftrightarrow \rho \leq 1$, we can state that these two ratios are certainly nondecreasing functions of x over the support of the corresponding p.f.

The concepts of *stochastically smaller in the usual sense* in the univariate and multivariate cases and *stochastically smaller in the hazard rate sense* in the univariate case can be found in (Shaked and Shanthikumar [9] pp. 3, 17, 266), whereas on p. 281 of this same reference the stochastic ordering of stochastic processes in the usual sense is defined.

Definition A4. *Let X and Y be two non-negative integer r.v., with p.f. P_X and P_Y and c.d.f. F_X and F_Y. Then: X is said to be stochastically smaller than Y in the usual sense $(X \leq_{st} Y)$ if*

$$\overline{F}_X(x) = 1 - F_X(x) \leq 1 - F_Y(x) = \overline{F}_Y(x), \quad x \in \mathbb{N}_0; \quad (A6)$$

X is said to be stochastically smaller than Y in the hazard rate sense $(X \leq_{hr} Y)$ in case

$$\lambda_X(x) = \frac{P_X(x)}{\overline{F}_X(x-1)} \geq \frac{P_Y(x)}{\overline{F}_Y(x-1)} = \lambda_Y(x), \quad x \in \mathbb{N}_0. \quad (A7)$$

The stochastic orders $\leq_{lr}$, $\leq_{hr}$, and $\leq_{st}$ can be related: $X \leq_{lr} Y \Rightarrow X \leq_{hr} Y \Rightarrow X \leq_{st} Y$ (see Shaked and Shanthikumar [9] pp. 18, 43, Theorems 1.B.1, 1.C.1). Moreover, $X \leq_{st} Y \Rightarrow E(X) \leq E(Y)$ provided that these expectations exist.

Lemma A4. *Let $G(p)$ be an r.v. with geometric distribution with parameter p. Then*

$$G(p') \leq_{st} G(p), \quad 0 < p \leq p' < 1. \quad (A8)$$

Equation (A8) follows in a straightforward manner: $P_{G(p)}(x) = (1-p)^x p$, for $x \in \mathbb{N}_0$; thus, $\overline{F}_{G(p')}(x) = (1-p')^{x+1} \leq (1-p)^{x+1} = \overline{F}_{G(p)}(x)$, for $x \in \mathbb{N}_0$, when $0 < p \leq p' < 1$.

Let $\underline{x} = (x_1, \ldots, x_m)$ and $\underline{y} = (y_1, \ldots, y_m)$ be two vectors in $\mathbb{R}^m$; then we write $\underline{x} \leq \underline{y}$ if $x_t \leq y_t$, for $t = 1, \ldots, m$. Additionally, recall that $\mathcal{U} \subseteq \mathbb{R}^m$ is called an upper set if $\underline{y} \in \mathcal{U}$ whenever $\underline{x} \leq \underline{y}$ and $\underline{x} \in \mathcal{U}$ (see Shaked and Shanthikumar [9] p. 266).

Definition A5. *Let $\underline{X}$ and $\underline{Y}$ be two m-dimensional random vectors. Then $\underline{X}$ is said to be smaller than $\underline{Y}$ in the usual sense if*

$$P(\underline{X} \in \mathcal{U}) \leq P(\underline{Y} \in \mathcal{U}), \quad (A9)$$

for every upper set $\mathcal{U}$ in $\mathbb{R}^m$. We write $\underline{X} \leq_{st} \underline{Y}$.

Definition A6. *Let $\{X_t : t \in \mathbb{N}_0\}$ and $\{Y_t : t \in \mathbb{N}_0\}$ be two discrete-time stochastic processes with a common state space S. Then $\{X_t : t \in \mathbb{N}_0\}$ is said to be stochastically smaller than $\{Y_t : t \in \mathbb{N}_0\}$ in the usual sense if*

$$(X_{t_1}, \ldots, X_{t_m}) \leq_{st} (Y_{t_1}, \ldots, Y_{t_m}), \quad (A10)$$

for every $m \in \mathbb{N}$ and $(t_1, \ldots, t_m) \in \mathbb{N}_0^m$. In this case, we write $\{X_t : t \in \mathbb{N}_0\} \leq_{st} \{Y_t : t \in \mathbb{N}_0\}$.

As a consequence of Definition A6, $\{X_t : t \in \mathbb{N}_0\} \leq_{st} \{Y_t : t \in \mathbb{N}_0\}$ implies that $X_t \leq_{st} Y_t$, for all $t \in \mathbb{N}_0$.

Appendix A.2. Run Length Related Performance Measures

The run length of the upper one-sided geometric chart, $RL^u \equiv RL^u(p, \rho)$, is the first passage time to the set of states $\{U+1, U+2, \ldots\}$, where $u \in \{0, 1 \ldots, U\}$. Thus, we can use the Markov chain approach proposed by Brook and Evans [23] and provide the expected value of RL^u,

$$E(RL^u) = \mathbf{e}_u^\top \times (\mathbf{I} - \mathbf{Q})^{-1} \times \mathbf{1}, \tag{A11}$$

where: $\mathbf{e}_u^\top$ is the $(u+1)$th vector of the orthogonal basis for $\mathbb{R}^{(U+1)}$; $\mathbf{I}$ represents an identity matrix with rank $(U+1)$; $\mathbf{Q} \equiv \mathbf{Q}(p, \rho) = [p_{ij}(p, \rho)]_{i,j \in \{0,1,\ldots,U\}}$ is the sub-stochastic matrix that governs the transitions between the states in $\{0, 1, \ldots, U\}$, with entries given by (3); $\mathbf{1}$ is a column-vector with $(U+1)$ ones.

We can also add the survival and hazard rate functions of RL^u are equal to

$$\bar{F}_{RL^u}(m) = \mathbf{e}_u^\top \times \mathbf{Q}^m \times \mathbf{1}, \tag{A12}$$

$$\lambda_{RL^u}(m) = \frac{P(RL^u = m)}{P(RL^u \geq m)} = \frac{\bar{F}_{RL^u}(m-1) - \bar{F}_{RL^u}(m)}{\bar{F}_{RL^u}(m-1)}, \tag{A13}$$

for $m \in \mathbb{N}$.

The overall ARL of the upper one-sided geometric chart is given by

$$E(RL^{X_1}) = 1 + \sum_{u=0}^{U} ARL^u \times P(X_1 = u) \tag{A14}$$

(see Weiß [22] Section 20.2.2) or Weiß and Testik [35]).

References

1. McKenzie, E. Some simple models for discrete variate time series. *Water Resour. Bull.* **1985**, *21*, 645–650. [CrossRef]
2. McKenzie, E. Autoregressive moving-average processes with negative-binomial and geometric marginal distributions. *Adv. Appl. Probab.* **1986**, *18*, 679–705. [CrossRef]
3. Alzaid, A.; Al-Osh, M. First-order integer-valued autoregressive (INAR (1)) process: Distributional and regression properties. *Stat. Neerl.* **1988**, *42*, 53–61. [CrossRef]
4. Steutel, F.W.; van Harn, K. Discrete analogues of self-decomposability and stability. *Ann. Probab.* **1979**, *7*, 893–899. [CrossRef]
5. Al-Osh, M.A.; Aly, E.-E.A.A. First order autoregressive time series with negative binomial and geometric marginals. *Commun. Stat.–Theory Methods* **1992**, *21*, 2483–2492. [CrossRef]
6. Ristić, M.M.; Bakouch, H.S.; Nastić, A.S. A new geometric first-order integer-valued autoregressive (NGINAR(1)) process. *J. Stat. Plan. Inference* **2009**, *139*, 2218–2226. [CrossRef]
7. Barreto-Souza, W. Zero-Modified geometric INAR(1) process for modelling count time series with deflation or inflation of zeros. *J. Time Ser. Anal.* **2015**, *36*, 839–852. [CrossRef]
8. Morais, M.C.; Pacheco, A. On hitting times for Markov time series of counts with applications to quality control. *REVSTAT–Stat. J.* **2016**, *4*, 455–479.
9. Shaked, M.; Shanthikumar, J.G. *Stochastic Orders*; Springer: New York, NY, USA, 2007.
10. Karlin, S.; Proschan, F. Pólya type distributions of convolutions. *Ann. Math. Stat.* **1960**, *31*, 721–736. [CrossRef]
11. Kulkarni, V.G. *Modeling and Analysis of Stochastic Systems*; Chapman and Hall: London, UK, 1995.
12. Kalmykov, G.I. On the partial ordering of one-dimensional Markov processes. *Theory Probab. Its Appl.* **1962**, *7*, 456–459. [CrossRef]
13. Weiß, C.H. SPC methods for time-dependent processes of counts—A literature review. *Cogent Math.* **2015**, *2*, 1–11. [CrossRef]
14. Weiß, C.H. *An Introduction to Discrete-Valued Time Series*; Wiley: Hoboken, NJ, USA, 2018.
15. Li, C.; Wang, D.; Zhu, F. Effective control charts for monitoring the NGINAR(1) process. *Qual. Reliab. Eng. Int.* **2016**, *32*, 877–888. [CrossRef]
16. Li, C.; Wang, D.; Sun, J. Control charts based on dependent count data with deflation or inflation of zeros. *J. Stat. Comput. Simul.* **2019**, *89*, 3273–3289. [CrossRef]
17. Xie, M.; Goh, T.N.; Kuralmani, V. *Statistical Models and Control Charts for High-Quality Processes*; Springer Science+Business Media, LLC: New York, NY, USA, 2002.

18. Assaf, D.; Shaked, M.; Shanthikumar, J.G. First-passage times with PF$_r$ densities. *J. Appl. Probab.* **1985**, *22*, 185–196. [CrossRef]
19. Kijima, M. *Markov Processes for Stochastic Modeling*; Chapman and Hall: London, UK, 1997.
20. Karlin, S. Total positivity, absorption probabilities and applications. *Trans. Am. Math. Soc.* **1964**, *11*, 33–107. [CrossRef]
21. Morais, M.J.C. Stochastic Ordering in the Performance Analysis of Quality Control Schemes. Ph.D. Thesis, Universidade Técnica de Lisboa, Lisbon, Portugal, 2002.
22. Weiß, C.H. Categorical Time Series Analysis and Applications in Statistical Quality Control. Ph.D. Thesis, Fakultät für Mathematik und Informatik der Universität Würzburg, Würzburg, Germany, 2009.
23. Brook, D.; Evans, D.A. An approach to the probability distribution of CUSUM run length. *Biometrika* **1972**, *59*, 539–549. [CrossRef]
24. Wolfram Research, Inc. *Mathematica*; Version 10.3; Wolfram Research, Inc.: Champaign, IL, USA, 2015.
25. Montgomery, D.C.; Mastrangelo, C.M. Some statistical process control methods for autocorrelated data. *J. Qual. Technol.* **1991**, *23*, 179–193. [CrossRef]
26. Weiß, C.H.; Testik, M.C. The Poisson INAR(1) CUSUM chart under overdispersion and estimation error. *IIE Trans.* **2011**, *43*, 805–818. [CrossRef]
27. Rakitzis, A.C.; Weiß, C.H.; Castagliola, P. Control charts for monitoring correlated counts with a finite range. *Appl. Stoch. Model. Bus. Ind.* **2017**, *33*, 733–749. [CrossRef]
28. Morais, M.C.; Pacheco, A. On stochastic ordering and control charts for traffic intensity. *Seq. Anal.* **2016**, *35*, 536–559. [CrossRef]
29. Daley, D.J. Stochastically monotone Markov chains. *Z. Wahrscheinlichkeitstheorie Werwandte Geb.* **1968**, *10*, 305–317. [CrossRef]
30. Whitt, W. Comparing counting processes and queues. *Adv. Appl. Probab.* **1981**, *13*, 207–220. [CrossRef]
31. Shaked, M.; Shanthikumar, J.G. *Stochastic Orders and Their Applications*; Academic Press: San Diego, CA, USA, 1994.
32. Guerrero, M.B.; Barreto-Souza, W.; Ombao, H. Integer-valued autoregressive processes with prespecified marginal and innovation distributions: A novel perspective. *Stoch. Models* **2022**, *38*, 70–90. [CrossRef]
33. Zhang, J.; Zhu, F.; Khan, N.M. A new INAR model based on Poisson-BE2 innovations. *Commun. Stat. Theory Methods* **2022**. [CrossRef]
34. Huang, J.; Zhu, F.; Deng, D. A mixed generalized Poisson INAR model with applications. *J. Stat. Comput. Simul.* **2022**. [CrossRef]
35. Weiß, C.H.; Testik, M.C. CUSUM monitoring of first-order integer-valued autoregressive processes of Poisson counts. *J. Qual. Technol.* **2009**, *41*, 389–400. [CrossRef]

Disclaimer/Publisher's Note: The statements, opinions and data contained in all publications are solely those of the individual author(s) and contributor(s) and not of MDPI and/or the editor(s). MDPI and/or the editor(s) disclaim responsibility for any injury to people or property resulting from any ideas, methods, instructions or products referred to in the content.

Article

Adaptation of Partial Mutual Information from Mixed Embedding to Discrete-Valued Time Series

Maria Papapetrou, Elsa Siggiridou and Dimitris Kugiumtzis *

Department of Electrical and Computer Engineering, Aristotle University of Thessaloniki, 54124 Thessaloniki, Greece
* Correspondence: dkugiu@auth.gr; Tel.: +30-23-1099-5955

Abstract: A causality analysis aims at estimating the interactions of the observed variables and subsequently the connectivity structure of the observed dynamical system or stochastic process. The partial mutual information from mixed embedding (PMIME) is found appropriate for the causality analysis of continuous-valued time series, even of high dimension, as it applies a dimension reduction by selecting the most relevant lag variables of all the observed variables to the response, using conditional mutual information (CMI). The presence of lag components of the driving variable in this vector implies a direct causal (driving-response) effect. In this study, the PMIME is appropriately adapted to discrete-valued multivariate time series, called the discrete PMIME (DPMIME). An appropriate estimation of the discrete probability distributions and CMI for discrete variables is implemented in the DPMIME. Further, the asymptotic distribution of the estimated CMI is derived, allowing for a parametric significance test for the CMI in the DPMIME, whereas for the PMIME, there is no parametric test for the CMI and the test is performed using resampling. Monte Carlo simulations are performed using different generating systems of discrete-valued time series. The simulation suggests that the parametric significance test for the CMI in the progressive algorithm of the DPMIME is compared favorably to the corresponding resampling significance test, and the accuracy of the DPMIME in the estimation of direct causality converges with the time-series length to the accuracy of the PMIME. Further, the DPMIME is used to investigate whether the global financial crisis has an effect on the causality network of the financial world market.

Keywords: Granger causality; conditional mutual information; mixed embedding; symbol sequences; discrete-valued time series; financial complex network

1. Introduction

A challenge in many domains of science and engineering is to study the causality of observed variables in the form of multivariate time series. Granger causality has been the key concept for this, where Granger causality from one variable to another suggests that the former improves the prediction ahead in time of the latter. Many methods have been developed based on the Granger causality idea to identify directional interactions among variables from their time series (see [1] for a recent comparative study of many Granger causality measures) and have been applied in various fields, such as economics [2], medical sciences [3], and earth sciences [4]. Of particular interest are measures of direct Granger causality that estimate the causal effect of the driving to the response variable that cannot be explained by the other observed variables. The estimated direct causal effects can then be used to form connections between the nodes being the observed variables in a causality network that estimates the connectivity (coupling) structure of the underlying system.

The studies on the Granger causality typically regard the continuous-valued time series, and often the number K of the time series is relatively high, and the underlying system is complex [5,6]. However, in some applications, the observations are discrete valued, e.g., the sign of the financial index return, the levels of precipitation, the counts of spikes in the electroencephalogram, and the counts of significant earthquake occurrences within successive time intervals. In this work, we propose an appropriate method to estimate the direct Granger causality on discrete-valued multivariate time series.

In the analysis of discrete-valued multivariate time series, termed symbol sequences when no order of discrete values is assumed, the causality effects are estimated typically in terms of the fitted model. Different approaches have been proposed for the form of the probability distribution models based on strong assumptions about the structure of the system, i.e., a multivariate Markov chain. One of the most known approaches giving a model of reduced form is the mixture transition distribution (MTD) [7], which restricts the initially large number of parameters, assuming that there is only an effect of each lag variable separately [8,9]. Recently, the MTD was adapted for the causality estimation in [10,11] and discussed in the review in [12]. Other approaches assume the Poisson distribution [13,14] and the negative binomial distribution [15,16]. In the category of autoregressive models are Pegram's autoregressive models [17,18] and multivariate integer-valued autoregressive models (MINAR) [19,20]. Another proposed method, which, however, simplifies the problem to a linear one, is the so-called CUTE method [21]. Having obtained the model under given restrictions, one can then identify the causality of a driving variable to the response variable from the existence of lag terms of the driving variable in the model form.

Here, we follow a different approach and estimate the causality relationships directly using the information measures of mutual information (MI) and conditional MI (CMI). These measures have been employed to estimate causality and derive causality networks from continuous-valued time series [1,22,23]. For discrete-valued time series, MI and CMI have been used, e.g., for the estimation of the Markov chain order [24] and the estimation of autocorrelation in conjunction with Pegram's autoregressive models [25]. They have also been used on discrete data derived from continuous-valued time series, either as ranks of components of embedding vectors [26,27] or as ordinal patterns [28–30]. However, we are not aware of any work on using information measures for a causality analysis of discrete-valued multivariate time series or symbol sequences (there is a reference to this in the supplementary material in [31]).

The framework of the proposed analysis is the estimation of the direct causality of a discrete driving variable X to a discrete response variable Y from the symbol sequences of K observed discrete variables, where X and Y are two of them. The direct causality implies the dependence of Y at one time step ahead, Y_{t+1}, on X at some lag $\tau \geq 0$, $X_{t-\tau}$ that cannot be explained by any other variable at any lag. In the model setting, the direct causality is identified by the presence of the term $X_{t-\tau}$ in the model for Y_{t+1}. In the information theory setting, it is identified by the presence of significant information of $X_{t-\tau}$ for the response Y_{t+1} that cannot be explained by other lag variables, which is quantified by the CMI of $X_{t-\tau}$ and Y_{t+1} given the other lag variables. We develop this idea in a progressive algorithm that builds a set of the most informative lag variables for Y_{t+1}, called the discrete partial conditional mutual information from mixed embedding (DPMIME), based on a similar measure called the PMIME for continuous variables [32,33]. The presence of lag variables $X_{t-\tau}$ (for one or more different lags τ) in the derived set, the so-called mixed embedding vector, identifies the existence of the direct causality from X to Y, and the relative contribution of the lag variables of X in explaining Y_{t+1} conditioned on the other components of the mixed embedding vector (regarding the other $K-1$ variables) quantifies the strength of this relationship. Further, we develop a parametric significance test for the CMI of the selected lag variable and Y_{t+1} at each step of the DPMIME algorithm, which does not have an analogue in the PMIME regarding continuous variables.

In the evaluation of the DPMIME with Monte Carlo simulations, we compare the DPMIME to PMIME on discretized time series from continuous-valued systems and also discrete-valued time series generated by multivariate sparse Markov chains and MTD and MINAR systems, with a predefined coupling structure. We also compare the parametric significance test to the resampling significance test in the DPMIME. Further, we form the causality networks of five capital markets from the DPMIME, using the sign of the change in the respective daily indices, as well as other causality measures (computed on the values of the indices), and we compare the networks from each measure before and after the global financial crisis of 2008.

The structure of the paper is as follows. First, in Section 2, we present the proposed measure DPMIME along with the resampling and parametric test for the CMI. In Section 3, we assess the efficiency of the proposed DPMIME measure with a resampling and parametric test and compare the DPMIME to the PMIME in a simulation study. The results of the application regarding the global financial crisis are presented in Section 4, and finally, in Section 5, the main conclusions are drawn.

2. Discrete Partial Mutual Information from Mixed Embedding

In this section, we present the measure of discrete partial mutual information from mixed embedding (DPMIME), the parametric significance test, and the resampling significance test used in the DPMIME. We also present performance indices for the causality measure when all the $K(K-1)$ causal effects are estimated for all possible directed pairs of the K observed discrete variables.

2.1. Iterative Algorithm for the Computation of DPMIME

Let $\{x_{1,t}, x_{2,t}, \ldots, x_{K,t}\}, t = 1, 2, \ldots, n$, be the observations of a stochastic process on K discrete random variables $X_1, X_2, \ldots, X_K$, typically a multivariate Markov chain. The discrete variables can be nominal or ordinal, and for convenience hereafter, we refer to the data as multivariate symbol sequence.

We are interested in defining a measure for the direct causality from X to Y, where X and Y are any of the K observed discrete variables. For a sufficiently large number of lags L, we formulate the set W_t of candidate lag variables that may have information explaining the response Y at one time step ahead, Y_{t+1}. The set W_t has $K \cdot L$ components ('·' denotes multiplication), $X_{i,t-\tau}, i = 1, \ldots, K, \tau = 0, \ldots, L-1$. The algorithm DPMIME aims to build up progressively the so-called mixed embedding vector, i.e., a subset $\mathbf{w}_t$ of W_t of the most informative lag variables explaining Y_{t+1}.

In the first step, the first lag variable to enter $\mathbf{w}_t$ is the one that maximizes the MI with Y_{t+1},

$$w_1 = \arg\max_{w \in W_t} I(Y_{t+1}; w) \tag{1}$$

and $\mathbf{w}_t = \mathbf{w}_t^1 = [w_1]$ (the superscript denotes the iteration, equal to the cardinality of the set). The MI of two variables X and Y is defined in terms of entropy and probability mass functions (pmfs) as [34]

$$I(X;Y) = H(X) + H(Y) - H(X,Y) = \sum_{x,y} p(x,y) \log \frac{p_{X,Y}(x,y)}{p_X(x)p_Y(y)},$$

where $H(X)$ is the entropy of X, the sum is over all values x and y of X and Y, $p_{X,Y}(x,y)$ is the joint pmf of (X,Y), and $p_X(x)$ is the pmf of X. The pmfs are assumed to regard the multinomial probability distribution and are estimated by the maximum likelihood estimate, where the probability for each value or pair of values is simply given by the relative frequency of occurrence in the sample (the multivariate symbol sequence). In the subsequent steps, the CMI instead of the MI is used to find the new component to enter $\mathbf{w}_t$. Suppose that at step j, the j most relevant lag variables to Y_{t+1} are found forming $\mathbf{w}_t = \mathbf{w}_t^j$. The next component to be added to $\mathbf{w}_t^j$ is one of the components in $W_t \setminus \mathbf{w}_t^j$ (the $K \cdot L$ components except the j components already selected) that maximizes the CMI to Y_{t+1}, i.e., the mutual information of the candidate w and Y_{t+1} conditioned on the components in $\mathbf{w}_t^j$

$$w_{j+1} = \arg\max_{w \in W_t \setminus \mathbf{w}_t^j} I(Y_{t+1}; w | \mathbf{w}_t^j). \tag{2}$$

The CMI of two variables X and Y given a third variable Z is defined in terms of entropy and pmfs as [34]

$$I(X;Y|Z) = -H(X,Y,Z) + H(X,Z) + H(Y,Z) - H(Z)$$
$$= \sum_{x,y,z} p_{X,Y,Z}(x,y,z) \log \frac{p_{X,Y,Z}(x,y,z)p_Z(z)}{p_{X,Z}(x,z)p_{Y,Z}(y,z)}.$$

At each step, when the lag variable is selected, using (1) for the first step and (2) for the subsequent steps, a significance test is run for the MI in (1) and the CMI in (2). The parametric and resampling significance tests are presented in detail later in this section. For the step $j+1$, where w_{j+1} is found in (2), if the CMI $I(Y_{t+1}; w_{j+1}|\mathbf{w}_t^j)$ is found statistically significant by the parametric or resampling test, the $\mathbf{w}_t$ is augmented as $\mathbf{w}_t = \mathbf{w}_t^{j+1} = [\mathbf{w}_t^j, w_{j+1}]$. Otherwise, there is no significant lag variable to be added to the mixed embedding vector and the algorithm terminates, giving the mixed embedding vector $\mathbf{w}_t = \mathbf{w}_t^j$.

The components of the mixed embedding vector $\mathbf{w}_t$ obtained upon termination of the algorithm are grouped in lag variables of the driving variable X, $\mathbf{w}_t^X$, the response variable Y, $\mathbf{w}_t^Y$, and all other $K-2$ variables, $\mathbf{w}_t^Z$, expressed as $\mathbf{w}_t = [\mathbf{w}_t^X, \mathbf{w}_t^Y, \mathbf{w}_t^Z]$. If $\mathbf{w}_t^X$ is empty, i.e., no-lag variable $X_{t-\tau}$ has information to explain Y_{t+1} in view of the other lag variables, there is no direct causality from X to Y. Otherwise, we quantify the direct causality from X to Y as the proportion of the information of Y_{t+1} explained by the lag variables of X. The measure DPMIME is thus defined as

$$\text{DPMIME}_{X \to Y} = \begin{cases} 0, & \text{if } \mathbf{w}_t^X = \emptyset. \\ \frac{I(Y_{t+1}; \mathbf{w}_t^X | \mathbf{w}_t^Y, \mathbf{w}_t^Z)}{I(Y_{t+1}; \mathbf{w}_t)}, & \text{otherwise.} \end{cases} \quad (3)$$

In the following, we present the resampling test and the parametric test for the significance of the CMI of the response Y_{t+1} and the selected component w_{j+1} given the components already selected in $\mathbf{w}_t^j$, $I(Y_{t+1}; w_{j+1}|\mathbf{w}_t^j)$.

2.2. Randomization Test for the Significance of CMI

First, we do not assume any asymptotic parametric distribution of the estimate of $I(Y_{t+1}; w_{j+1}|\mathbf{w}_t^j)$ under the null hypothesis $H_0 : I(Y_{t+1}; w_{j+1}|\mathbf{w}_t^j) = 0$. Thus, the empirical distribution of the estimate of $I(Y_{t+1}; w_{j+1}|\mathbf{w}_t^j)$ is formed by resampling on the initial sample of the variables Y_{t+1}, w_{j+1} and $\mathbf{w}_t^j$. For this, we follow the resampling scheme of the so-called time-shifted surrogates for the significance test for correlation or causality [35,36]. The resampling is actually applied only to w_{j+1}. To retain both the marginal distribution and intra-dependence (autocorrelation) of w_{j+1} and destroy any inter-dependence to Y_{t+1} and $\mathbf{w}_t^j$, we shift cyclically the symbol sequence of w_{j+1} by a random step k [35] (We do not consider here the case of periodic or periodic-like symbol sequences, where this randomization scheme is problematic, as it is likely that the generated surrogate symbol sequence is similar to the original symbol sequence.). Thus, for the original symbol sequence $\{w_{j+1,1}, w_{j+1,2}, \ldots, w_{j+1,n}\}$ of w_{j+1}, a randomized (surrogate) symbol sequence for the random step k is

$$\{w_{j+1,1}^*, w_{j+1,2}^*, \ldots, w_{j+1,n}^*\} = \{w_{j+1,k+1}, \ldots, w_{j+1,n}, w_{j+1,1}, \ldots, w_{j+1,k}\}.$$

We derive a number Q of such randomized symbol sequences and compute for each of them the corresponding estimates of $I(Y_{t+1}; w_{j+1}|\mathbf{w}_t^j)$ under the H_0, denoted

$$I(Y_{t+1}; w_{j+1}^{*1}|\mathbf{w}_t^j), I(Y_{t+1}; w_{j+1}^{*2}|\mathbf{w}_t^j), \ldots, I(Y_{t+1}; w_{j+1}^{*Q}|\mathbf{w}_t^j).$$

These Q values form the empirical null distribution of the estimate of $I(Y_{t+1}; w_{j+1}|\mathbf{w}_t^j)$. The H_0 is rejected if the estimate of $I(Y_{t+1}; w_{j+1}|\mathbf{w}_t^j)$ on the original data is at the right tail of the empirical null distribution. To assess this, we use rank ordering, where r^0 is the rank of the estimate of $I(Y_{t+1}; w_{j+1}|\mathbf{w}_t^j)$ in the ordered list of the $Q+1$ values, assuming ascending order. The p-value of the one-sided test is $1 - (r^0 - 0.326)/(Q + 1 + 0.348)$ (using the correction in [37] to avoid extreme values such as $p = 0$ when the original value is last in the ordered list, which is formally not correct). The DPMIME measure in Equation (3) derived using resampling test of CMI is denoted DPMIMErt.

2.3. Parametric Test for the Significance of CMI

Entropy and MI on discrete variables are well-studied quantities and there is rich literature about the statistical properties and distribution of their estimates. For the significance test for the CMI $I(X;Y|Z)$ for three discrete scalar or vector variables X, Y, and Z, the most prominent of the parametric null distribution approximations are worked out in [38], namely the Gaussian and Gamma distributions. For the Gamma null distribution, following the work in [39], it turns out that $\hat{I}(X,Y)$ follows approximately the Gamma distribution

$$\hat{I}(X,Y) \sim \Gamma\left(\frac{(P_X - 1)(P_Y - 1)}{2}, \frac{1}{n \ln 2}\right),$$

where n is the sample size and P_X is the number of the possible discrete values of X. Further, it follows that $\hat{I}(X,Y|Z)$ is also approximately Gamma distributed

$$\hat{I}(X,Y|Z) \sim \Gamma\left(\frac{P_Z}{2}(P_X - 1)(P_Y - 1), \frac{1}{n \ln 2}\right). \tag{4}$$

We use the Gamma distribution to approximate the null distribution of the estimate of $I(Y_{t+1}; w_{j+1}|\mathbf{w}_t^j)$ for the significance test of CMI, setting Y_{t+1}, w_{j+1} and $\mathbf{w}_t^j$ as X, Y, and Z, respectively, in Equation (4). The parametric significance test is right-sided, as is for the resampling significance test, and the p-value is the complementary of the Gamma cumulative density function for the value of the estimate of $I(Y_{t+1}; w_{j+1}|\mathbf{w}_t^j)$. The DPMIME measure in Equation (3) derived using the parametric test of CMI is denoted DPMIMEpt. Both tests in the computation of DPMIMErt and DPMIMEpt are performed at the significance level $\alpha = 0.05$.

2.4. Statistical Evaluation of Method Accuracy

For a system of K variables, there are $K(K-1)$ ordered pairs of variables to estimate causality. In the simulations of known systems, we know the true interactions between the system variables from the system equations. We further assume that the causal effects in each realization of the system match the designed interactions. Though this cannot be established analytically, former simulations have shown that for weak coupling, below the limit of generalized synchronization, the match holds [1]. Thus, we can assess the match of the $K(K-1)$ estimated causal effects to the true causal dependencies using performance indices. Here, we consider the indices of specificity, sensitivity, Matthews correlation coefficient, F-measure, and Hamming distance. All the indices refer to binary entries, i.e., there is causal effect or not, so we do not use the magnitude of DPMIME in (3), but only if it is positive or not.

The sensitivity is the proportion of the true causal effects (true positives, TPs) correctly identified as such, given as sens = TP/(TP + FN), where FN (false negative) denotes the number of pairs having true causal effects but have gone undetected. The specificity is the proportion of the pairs correctly not being identified as having causal effects (true negatives, TNs), given as spec = TN/(TN + FP), where FP (false positive) denotes the number of pairs found falsely to have causal effects. For the perfect match of estimated and true causality, sensitivity and specificity are one. The Matthews correlation coefficient (MCC) weighs sensitivity and specificity [40]

$$\text{MCC} = \frac{\text{TP} \cdot \text{TN} - \text{FP} \cdot \text{FN}}{\sqrt{(\text{TP} + \text{FP}) \cdot (\text{TP} + \text{FN}) \cdot (\text{TN} + \text{FP}) \cdot (\text{TN} + \text{FN})}}$$

MCC ranges from -1 to 1. If MCC = 1, there is perfect identification of the pairs of true and no causality; if MCC = -1, there is total disagreement and pairs of no causality are identified as pairs of causality and vice versa, whereas MCC at the zero level indicates random assignment of pairs to causal and non-causal effects. The F-measure is the harmonic mean of precision and sensitivity. The precision, also called positive predictive value, is the number of detected true causal effects divided by the total number of detected casual effects, F = TP/(TP + FP). The F-measure (FM) ranges from 0 to 1. If FM = 1, there is perfect identification of the pairs of true causality, whereas if FM = 0, no true coupling is detected. The Hamming distance (HD) is the sum of false positives (FPs) and false negatives (FNs). Thus, HD obtains non-negative integer values bounded below by zero (perfect identification) and above by $K(K-1)$ if all pairs are misclassified.

3. Simulations

One of the aims of the simulation study is to assess whether and how the DPMIME on discrete-valued time series attains the causality estimation accuracy of PMIME on the respective continuous-valued time series. Therefore, we generate discrete-valued time series on the basis of the causality structure of a continuous-valued time series. The continuous-valued time series is generated by a known dynamical system so that the original causal interactions are given by the system equations. In the simulation study, we consider four different ways to generate discrete-valued time series aiming at having the original causal interactions, as presented below.

1. *Continuous to Discrete by quantization (Con2Dis)*: The multivariate symbol sequence of a predefined number of symbols M is directly derived by quantization of the values of the multivariate continuous-valued time series of K observed variables. The range of values of each variable is partitioned to M equiprobable intervals and each interval is assigned to one of the M symbols.
2. *Realization of estimated sparse Markov Chain (SparseMC)*: The multivariate symbol sequence is generated as a realization of a Markov chain of reduced form estimated on the Con2Dis multivariate symbol sequence (as derived above from the continuous-valued multivariate time series). First, the transition probability matrix of a Markov chain of predefined order L is estimated on the Con2Dis multivariate symbol sequence. An entry in this matrix regards the probability of a symbol of the response variable conditioned on the 'word' of size $K \cdot L$ of L last symbols of all K variables. For M discrete symbols, the size of the transition probability matrix for one of the K response variables is $M^{K \cdot L} \times M$. The causal interactions in the original dynamical system assign zero transition probabilities to words that contain non-existing causal interactions so that the Markov chain has a reduced form as the lag variables are less (or much less for a sparse causality network) than $K \cdot L$. For example, let us assume the case of $K = 3$, $L = 2$, and $M = 2$ and the true lag causal relationships for the response $X_{1,t+1}$ are $X_{1,t}$, $X_{1,t-1}$, and $X_{2,t}$. The full form of the Markov chain comprises $2^{3 \cdot 2} = 64$ conditioned probabilities for each of the two symbols of $X_{1,t+1}$, but we estimate only the $2^3 = 8$ probabilities as the lag variables $X_{2,t-1}$, $X_{3,t}$, and $X_{3,t-1}$ are not considered to have any causal effect on $X_{1,t+1}$. Even for a sparse causality network (few true lag causal relationships), the multivariate Markov chain can only be estimated for relatively small values of K, L, and M. Once the sparse transition probability matrix is formed, the generation of a multivariate symbol sequence of length n goes as follows. The first L symbols for each of the K variables are chosen randomly, and they assign to the initial condition. Then, for times $t + 1$, $t = L + 1, ..., n + T$, the new symbol of

each of the K variables is drawn according to the estimated conditioned probabilities. Finally, the first T symbols for each variable are assigned to a transient period and are omitted to form the SparseMC multivariate symbol sequence of length n.

3. Realization of estimated mixture transition distribution model (MTD): Instead of determining the multivariate Markov chain of reduced form in SparseMC, a specific and operatively more tractable form called mixture transition distribution model (MTD) has been proposed [41]. In essence, instead of determining the transition probability from the word of the causal lag variables to the response variable, the MTD determines the transition probability from each causal lag variable to the response variable. Here, as lag variable, we consider any lag of the driving variable to the response in the true dynamical system, e.g., for the example above for the driving variable X_2 to the response X_1, we consider both lags of X_2 (assuming $L = 2$) and not only the true lag one. In its full form, the MTD assumes that the state probability distribution of the j-th variable at time $t + 1$ (response variable) depends on the state probability distribution of all K variables at the last L times as

$$X_{j,t+1} = \sum_{i=1}^{K}\sum_{l=1}^{L} \lambda_{j,i,l} P_{j,i,l} X_{i,t-l+1}, \quad i = 1,2,\ldots,K, \quad t = L, L+1, \ldots,$$

where $P_{j,i,l}$ is the transition probability from $X_{i,t-l+1}$ to $X_{j,t+1}$, and $\lambda_{j,i,l}$ is a parameter giving the weight on $X_{i,t-l+1}$ in determining $X_{j,t+1}$, and for $j = 1, 2, \ldots, K$, the following holds $\sum_{i=1}^{K}\sum_{l=1}^{L} \lambda_{j,i,l} = 1$. We restrict the full form of MTD by dropping from the sum the variables that are non-causal to X_j, preserving that the remaining $\lambda_{j,i,l}$ sum up to one. Thus, $\lambda_{j,i,l}$ denotes the strength of lag causality from $X_{i,t-l+1}$ to $X_{j,t+1}$. Further, after a simulation study for the optimal tolerance threshold λ_0, we determine $\lambda_0 = 0.01$, and if $\lambda_{j,i,l} < \lambda_0$, we set $\lambda_{j,i,l} = 0$ to omit terms having small coefficients. In this way, we attempt to retain only significant dependencies of the response on the lag variables. We use the estimated MTD model as the generating process and generate a multivariate symbol sequence. To fit MTD to the Con2Dis multivariate symbol sequence, we use the package **markovchain** package in R language [42], implementing the fitting of higher-order multivariate Markov chains as described in [43,44].

4. Realization of estimated multivariate integer-autoregressive system (MINAR): Another simplified form of the multivariate Markov chain is given by the multivariate integer-autoregressive systems (MINAR) [19]. Here, we do not estimate MINAR from the Con2Dis multivariate symbol sequence, as done for the sparse multivariate Markov chain (SparseMC) and the MTD process, but define the MINAR of order one, MINAR(1), by setting to zero the coefficients that regard no-lag causality in the original dynamical system. Therefore, the j-th variable at time t is given as $X_{j,t} = \sum_{i=1}^{K} \alpha_{i,j} \circ X_{i,t-1} + R_{j,t}$ for $j = 1, \ldots, K$, where $\alpha_{i,j} \in [0,1]$ are the coefficients of MINAR(1) (set to zero if the corresponding driver–response relationship does not exist in the original dynamical system), $\circ$ denotes the thinning operator (The thinning operator defines that $a \circ x$ is the sum of x Bernoulli outcomes of probability a.), and $R_{j,t}$ is a random variable taking integer values from a given distribution (here, we set the discrete uniform of two symbols). We note that the way the integer-valued sequence is generated does not determine a fixed number of integer values for each of the K variables so that the generated multivariate symbol sequence does not have a predefined number M of symbols.

The multivariate symbol sequences of all four types are generated under the condition of preserving the coupling structure of the original continuous-valued system. However, only the first type Con2Dis directly preserves the original coupling structure, as the Con2Dis multivariate symbol sequence is directly converted from the continuous-valued realization of the original system. For the other three types, a restricted model is first fitted to the Con2Dis multivariate symbol sequence, which is then used to generate a multivariate symbol sequence. Among the three models, the sparse Markov chain (SparseMC) is best constrained to preserve the original coupling structure. The other two models, the MTD and MINAR, are included in the study as there are known models for discrete-valued time series, adapted here to the given coupling structure. However, the MTD does not preserve the exact lag coupling structure of the original system and the MINAR generates multivariate symbol sequences of varying number of symbols at each realization so that the estimation of the causality structure with the DPMIME on the MTD and MINAR multivariate symbol sequences is not expected to be accurate.

We compute the DPMIME on each multivariate symbol sequence and evaluate the statistical accuracy of the DPMIME to estimate the true variable interactions and subsequently the true coupling network. Further, we compute also the PMIME on the initial continuous-valued time series and examine whether DPMIME can attain the accuracy of PMIME.

3.1. The Simulation Setup

In the simulation study, we use as the original dynamical system the coupled Hénon maps [33,45] and consider four settings regarding different connectivity structures for $K = 5$ (here, the K variables constitute the K subsystems being coupled). We also consider a vector stochastic process as a fifth generating system.

The first system (S1) has an open-chain structure of $K = 5$ coupled Hénon maps, as shown in Figure 1a, defined as

$$\begin{aligned}
X_{1,t+1} &= 1.4 - X_{1,t}^2 + 0.3 X_{1,t-1} \\
X_{2,t+1} &= 1.4 - (0.5C(X_{1,t} + X_{3,t}) + (1-C)X_{2,t})^2 + 0.3 X_{2,t-1} \\
X_{3,t+1} &= 1.4 - (0.5C(X_{2,t} + X_{4,t}) + (1-C)X_{3,t})^2 + 0.3 X_{3,t-1} \\
X_{4,t+1} &= 1.4 - (0.5C(X_{3,t} + X_{5,t}) + (1-C)X_{4,t})^2 + 0.3 X_{4,t-1} \\
X_{5,t+1} &= 1.4 - X_{5,t}^2 + 0.3 X_{5,t-1}
\end{aligned} \quad (5)$$

The first and last variable in the chain of $K = 5$ variables drives its adjacent variable and each of the other variables drive the adjacent variable to its left and right. The coupling strength C is set to 0.2 regarding weak coupling.

The second system (S2) has a randomly chosen structure, as shown in Figure 1b, and it is defined as

$$\begin{aligned}
X_{1,t+1} &= 1.4 - X_{1,t}((1-C)X_{1,t} + CX_{3,t}) + 0.3 X_{1,t-1} \\
X_{2,t+1} &= 1.4 - X_{2,t}^2 + 0.3 X_{2,t-1} \\
X_{3,t+1} &= 1.4 - X_{3,t}(0.5CX_{2,t} + (1-C)X_{3,t} + 0.5CX_{5,t}) + 0.3 X_{3,t-1} \\
X_{4,t+1} &= 1.4 - X_{4,t}(0.5CX_{2,t} + 0.5CX_{3,t} + (1-C)X_{4,t}) + 0.3 X_{4,t-1} \\
X_{5,t+1} &= 1.4 - X_{5,t}(CX_{4,t} + (1-C)X_{5,t}) + 0.3 X_{5,t-1}
\end{aligned} \quad (6)$$

The coupling strength C is set to 0.5. There is no predefined pattern for the interactions of the variables, other than the number of interactions being six, as for S1.

The other two systems, S3 and S4, also have a randomly chosen structure similar to S1 (see Figure 1c,d). S3 is defined as

$$\begin{aligned}
X_{1,t+1} &= 1.4 - X_{1,t}((1-C)X_{1,t} + CX_{5,t}) + 0.3 X_{1,t-1} \\
X_{2,t+1} &= 1.4 - X_{2,t}^2 + 0.3 X_{2,t-1} \\
X_{3,t+1} &= 1.4 - X_{3,t}((1-C)X_{3,t} + CX_{5,t}) + 0.3 X_{3,t-1} \\
X_{4,t+1} &= 1.4 - X_{4,t}((1-C)X_{4,t} + CX_{5,t}) + 0.3 X_{4,t-1} \\
X_{5,t+1} &= 1.4 - X_{5,t}\left(\frac{1}{3}CX_{1,t} + \frac{1}{3}CX_{2,t} + \frac{1}{3}CX_{3,t} + (1-C)X_{5,t}\right) + 0.3 X_{5,t-1}
\end{aligned} \quad (7)$$

and S4 is defined as

$$
\begin{aligned}
X_{1,t+1} &= 1.4 - X_{1,t}^2 + 0.3 X_{1,t-1} \\
X_{2,t+1} &= 1.4 - X_{2,t}(CX_{1,t} + (1-C)X_{2,t}) + 0.3 X_{2,t-1} \\
X_{3,t+1} &= 1.4 - X_{3,t}(CX_{2,t} + (1-C)X_{3,t}) + 0.3 X_{3,t-1} \\
X_{4,t+1} &= 1.4 - X_{4,t}(CX_{3,t} + (1-C)X_{4,t}) + 0.3 X_{4,t-1} \\
X_{5,t+1} &= 1.4 - X_{5,t}(0.5 C X_{1,t} + 0.5 C X_{4,t} + (1-C)X_{5,t}) + 0.3 X_{5,t-1}
\end{aligned}
\tag{8}
$$

The coupling strength C is set to 0.5 for S3 and 0.4 for S4. System S3 has node 5 as a hub (three in-coming and three out-going connections) and system S4 has a causal chain from node 1, to 2, to 3, to 4.

The fifth system (S5) is a vector autoregressive process on $K = 5$ variables (model 1 in [46]), and it is defined as

$$
\begin{aligned}
X_{1,t+1} &= 0.4 X_{1,t} - 0.5 X_{1,t-1} + 0.4 X_{5,t} + u_{1,t+1} \\
X_{2,t+1} &= 0.4 X_{2,t} - 0.3 X_{1,t-3} + 0.4 X_{5,t-1} + u_{2,t+1} \\
X_{3,t+1} &= 0.5 X_{3,t} - 0.7 X_{3,t-1} - 0.3 X_{5,t-2} + u_{3,t+1} \\
X_{4,t+1} &= 0.8 X_{4,t-2} + 0.4 X_{1,t-1} + 0.3 X_{2,t-1} + u_{4,t+1} \\
X_{5,t+1} &= 0.7 X_{5,t} - 0.5 X_{5,t-1} - 0.4 X_{4,t} + u_{5,t+1}
\end{aligned}
\tag{9}
$$

The terms $u_{j,t+1}$ are white noise with zero mean. The connectivity structure of S5 is shown in Figure 1e.

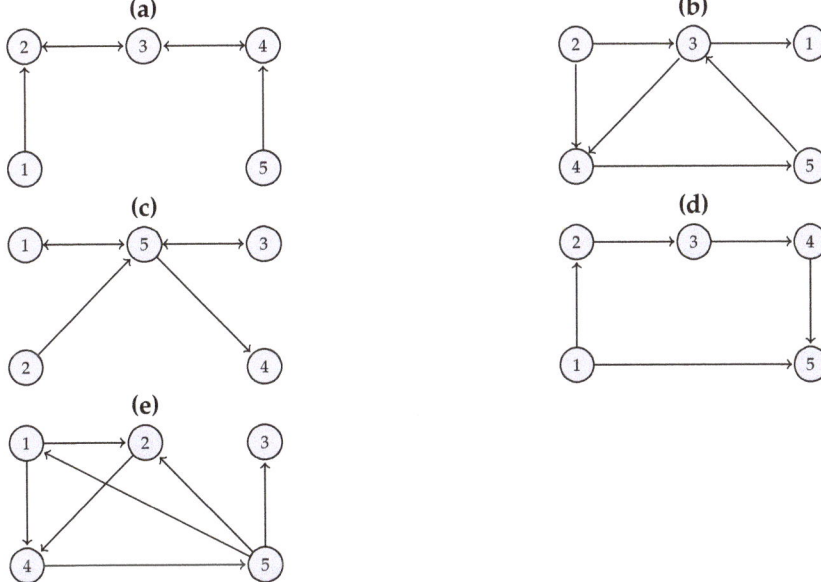

Figure 1. The graphs of the connectivity structure of the simulated systems: (**a**) open-chain structure (S1), (**b**–**d**) randomly chosen structure for (S2)–(S4), respectively, and (**e**) vector autoregressive process (S5).

To derive statistically stable results, we generate 100 realizations for each system and for different time-series lengths n. The number of symbols (M) for the discretization of the continuous-valued multivariate time series is 2 and 4, respectively. For the discretization, an equiprobable partition is used so that when $M = 2$, all values of the time series larger than the median are set to 0 and the rest to 1, and when $M = 4$, the quartiles of the time series define the four symbols.

3.2. An Illustrative Example

The performance of the DPMIME is first illustrated with a specific example, focusing on the first two equations of system S1 and thus considering as a response in the DPMIME (and PMIME) only the first and second variable. We consider only the first type (Con2Dis) for the generation of the multivariate symbol sequence with $M = 2$ symbols and length $n = 1024$. The parametric test for the significance of each component to be added to the mixed embedding vector is used (DPMIMEpt), and the maximum lag is $L = 5$. The PMIME is computed for the same $L = 5$ on the continuous-valued time series (before discretization). Table 1 shows the frequency of occurrence of any of the 25 lag terms in the mixed embedding vector of DPMIMEpt and PMIME for the response X_1 and X_2 in 100 Monte Carlo realizations. For the true lag terms, i.e., the terms that occur in the system equations, the frequencies are highlighted.

Table 1. Each cell in columns 2–5 has the frequency of occurrence over 100 realizations of the lag variable (first column) in the mixed embedding vector for DPMIMEpt and PMIME, where the response is the first or the second variable of system S1 and for $n = 1024$, $L = 5$, and $M = 2$. The frequencies of the lag variables occurring in the system equations are highlighted.

	$X_{1,t+1}$		$X_{2,t+1}$	
	DPMIMEpt	PMIME	DPMIMEpt	PMIME
$X_{1,t}$	100	100	6	1
$X_{1,t-1}$	100	100	84	99
$X_{1,t-2}$	100	5	3	0
$X_{1,t-3}$	100	1	2	0
$X_{1,t-4}$	92	0	1	0
$X_{2,t}$	0	0	100	100
$X_{2,t-1}$	0	0	91	100
$X_{2,t-2}$	0	0	72	5
$X_{2,t-3}$	0	0	8	1
$X_{2,t-4}$	0	0	31	0
$X_{3,t}$	1	0	32	44
$X_{3,t-1}$	0	0	39	53
$X_{3,t-2}$	0	0	15	6
$X_{3,t-3}$	1	0	10	1
$X_{3,t-4}$	1	0	5	0
$X_{4,t}$	0	0	0	0
$X_{4,t-1}$	0	0	1	0
$X_{4,t-2}$	0	0	0	0
$X_{4,t-3}$	0	0	0	0
$X_{4,t-4}$	0	0	0	0
$X_{5,t}$	0	0	0	0
$X_{5,t-1}$	0	0	0	0
$X_{5,t-2}$	0	0	0	0
$X_{5,t-3}$	0	0	0	0
$X_{5,t-4}$	0	0	0	0

The variable $X_{1,t+1}$ depends on the variables $X_{1,t}$ and $X_{1,t-1}$, which are selected by both algorithms of DPMIMEpt and PMIME in all realizations (frequency 100). The DPMIMEpt selects also the lag terms $X_{1,t-2}$, $X_{1,t-3}$, and $X_{1,t-4}$ of the response variable, but their inclusion in the mixed embedding vector does not result in any false causal effects. (It turns out that it is hard to find the exact lag components of the driving variable in the case of discrete-valued time series, which questions the use of DPMIME for building the input of a regression model to the response.) No other lag terms are found (the maximum frequency is one).

The second equation of S1 defines the dependence of $X_{2,t+1}$ on $X_{1,t}$, $X_{2,t}$, $X_{2,t-1}$, and $X_{3,t}$. For X_2 as response, both algorithms do not include the lag term $X_{1,t}$ in the mixed embedding vector (frequency 6 for DPMIMEpt and 1 for PMIME) but include instead $X_{1,t-1}$ (frequency 84 for DPMIMEpt and 99 for PMIME) so that the variable X_1 is represented in the mixed embedding vector and the correct causal effect from X_1 to X_2 is established. The lag terms $X_{2,t}$ and $X_{2,t-1}$ are always present in the mixed embedding vector for both algorithms ($X_{2,t}$ occurs less frequently at 91% for DPMIMEpt) and terms of larger lag of X_2 occur for DPMIMEpt at a smaller frequency. The representation of X_3 in the mixed embedding vector is spread over the two first lags for PMIME and to the first four lags for DPMIMEpt so that though the true lag term $X_{3,t}$ is not well identified (frequency 32 and 44 for DPMIMEpt and PMIME, respectively), the causal effect from X_3 to X_2 is well established. The variables X_4 and X_5 are not represented in the mixed embedding vector, and thus both DPMIMEpt and PMIME correctly find no causal effect from these variables to X_2.

The example shows that the two algorithms have a similar performance, with DPMIMEpt tending to include more lag terms of the causal variables, but both algorithms do not include lag terms of variables that have no causal effect on the response variable.

3.3. System 1

The example above is for the first two variables of S1, and in the following, we compute DPMIMEpt and PMIME for all $K = 5$ response variables of S1 and detect the presence of causal effects by the presence of a lag term (or terms) of the driving variable in the mixed embedding vector for the response variable. The true causal effects as derived by the equations of S1 are $X_1 \to X_2$, $X_2 \to X_3$, $X_3 \to X_2$, $X_3 \to X_4$, $X_4 \to X_3$, and $X_5 \to X_4$. The distribution of the DPMIMEpt and PMIME (in the form of boxplots) and the rate of detection of causal effects (numbers under the boxplots) for all 20 directed variable pairs are shown in Figure 2.

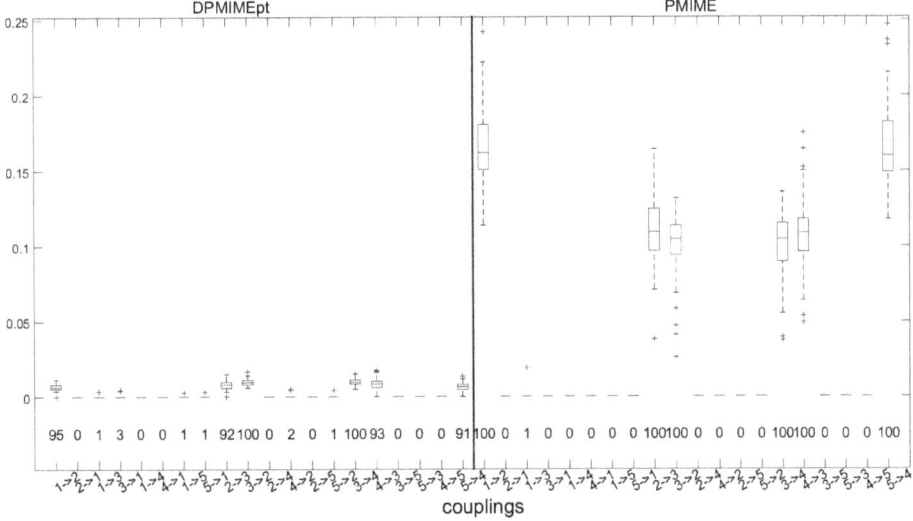

Figure 2. Boxplots of DPMIMEpt ($M = 2$) and PMIME for all variable pairs of S1, for 100 realizations of the system S1, using $L = 5$ and $n = 1024$. At each panel, the number of times the causal effect is detected is displayed below each boxplot.

Both measures perfectly define the non-existent causal effects with a percentage of detection less than 3%. The DPMIMEpt detects the true causal effects in high percentages, approaching the perfect identification achieved by PMIME. However, as seen by the size of the boxplots, the DPMIME obtains smaller values than the PMIME. Though both are defined by the same CMI to MI ratio in (3), this ratio is smaller for the DPMIME.

To quantify the performance of the DPMIMEpt and PMIME at each realization of S1, we calculate the performance indices sens, spec, MCC, FM, and HD on the 20 binary directed connections, where six of them are true. In Table 2, the average indices over the 100 realizations of S1 for $n = 1024$ are shown for both measures.

Table 2. Average of sensitivity (sens), specificity (spec), MCC, F-measure (FM), and Hamming distance (HD) over 100 realizations of system S1 for the causality measures DPMIMEpt ($M = 2$) and PMIME, using $L = 5$ and $n = 1024$.

	DPMIMEpt	PMIME
sens	0.952	1
spec	0.994	0.999
MCC	0.956	0.999
FM	0.956	0.999
HD	0.380	0.010

For the DPMIMEpt using $M = 2$, both sensitivity and specificity are very high, and the overall indices are high as well, e.g., the HD shows that more often none or less often one causal effect out of 20 causal effects is misclassified (average HD is 0.38). Thus, the performance of the DPMIMEpt is close to the almost perfect performance of the PMIME.

Next, we compare the parametric test (PT) and resampling test (RT) for the CMI used in the DPMIME as the criterion to terminate the algorithm building the mixed embedding vector. We consider the four types for generating multivariate symbol sequences (Con2Dis, SparseMC, MTD, and MINAR) and for $M = 2$ and $M = 4$ symbols. The latter does not apply to MINAR as the generated sequences have not a predefined number of symbols (integers). In the comparison, we again use as reference (gold standard) the PMIME, because this is computed directly on the continuous-valued time series, whereas the DPMIMEpt and DPMIMErt are computed on the discrete-valued time series. We also examine the performance of measures for different time-series lengths n. Here, we only report results for the performance index MCC in Table 3.

The DPMIMEpt and DPMIMErt fail to define the pairs with causal and non-causal effects when applied to the multivariate symbol sequences generated by the MTD. As already mentioned, the MTD model fails to preserve the causality of the original system and, in turn, the generated discrete-valued sequences do not allow for the estimation of the true causal effects. For example, for $M = 2$ when $n = 1024$, the performance indices sens, spec, MCC, FM, and HD are 0.36, 0.60, −0.03, 0.31, and 9.41, respectively, indicating a very low specificity. The DPMIME using either significance test on the Con2Dis sequences scores similarly in the MCC and at a lower level than the PMIME, converging to the highest level with the increase of n. This holds for both $M = 2$ and $M = 4$, but for $M = 4$, the performance of DPMIME is worse than that of PMIME and the difference decreases with n, indicating that for a larger number of symbols longer time series are needed. The accuracy of DPMIME on the SparseMC sequences is similar as for the Con2Dis sequences when $M = 4$, but for $M = 2$, the accuracy does not improve with n unlike in the case of Con2Dis. The DPMIME performs better on MINAR sequences than on MTD sequences, especially when the resampling significance test is used in DPMIME. In this particular case, the parametric test is not as accurate as the resampling test. The finding that DPMIMEpt and DPMIMErt (except in the case of MINAR) perform similarly has practical importance because we can rely on DPMIMEpt and save computation time, which for long time series and many observed variables, DPMIMErt would be computationally very intensive.

Table 3. Average MCC over 100 realizations of system S1 for the causality measures DPMIME using $L = 5$, number of symbols $M = 2, 4$ (column 1), the parametric test (PT), and the resampling test (RT) (column 2) on multivariate sequences of type Con2Dis, SparseMC, MTD, and MINAR, as well as PMIME (colum 3), and for $n = 512, 1024, 2048, 4096$ (columns 4–7, respectively).

			$n = 512$	$n = 1024$	$n = 2048$	$n = 4096$
$M = 2$	PT	Con2Dis	0.78	0.96	1	1
	RT	Con2Dis	0.76	0.95	1	1
	PT	SparseMC	0.72	0.79	0.79	0.80
	RT	SparseMC	0.71	0.82	0.83	0.82
	PT	MTD	−0.02	−0.03	−0.05	−0.05
	RT	MTD	0.00	−0.01	−0.04	−0.03
$M = 4$	PT	Con2Dis	0.49	0.70	0.99	1
	RT	Con2Dis	0.39	0.70	0.99	1
	PT	SparseMC	0.54	0.72	1	1
	RT	SparseMC	0.43	0.72	1	1
	PT	MTD	0.00	−0.02	−0.03	−0.06
	RT	MTD	0.00	−0.02	−0.02	0.03
	PT	MINAR	0.08	0.12	0.25	0.43
	RT	MINAR	0.41	0.61	0.75	0.81
		PMIME	0.98	1	1	1

Similar results as for MCC in Table 3 are obtained using the performance index HD, as shown in Figure 3a. The misclassification is larger when the time series gets smaller (from 4096 to 512) for the same number of symbols M, and when M gets larger (from 2 to 4) for the same n. However, the HD is at the same level for all these settings for DPMIMEpt and DPMIMErt, and for both measures, it converges to zero (no misclassification of all 20 variable pairs) for $n \geq 2048$, as does PMIME even for small n.

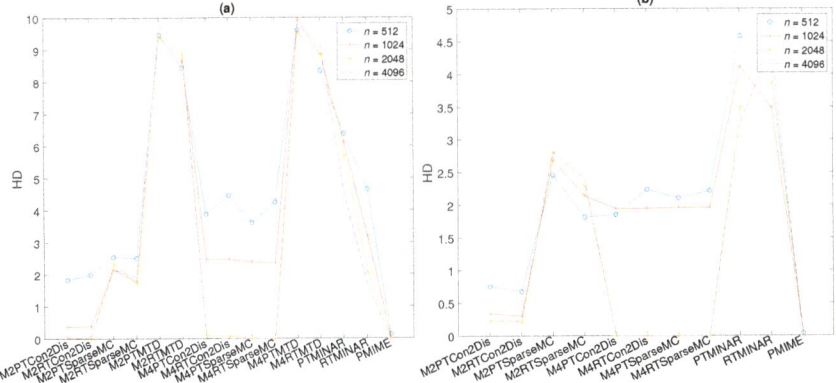

Figure 3. Average HD over 100 realizations of system S1 in (**a**) and S2 in (**b**) for the causality measures DPMIME ($L = 5$) using symbols $M = 2, 4$ (denoted with M and the number 2 or 4 in the beginning of each word label in the abscissa), the parametric test, and the resampling test (given by PT or RT after the symbol notation of each word label) on multivariate sequences of type Con2Dis, SparseMC, MTD (present only in (**a**)), and MINAR, as well as PMIME (the acronym is at the end of each word label), and for $n = 512, 1024, 2048, 4096$, as given in the legend.

3.4. System 2

System S2 differs from S1 in that it has a randomly chosen coupling structure. We show the summary results of the performance index MCC in Table 4 and the HD in Figure 3b. The performance of DPMIMEpt and DPMIMErt on the Con2Dis and SparseMC sequences is similar to that on system S1 for the different settings of time-series length n and number of symbols M (the MTDs are not included in the results due to their poor performance in the previous system). The accuracy in detecting the true causal effects is better for smaller M when n is small, converging to the highest performance level with n and faster for $M = 4$ (MCC = 1 and HD = 0). The highest level is again attained by PMIME even for the smallest time-series length $n = 512$. For the smallest $n = 512$, the performance of DPMIMEpt and DPMIMErt is better for system S2 than for system S1. Another difference to system S1 is that for the largest tested $n = 4096$, the DPMIMEpt and DPMIMErt reach the highest level for $M = 4$ but not for $M = 2$, indicating that once there is enough data, the use of the largest number of symbols allows for a better detection of the causal effects. For system S2, the DPMIMEpt and DPMIMErt on the MINAR sequences give similar MCC scores that do not tend to get higher with n, unlike the respective scores for system S1. This lack of improvement with n in the causality estimation on MINAR sequences is attributed to the varying number of integers of the generated time series increasing with n so that the number of symbols M is relatively large compared to the length of time series n.

Table 4. As for Table 3 but for system S2 (MTD not included).

			$n = 512$	$n = 1024$	$n = 2048$	$n = 4096$
$M = 2$	PT	Con2Dis	0.92	0.97	0.98	0.97
	RT	Con2Dis	0.93	0.97	0.98	0.98
	PT	SparseMC	0.76	0.76	0.75	0.75
	RT	SparseMC	0.82	0.80	0.79	0.78
$M = 4$	PT	Con2Dis	0.78	0.77	1	1
	RT	Con2Dis	0.73	0.77	1	1
	PT	SparseMC	0.75	0.77	1	1
	RT	SparseMC	0.74	0.77	1	1
	PT	MINAR	0.29	0.38	0.49	0.54
	RT	MINAR	0.42	0.53	0.49	0.43
		PMIME	1	1	1	1

3.5. System 3 and System 4

Systems S3 and S4 also have a randomly chosen structure, as with system S2. The summary results of the performance index MCC are shown in Table 5 for S2 and Table 6 for S3.

For the different settings of both S2 and S3, the generation of symbol sequences by Con2Dis and SparseMC, the number of symbols $M = 2$ and $M = 4$, and the time-series lengths n, the DPMIMEpt and DPMIMErt always perform similarly and less accurately than the PMIME. There are however differences in the DPMIME in Con2Dis and SparseMC with respect to M. As for S1 and S2, for both S3 and S4, the DPMIME on SparseMC symbol sequences tends to perform better for $M = 4$ than for $M = 2$, and this occurs more consistently for a larger n. On the other hand, the DPMIME on Con2Dis symbol sequences tends to perform better for $M = 2$ than for $M = 4$, particularly for a smaller n. For a larger n, for S3, the best performance is observed for Con2Dis and $M = 2$, and for S4, SparseMC and $M = 4$.

Table 5. As for Table 3 but for system S3 ($L = 5$, MTD and MINAR not included).

			$n = 512$	$n = 1024$	$n = 2048$	$n = 4096$
$M = 2$	PT	Con2Dis	0.8	0.93	0.93	0.95
	RT	Con2Dis	0.78	0.92	0.94	0.95
	PT	SparseMC	0.73	0.75	0.73	0.76
	RT	SparseMC	0.75	0.79	0.77	0.79
$M = 4$	PT	Con2Dis	0.68	0.76	0.81	0.88
	RT	Con2Dis	0.68	0.76	0.81	0.88
	PT	SparseMC	0.65	0.76	0.81	0.88
	RT	SparseMC	0.65	0.76	0.80	0.88
		PMIME	0.98	0.99	1	1

Table 6. As for Table 3 but for system S4 ($L = 5$, MTD and MINAR not included).

			$n = 512$	$n = 1024$	$n = 2048$	$n = 4096$
$M = 2$	PT	Con2Dis	0.94	0.98	0.99	0.94
	RT	Con2Dis	0.93	0.98	0.99	0.96
	PT	SparseMC	0.68	0.73	0.70	0.72
	RT	SparseMC	0.74	0.79	0.75	0.75
$M = 4$	PT	Con2Dis	0.82	0.81	0.78	0.70
	RT	Con2Dis	0.86	0.85	0.80	0.72
	PT	SparseMC	0.79	0.87	0.98	1
	RT	SparseMC	0.77	0.87	0.97	1
		PMIME	0.99	0.99	1	

3.6. System 5

System S5 is a five-dimensional vector autoregressive process of order 4. This system is chosen in order to examine the performance of the causality measures in a linear stochastic system. The summary results of the performance index MCC are presented in Table 7. First, it is worth noting that the PMIME does not reach the highest accuracy level as for the nonlinear deterministic systems S1–S4, but the MCC ranges from 0.76 to 0.78 for the different n. The highest accuracy level is attained by the DPMIMEpt for a smaller n and the DPMIMEpt for a larger n on the Con2Dis symbol sequences when $M = 4$. For $M = 4$, the randomization test tends to outperform the parametric test for a larger n and attains the maximum MCC = 1. On the other hand, when $M = 2$, the accuracy of both tests is at the same level and does not improve significantly with the increase of n as does for PMIME.

Table 7. As for Table 3 but for system S5 ($L = 8$, MTD and MINAR not included).

			$n = 512$	$n = 1024$	$n = 2048$	$n = 4096$
$M = 2$	PT	Con2Dis	0.61	0.71	0.72	0.75
	RT	Con2Dis	0.66	0.74	0.74	0.76
	PT	SparseMC	0.60	0.65	0.68	0.67
	RT	SparseMC	0.66	0.68	0.71	0.70
$M = 4$	PT	Con2Dis	0.81	0.98	0.88	0.86
	RT	Con2Dis	0.75	0.98	1	0.98
	PT	SparseMC	0.82	0.98	0.88	0.86
	RT	SparseMC	0.75	0.98	1	0.98
		PMIME	0.77	0.78	0.78	0.76

3.7. Effect of Observational Noise

In the last part of the simulation study, we investigate the effect of observational noise, restricting to observational noise on the original continuous-valued time series. We consider system S1 and add to each of the five generated time series white Gaussian noise with standard deviation (SD) being a given percentage of the SD of the time series. The Con2Dis approach with $M = 2$ is then applied to derive the symbol sequences of different lengths n and the DPMIME is applied using the parametric test (DPMIMEpt) and randomization test (DPMIMErt). In Table 8, the results are presented, including the PMIME measure as well. The type of test does not seem to affect the performance of the DPMIME for all different noise levels. For noise levels up to 10%, the MCC is rather stable and effectively the same as for the noise-free symbol sequences and decreases with a further increase in noise level (20% and 40%) for all different n. However, even for the high noise level of 40% when $n = 4096$, the MCC is 0.8 for DPMIMEpt and 0.85 for DPMIMErt and close to the MCC for PMIME at 0.89. Overall, a smooth decrease in the accuracy of the DPMIME is observed with the increase in the level of observational noise, which suggests the appropriateness of DPMIME for real-world symbol sequences.

Table 8. Average MCC over 100 realizations of system S1 for the causality measures DPMIMEpt and DPMIMErt on the Con2Dis symbols sequences ($M = 2$) and PMIME on the original continuous-valued time series (column 2), where noise of different levels is added (column 1), and the time-series length is $n = 512, 1024, 2048, 4096$ (columns 3–6). The added white noise is Gaussian with standard deviation given by the percentage of the standard deviation of the data.

Noise	Measure	$n = 512$	$n = 1024$	$n = 2048$	$n = 4096$
0%	PMIMEpt	0.78	0.96	1	1
	PMIMErt	0.76	0.95	1	1
	PMIME	0.98	1	1	1
5%	PMIMEpt	0.81	0.95	0.98	0.99
	PMIMErt	0.80	0.94	0.98	0.99
	PMIME	0.99	1	1	1
10%	PMIMEpt	0.76	0.95	0.99	0.99
	PMIMErt	0.78	0.95	1	0.99
	PMIME	0.95	0.99	1	1
20%	PMIMEpt	0.72	0.87	0.93	0.93
	PMIMErt	0.71	0.88	0.94	0.94
	PMIME	0.92	0.96	0.97	0.97
40%	PMIMEpt	0.46	0.67	0.76	0.80
	PMIMErt	0.48	0.70	0.80	0.85
	PMIME	0.70	0.84	0.88	0.89

4. Application to Real Data

We consider a real-world application to compare DPMIME to other causality measures. These are the linear direct causality measure called the conditional Granger causality index (CGCI) [47,48], the information-based direct causality measure of partial transfer entropy (PTE) [49], and finally, the original partial mutual information from mixed embedding (PMIME).

The dataset is the Morgan Stanley Capital International (MSCI) market capitalization weighted index of five selected markets in Europe and South America: Greece, Germany, France, UK, and USA. Specifically, we consider two datasets: the first one is in the time period 1 January 2004 to 31 January 2008 and the second one in the period from 3 March 2008 to 30 March 2012. The separation was made with regard to the occurrence of the global financial crisis (GFC), also referred to as the Great Recession, dated from the beginning of year 2008 to year 2013 [50]. The two selected periods are therefore called preGFC and postGFC. The interest here is to study whether and how each of the causality measures detects changes in the connectivity structure in the system of the five markets from preGFC to postGFC. Each dataset comprises $n=1065$ observations, which correspond to daily returns (first differences in the logarithms of the indices). For DPMIME, the data were discretized to two symbols: 1 if the return is positive and 0 otherwise. For consistency, the amount of past information denoted L is the same for all causality measures and set to $L = 2$, where for DPMIME and PMIME L stands for the maximum lag, for PTE it stands for the embedding dimension, and for CGCI it stands for the order of the (restricted and unrestricted) vector autoregressive (VAR) model.

The causality measures DPMIMEpt, PMIME, CGCI, and PTE are computed for each pair of national markets in the preGFC and postGFC periods. While DPMIME and PMIME assign zero to the non-significant causal relationships, CGCI and PTE require a threshold, here given by the parametric significance test for CGCI and the resampling significance test for PTE (the time-shifted surrogates as for the significance of CMI in DPMIMErt). Then, the causality networks are formed drawing weighted connections with weights being the value of the significant measure, and the networks are shown in Figure 4 for the preGFC and postGFC periods.

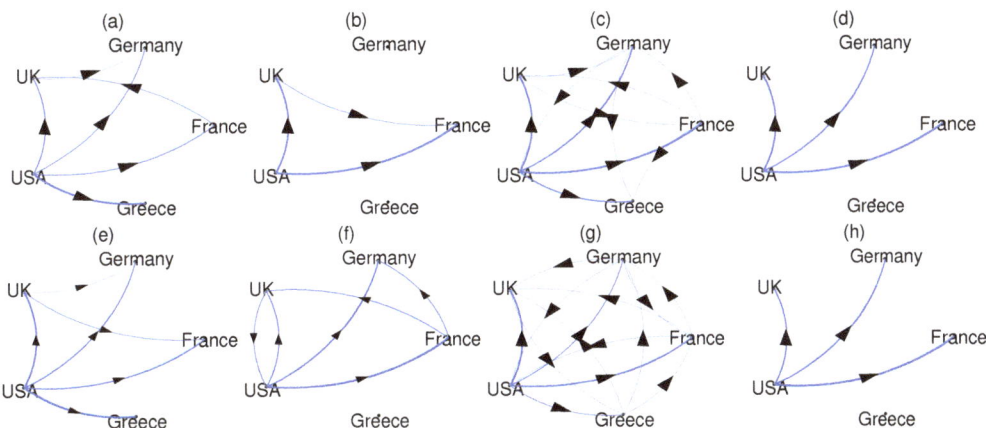

Figure 4. The causality networks of weighted connections for the preGFC period using the measures DPMIMEpt in (**a**), PMIME in (**b**), CGCI in (**c**), and PTE in (**d**), and respectively for the postGFC period in (**e**–**h**).

All causality measures suggest the USA market has a causal effect on many other markets before and after the GFC. In the DPMIMEpt networks (Figure 4a,e), there is in additional causal effect from UK to Germany in both the preGFC and postGFC periods, while the driving from France to UK in preGFC reverses in the postGFC period. Regarding the latter, the PMIME networks show the opposite, UK to France in preGFC and France to UK in postGFC (Figure 4b,f). The PMIME networks show no causal effect of USA on Greece in both periods, which has no straightforward interpretation. In the postGFC period, the PMIME finds a bidirectional causal relationship for the USA and UK. The CGCI measure gives almost full networks in both periods (Figure 4c,g), failing to reveal any particular connectivity structure in the system of the five national markets. On the other hand, the PTE turns out to be the most conservative measure, giving only the causal effect of the USA to UK, Germany and France (not Greece) in both periods (Figure 4d,h).

The DPMIMErt gave similar results to DPMIMEpt (not shown here). We repeated the same analysis for the DPMIME and $L = 1$, and the results were stable. Overall, the DPMIME estimates reasonable causal relationships, the USA to all four other markets in both periods, whereas the PMIME and PTE exclude Greece, and the UK and France causal relationship changes direction from preGFC to postGFC.

5. Discussion

In this study, we propose a Granger causality measure for discrete-valued multivariate time series or multivariate symbol sequences based on partial mutual information from a mixed embedding named DPMIME. The rationale is to build the so-called mixed embedding vector that has as components the lag terms of the observed variables that best explain the response ahead in time. To quantify the causality of a driving variable to a response variable in view of all the observed variables, we first check whether the lag terms of the driving variable are included in the mixed embedding vector. If there are not any, then the measure is zero and there is no causal effect, whereas if there are, then the proportion of the information on the response explained by these lag terms determines the strength of the causal effect from the driving to the response. For the termination of the algorithm building the mixed embedding vector, we develop a parametric test using a Gamma approximation of the asymptotic null distribution of the conditional mutual information, CMI (information of the tested lag term and the response given the other lag terms already included in the mixed embedding vector). This is a main difference to the PMIME, the analogue of the same algorithm already developed by our team for continuous-valued time series. The PMIME employs a resampling significance test as there is no parametric approximation of the null distribution of the CMI for continuous variables. Another main difference to the PMIME is that for discrete-valued time series, we use a different estimate for the information measures of the mutual information, MI, and CMI used in the algorithm, i.e., we use the maximum likelihood estimate for the probabilities of all discrete probability distributions involved in the definition of the MI and CMI, whereas in the PMIME, the nearest neighbor estimate [51] is used for the entropies involved in the definition of the MI and CMI. We develop two versions of the DPMIME, one using the parametric significance test for the termination criterion, denoted DPMIMEpt, and another using the resampling significance test, denoted DPMIMErt, as for the PMIME.

The previous studies of our research team have showed that the PMIME is one of the most appropriate measures to estimate direct causality in multivariate time series and particularly in the setting of high-dimensional time series (many observed variables) [1,27,52]. Therefore, to evaluate the proposed measure DPMIME for the causality of discrete-valued multivariate time series and multivariate symbol sequences, we compare it to the PMIME. For the simulations, dynamical systems of continuous-valued variables were used to generate multivariate time series and compute on them the PMIME. Then, the discrete-valued time series were generated by discretizing the continuous-valued time series, denoted Con2Dis. Moreover, systems for the generation of discrete-valued time series were fitted to the Con2Dis multivariate sequence: the sparse Markov Chain (SparseMC), the mixture transition distribution models (MTD), and the multivariate integer-autoregressive systems (MINAR). The simulations showed that the MTD cannot preserve the original coupling structure, whereas the varying number of integers (assigned to symbols) of the MINAR sequences complicates the use of the DPMIME on these sequences. The SparseMC sequences could preserve sufficiently well the coupling structure in the discrete-valued (Con2Dis) and continuous-valued time series, as the DPMIME could detect the original causality relationships almost as well in the SparseMC sequences as in the Con2Dis sequences. Thus, the main focus in the simulation study was on the performance of the DPMIMEpt and DPMIMErt (for the parametric and resampling significance test) on the Con2Dis multivariate symbol sequences, as compared to the PMIME, where the latter has the role of golden standard as it is computed on the complete available information from the system, i.e., the continuous-valued time series. Further, we assess whether the DPMIMEpt can perform as well as the DPMIMErt, as the DPMIMEpt is much faster to compute and would be preferred in applications.

The simulation systems are coupled Hénon maps of five subsystems, one with an open-chain coupling structure and the other three with a different randomly chosen coupling structure, as well as a vector autoregressive process (VAR) on five variables. The performance indices were computed on binary causality estimates (presence or absence of causal effect) for all pairs of variables (subsystems). The average of the performance indices over 100 realizations for each setting of the time-series length n and number of symbols M from the discretization were reported. The results on all the simulated systems showed that the DPMIMEpt scores lower than the PMIME, as expected, but converges to the performance level of the PMIME with an increasing n, except for the VAR system where the accuracy of the DPMIMEpt in detecting the true causal effects is better when $n = 1024$ and similar to the PMIME when $n \geq 2048$. The difference to the PMIME is larger for a small n and larger M (going from 2 to 4 symbols), which is anticipated as the discretization smooths out information in the time series about the evolution of the underlying system. However, the convergence of the DPMIME to the PMIME for a data size of $n \geq 2048$ indicates that the proposed measure can be used in applications with a moderate length of the discrete-valued time series that can have an even high dimension (we tested here for five subsystems).

The finding that the DPMIMEpt and DPMIMErt perform similarly has high practical relevance. The DPMIME is based on multiple computations of the CMI on progressively higher dimensions that are computationally intensive. If we had to rely on the DPMIMErt, the computation time at each iteration of the algorithm would be multiplied with the number of the resampled data used for the resampling significance test. In applications on long sequences of many symbols, the computation time may be prohibitively long using the DPMIMErt with, say, 100 resampling sequences for each test, but it would be approximately 100 times less when using the DPMIMEpt. Thus, the DPMIMEpt is an appropriate measure to estimate the direct causality in many symbol sequences.

The DPMIME was further applied and compared to other causality measures (PMIME, CGCI, and PTE) in one real-world application. We used data from the Morgan Stanley Capital International market capitalization weighted index of five national markets to examine the causality structure of the system of the five markets before and after the start of the global financial crisis. The proposed measure DPMIME detects the crucial role of the US market before and after the start of the global financial crisis without being as conservative as the PTE and without giving full networks as the CGCI.

Author Contributions: Conceptualization, D.K.; Data curation, M.P. and E.S.; Formal analysis, M.P., E.S. and D.K.; Funding acquisition, D.K.; Methodology, M.P., E.S. and D.K.; Project administration, D.K.; Supervision, D.K.; Visualization, M.P. and E.S.; Writing—original draft, M.P. and E.S.; Writing—review & editing, D.K. All authors have read and agreed to the published version of the manuscript.

Funding: The research was funded by the Research Project No. 5047900 of the Program 'Development of Human Potential, Education and Lifelong Learning 2014-2020' of the Ministry of Development and Investments of the Hellenic Republic.

Institutional Review Board Statement: Not applicable

Data Availability Statement: Not applicable.

Conflicts of Interest: The authors declare no conflict of interest.

References

1. Siggiridou, E.; Koutlis, C.; Tsimpiris, A.; Kugiumtzis, D. Evaluation of Granger causality measures for constructing networks from multivariate time series. *Entropy* **2019**, *21*, 1080. [CrossRef]
2. Billio, M.; Getmansky, M.; Lo, A.W.; Pelizzon, L. Econometric measures of connectedness and systemic risk in the finance and insurance sectors. *J. Financ. Econ.* **2012**, *104*, 535–559. [CrossRef]
3. Porta, A.; Faes, L. Wiener-Granger Causality in network physiology with applications to cardiovascular control and neuroscience. *Proc. IEEE* **2016**, *104*, 282–309. [CrossRef]
4. Fan, J.; Meng, J.; Ludescher, J.; Chen, X.; Ashkenazy, Y.; Kurths, J.; Havlin, S.; Schellnhuber, H.J. Statistical physics approaches to the complex earth system. *Phys. Rep. Rev. Sect. Phys. Lett.* **2021**, *896*, 1–84.

5. Fieguth, P. *An Introduction to Complex Systems: Society, Ecology and Nonlinear Dynamics*; Springer: Cham, Switzerland, 2017.
6. Thurner, S.; Hanel, R.; Klimek, P. *Introduction to the Theory of Complex Systems*; Oxford University Press: Oxford, UK, 2018.
7. Raftery, A. A model for high order Markov chains. *J. R. Stat. Soc.* **1985**, *47*, 528–539. [CrossRef]
8. Nicolau, J. A new model for multivariate Markov chains. *Scand. J. Stat.* **2014**, *41*, 1124–1135. [CrossRef]
9. Zhou, G.; Ye, X. High-order interacting multiple model filter based on mixture transition distribution. In Proceedings of the International Conference on Radar Systems, Belfast, Ireland, 23–26 October 2017; pp. 1–4.
10. Tank, A.; Fox, E.; Shojaie, A. Granger causality networks for categorical time series. *arXiv* **2017**, arXiv:1706.02781.
11. Tank, A.; Li, X.; Fox, E.B.; Shojaie, A. The convex mixture distribution: Granger causality for categorical time aeries. *SIAM J. Math. Data Sci.* **2021**, *3*, 83–112. [CrossRef]
12. Shojaie, A.; Fox, E.B. Granger Causality: A review and recent advances. *Annu. Rev. Stat. Its Appl.* **2022**, *9*, 289–319. [CrossRef]
13. Fokianos, K.; Rahbek, A.; Tjøstheim, D. Poisson autoregression. *J. Am. Stat. Assoc.* **2009**, *104*, 1430–1439. [CrossRef]
14. Neumann, M. Absolute regularity and ergodicity of Poisson count processes. *Bernoulli* **2011**, *17*, 1268–1274. [CrossRef]
15. Davis, A.; Wu, R. A negative binomial model for time series of counts. *Biometrika* **2009**, *96*, 735–749. [CrossRef]
16. Christou, V.; Fokianos, K. On count time series prediction. *J. Stat. Comput. Simul.* **2015**, *82*, 357–373. [CrossRef]
17. Song, P.; Freeland, R.; Biswas, A.; Zhang, S. Statistical analysis of discrete-valued time series using categorical ARMA models. *Comput. Stat. Data Anal.* **2013**, *57*, 112–124. [CrossRef]
18. Angers, J.; Biswas, A.; Maiti, R. Bayesian forecasting for time series of categorical data. *J. Forecast.* **2017**, *36*, 217–229. [CrossRef]
19. Pedeli, X.; Karlis, D. Some properties of multivariate INAR(1) processes. *Comput. Stat. Data Anal.* **2013**, *67*, 213–225. [CrossRef]
20. Scotto, M.; Weiss, C.; Gouveia, S. Thinning-based models in the analysis of integer-valued time series: A review. *Stat. Model.* **2015**, *15*, 590–618. [CrossRef]
21. Budhathoki, K.; Vreeken, J. Causal inference on event sequences. In Proceedings of the SIAM International Conference on Data Mining, San Diego, CA, USA, 3–5 May 2018; pp. 55–63.
22. Schreiber, T. Measuring information transfer. *Phys. Rev. Lett.* **2000**, *85*, 461–464. [CrossRef]
23. Paluš, M. Coupling in complex systems as information transfer across time scales. *Philos. Trans. R. Soc. A Math. Phys. Eng. Sci.* **2019**, *377*, 20190094. [CrossRef]
24. Papapetrou, M.; Kugiumtzis, D. Markov chain order estimation with conditional mutual information. *Phys. A* **2013**, *392*, 1593–1601. [CrossRef]
25. Biswas, A.; Guha, A. Time series analysis of categorical data using auto-mutual information. *J. Stat. Plan. Inference* **2009**, *139*, 3076–3087. [CrossRef]
26. Staniek, M.; Lehnertz, K. Symbolic transfer entropy. *Phys. Rev. Lett.* **2008**, *100*, 158101. [CrossRef] [PubMed]
27. Kugiumtzis, D. Partial transfer entropy on rank vectors. *Eur. Phys. J. Spec. Top.* **2013**, *222*, 401–420. [CrossRef]
28. Amigó, J. *Permutation Complexity in Dynamical Systems Ordinal Patterns, Permutation Entropy and All That*; Springer Science & Business: Berlin, Germany, 2010.
29. Herrera-Diestra, J.L.; Buldu, J.M.; Chavez, M.; Martinez, J.H. Using symbolic networks to analyse dynamical properties of disease outbreaks. *Proc. R. Soc. A Math. Phys. Eng. Sci.* **2020**, *476*, 20190777. [CrossRef]
30. Weiss, C.H.; Marin, M.R.; Keller, K.; Matilla-Garcia, M. Non-parametric analysis of serial dependence in time series using ordinal patterns. *Comput. Stat. Data Anal.* **2022**, *168*, 107381. [CrossRef]
31. Runge, J.; Nowack, P.; Kretschmer, M.; Flaxman, S.; Sejdinovic, D. Detecting and quantifying causal associations in large nonlinear time series datasets. *Sci. Adv.* **2019**, *5*, 11. [CrossRef]
32. Vlachos, I.; Kugiumtzis, D. Non-uniform state space reconstruction and coupling detection. *Phys. Rev. E* **2010**, *82*, 016207. [CrossRef]
33. Kugiumtzis, D. Direct-Coupling Information Measure from Nonuniform Embedding. *Phys. Rev. E* **2013**, *87*, 062918. [CrossRef]
34. Cover, T.; Thomas, J. *Elements of Information Theory*; John Wiley and Sons: New York, NY, USA, 1991.
35. Quian Quiroga, R.; Kraskov, A.; Kreuz, T.; Grassberger, P. Performance of different synchronization measures in real data: A case study on electroencephalographic signals. *Phys. Rev. E* **2002**, *65*, 041903. [CrossRef]
36. Lancaster, G.; Iatsenko, D.; Pidde, A.; Ticcinelli, V.; Stefanovska, A. Surrogate data for hypothesis testing of physical systems. *Phys. Rep.* **2018**, *748*, 1–60.
37. Yu, G.H.; Huang, C.C. A distribution free plotting position. *Stoch. Environ. Res. Risk Assess.* **2001**, *15*, 462–476. [CrossRef]
38. Papapetrou, M.; Kugiumtzis, D. Markov chain order estimation with parametric significance tests of conditional mutual information. *Simul. Model. Pract. Theory* **2016**, *61*, 1–13. [CrossRef]
39. Goebel, B.; Dawy, Z.; Hagenauer, J.; Mueller, J. An approximation to the distribution of finite sample size mutual information estimates. *IEEE Int. Conf. Commun.* **2005**, *2*, 1102–1106.
40. Matthews, B.W. Comparison of the predicted and observed secondary structure of T4 Phage Lysozyme. *Biochim. Biophys. Acta* **1975**, *405*, 442–451. [CrossRef]
41. Berchtold, A.; Raftery, A. The mixture transition distribution model for high-order Markov chains and non-Gaussian time series. *Stat. Sci.* **2002**, *17*, 328–356. [CrossRef]
42. Spedicato, G. Discrete Time Markov Chains with R. *R J.* **2017**, *9*, 84–104. [CrossRef]
43. Ching, W.; Ng, M.; Fung, E. Higher-order multivariate Markov chains and their applications. *Linear Algebra Its Appl.* **2008**, *428*, 492–507. [CrossRef]

44. Ching, W.; Huang, X.; Ng, M.; Siu, T. *Markov Chains: Models, Algorithms and Applications*; Springer Nature: New York, NY, USA, 2013.
45. Politi, A.; Torcini, A. Periodic orbits in coupled henon maps: Lyapunov and multifractal analysis. *Chaos* **1992**, *2*, 293–300. [CrossRef]
46. Schelter, B.; Winterhalder, M.; Hellwig, B.; Guschlbauer, B.; Lücking, C.H.; Timmer, J. Direct or indirect? Graphical models for neural oscillators. *J. Physiol.* **2006**, *99*, 37–36. [CrossRef]
47. Geweke, J.F. Measures of conditional linear dependence and feedback between time series. *J. Am. Stat. Assoc.* **1984**, *79*, 907–915. [CrossRef]
48. Guo, S.; Seth, A.K.; Kendrick, K.; Zhou, C.; Feng, J. Partial Granger causality—Eliminating exogenous inputs and latent variables. *J. Neurosci. Methods* **2008**, *172*, 79–93. [CrossRef] [PubMed]
49. Papana, A.; Kugiumtzis, D.; Larsson, P.G. Detection of direct causal effects and application in the analysis of electroencephalograms from patients with epilepsy. *Int. J. Bifurc. Chaos* **2012**, *22*, 1250222. [CrossRef]
50. Xu, H.; Couch, K.A. The business cycle, labor market transitions by age, and the Great Recession. *Appl. Econ.* **2017**, *49*, 5370–5396. [CrossRef]
51. Kraskov, A.; Stögbauer, H.; Grassberger, P. Estimating mutual information. *Phys. Rev. E* **2004**, *69*, 066138. [CrossRef]
52. Koutlis, C.; Kugiumtzis, D. Discrimination of coupling structures using causality networks from multivariate time series. *Chaos* **2016**, *26*, 093120. [CrossRef]

MDPI
St. Alban-Anlage 66
4052 Basel
Switzerland
www.mdpi.com

Entropy Editorial Office
E-mail: entropy@mdpi.com
www.mdpi.com/journal/entropy

Disclaimer/Publisher's Note: The statements, opinions and data contained in all publications are solely those of the individual author(s) and contributor(s) and not of MDPI and/or the editor(s). MDPI and/or the editor(s) disclaim responsibility for any injury to people or property resulting from any ideas, methods, instructions or products referred to in the content.